BD 027340501 2

19 Advances in Biochemical Engineering

Managing Editor: A. Fiechter

Reactors and Reactions

With Contributions by
H. W. Blanch, P. J. Bottino, S. Fukui,
O. L. Gamborg, M. Moo-Young, H.-J. Rehm,
I. Reiff, K. Schügerl, A. Tanaka

With 142 Figures and 37 Tables

Springer-Verlag
Berlin Heidelberg New York 1981

ISBN 3-540-10464-X Springer-Verlag Berlin Heidelberg New York
ISBN 0-387-10464-X Springer-Verlag New York Heidelberg Berlin

This work is subject to copyright. All rights are reserved, whether the whole or part of the material is concerned, specifically those of translation, reprinting, re-use of illustrations, broadcasting, reproduction by photocopying machine or similar means, and storage in data banks. Under § 54 of the German Copyright Law where copies are made for other than private use, a fee is payable to „Verwertungsgesellschaft Wort", Munich.

© by Springer-Verlag Berlin · Heidelberg 1981
Library of Congress Catalog Card Number 72-152360
Printed in GDR

The use of registered names, trademarks, etc. in this publication does not imply, even in the absence of a specific statement, that such names are exempt from the relevant protective laws and regulations and therefore free for general use.

2152/3020-543210

Managing Editor

Professor Dr. A. Fiechter
Eidgenössische Technische Hochschule,
Hönggerberg, CH-8093 Zürich

Editorial Board

Prof. Dr. *S. Aiba*	Department of Fermentation Technology, Faculty of Engineering, Osaka University, Yamada-Kami, Suita-Shi, Osaka 565, Japan
Prof. Dr. *B. Atkinson*	University of Manchester, Dept. Chemical Engineering, Manchester/England
Prof. Dr. *J. Böing*	Röhm GmbH, Chem. Fabrik, Postf. 4166, D-6100 Darmstadt
Prof. Dr. *E. Bylinkina*	Head of Technology Dept., National Institute of Antibiotika. 3a Nagatinska Str., Moscow M-105/USSR
Prof. Dr. *H. Dellweg*	Techn. Universität Berlin, Lehrstuhl für Biotechnologie, Seestraße 13, D-1000 Berlin 65
Prof. Dr. *A. L. Demain*	Massachusetts Institute of Technology, Dept. of Nutrition & Food Sc., Room 56-125, Cambridge, Mass. 02139/USA
Prof. Dr. *R. Finn*	School of Chemical Engineering, Olin Hall, Ithaca, NY 14853/USA
Prof. *S. Fukui*	Dept. of Industrial Chemistry, Faculty of Engineering, Sakyo-Ku, Kyoto 606, Japan
Prof. Dr. *K. Kieslich*	Wissenschaftl. Direktor, Ges. für Biotechnolog. Forschung mbH, Mascheroder Weg 1, D-3300 Braunschweig
Prof. Dr. *R. M. Lafferty*	Techn. Hochschule Graz, Institut für Biochem. Technol., Schlögelgasse 9, A-8010 Graz
Prof. Dr. *K. Mosbach*	Biochemical Div., Chemical Center, University of Lund, S-22007 Lund/Sweden
Prof. Dr. *H. J. Rehm*	Westf. Wilhelms Universität, Institut für Mikrobiologie, Tibusstraße 7—15, D-4400 Münster
Prof. Dr. *P. L. Rogers*	School of Biological Technology, The University of New South Wales. PO Box 1, Kensington, New South Wales, Australia 2033
Prof. Dr. *H. Sahm*	Institut für Biotechnologie, Kernforschungsanlage Jülich, D-5170 Jülich
Prof. Dr. *K. Schügerl*	Institut für Technische Chemie, Universität Hannover, Callinstraße 3, D-3000 Hannover
Prof. Dr. *H. Suomalainen*	Director, The Finnish State Alcohol Monopoly, Alko, P.O.B. 350, 00101 Helsinki 10/Finland
Prof. *G. T. Tsao*	Director, Lab. of Renewable Resources Eng., A. A. Potter Eng. Center, Purdue University, West Lafayette, IN 47907/USA

Table of Contents

Design of Biochemical Reactors
Mass Transfer Criteria for Simple and Complex Systems

M. Moo-Young
Dept. of Chemical Engineering, University of Waterloo, Waterloo, Ontario, Canada

H. W. Blanch
Dept. of Chemical Engineering, University of California, Berkeley, California, U.S.A.

Biochemical reactors are treated as heterogeneous catalytic reactors in which physical mass transfer completely or significantly controls the overall rate of the process being promoted in the reactor. The treatment used to develop basic design strategies takes into account the special constraints imposed by biological and biochemical phenomena on the systems.

By identifying the fundamental principles involved, generalized mass transfer criteria for biochemical reactors are developed for both inter-particle and intra-particle pathways in solid-fluid and fluid-fluid contacting systems for such diverse processes as aerobic fermentations, anaerobic fermentations, immobilized enzyme reactions, and insoluble substrate utilization. A wide range of practical operating conditions extending from rheologically simple non-viscous materials to complex viscous non-Newtonian and multiphase systems, and from geometrically simple bubble-column and packed-bed devices to complex stirred-tank and tubular-loop configurations are considered. Recent advancements in the development of correlations for mass transfer coefficients, interfacial areas, and related parameters are reviewed.

The processing energy required to induce and maintain the physical mass transfer pathways in the various reactor systems are also considered. It is shown that with the present state of the art, the application of engineering correlations to the scaling-up of biochemical reactors, especially stirred-tank reactor types, is more difficult than may be generally realized. Finally, attention is drawn to the areas of ignorance which need further exploration to help in the establisment of rational design and operation procedures for biochemical reactors.

1 Introduction

1.1 Reactor Types and Mass Transfer Implications

A biochemical reactor is a device in which materials are treated to promote biochemical transformation of matter by the action of living cells or cell-free enzyme systems. In the literature, the terms "biochemical reactor", "biological reactor" and "bioreactor" have been used interchangeably and indiscriminately. In this manuscript, we give preference to the first term: it is the most widely accepted of the three at present; it also reflects the fact that the design and operation of these reactor types are based on principles previously established for chemical reactors, analogous devices which are used to promote chemical transformations of matter[89)].

Biochemical reactors are widely employed in the food industries, in fermentation in waste treatment, and in many biomedical facilities. In industrial processes, they are invariably at the heart of the process (see Fig. 1). Broadly speaking, there are two types of biochemical reactors: microbial fermenters and enzyme (cell-free) reactors. Depending on the process requirements (aerobic, anaerobic, solid state, immobilized), numerous subdivisions of this classification are possible.

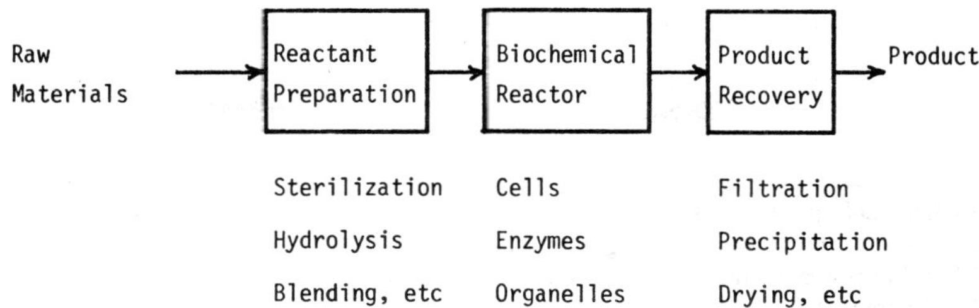

Fig. 1 Generalized outline of a biochemical process illustrating the central importance of the reactor

In the near future, it is possible that biochemical reactors which are based on cell-free organelles will also be developed.

In fermentors, cell-growth is promoted or maintained to allow formation of products such as metabolites (e.g. antibiotics, alcohols, citric acid), biomass (e.g. Baker's yeast, SCP), transformed substrates (e.g. physiologically active steroids) or purified solvents (e.g. in water reclamation). By contrast, cell-growth is destroyed or prevented in sterilizers so that undesirable metabolites cannot be produced (as in food preservation by heat treatment and clean air preparation by filtration). Systems based on macro-organism cultures (consisting of mammalian or plant cells) are usually referred to as "tissue cultures" while those based on dispersed non-tissue forming cultures of micro-organisms (bacteria, yeasts, fungi) are loosely referred to as "microbial" reactors. Because bioprocesses are usually promoted by microbes, the terms "fermentors" and "microbial reactors" are often used interchangeably, sometimes erroneously.

In enzyme reactors, substrate transformation is promoted without the life-support system of whole cells (e.g., enzymic saccharification of polysaccharides to make syrup). Frequently, these reactors employ "immobilized enzymes" where solid or semi-solid supports are used to internally entrap or externally attach the biocatalyst so that it is not lost as in "free-enzyme" systems, and may be re-used in a process[7)].

Virtually, all biochemical reactors of technological importance deal with heterogeneous systems involving one or more phases. Thus, to be effective in achieving the required biochemical changes, interphase mass (and heat) transfer must occur in these systems. Depending on the primary interphase-contacting requirement, there are two basic types of biochemical reactors: fluid-fluid contactors and solid-fluid contactors. Numerous subdivisions of this classification are possible, e.g. bubble columns, stirred tanks, trickle-bed filters, fluidized beds, air-lift towers, etc.

In this review, a comprehensive examination of the various aspects of physical mass transfer (and — by analogy heat transfer) which form the basis of the rational engineering design and operation of all types of biochemical reactors is given. Previous reviews[149, 188)] have generally been limited to special aspects such as fluid types (usually Newtonian systems) or reactor geometries (usually stirred-tanks or bubble columns), with little regard to the biological or biochemical constraints. We will consider mass transfer mechanisms for a wide range of materials and geometric properties of practical interest, including electrolyte solutions, simple low-viscosity Newtonian as well as complex high-viscosity non-Newtonian fluids, free-suspension as well as fixed or immobilized dispersed systems and a variety of reactor configurations including tubular and tank devices with mechanically-induced or pneumatically-induced agitation. Particular attention is given to gas-liquid contacting phenomena which usually determine the performance of aerobic fermentors, the most widely-used biochemical reactors.

In industrial practice, theoretical explanation frequently lags behind technological realization: many biochemical process developments are good examples of this apparent paradox. In this review, the basic mass transfer concepts, which determine biochemical reactor performance, are generalized so that the rationale for traditional empiricisms as well as for recent developments and potential innovations can be identified in terms of unifying fundamental principles. Basic correlations for both inter-particle and intra-particle mass transfer will be examined. Combined with knowledge of the interfacial area, which will also be discussed, these correlations lead to the prediction of the various mass transfer rates in biochemical reactors in terms of operating conditions. Empirical correlations relating overall mass transfer rates directly to process variables will also be given; however, it should be noted that although these correlations currently facilitate design calculations, their general applicability is questionable. Methods of calculating agitation power which affects the mass transfer processes will also be presented for both Newtonian and non-Newtonian systems. Finally, the utilization of mass transfer criteria in the scaling-up of biochemical reactors is briefly discussed.

1.2 Systems and Operating Constraints

The application of chemical engineering principles is useful in the analysis of the design and operation of biochemical reactors. However, classical approaches to the analysis are limited by the following special constraints:

a) The bulk densities of suspended microbial cells and substrate particles generally approach those of their liquid environments so that relative flow between the dispersed and continuous phases in normally low. This situation may be contrasted with the relatively heavy metallic catalyst particles generally used in chemical reactors.
b) The sizes of single microbial cells are very small (in the range of a few microns) compared to chemical catalyst particles; coupled with the above constraints, it is generally difficult to promote high particle Reynolds numbers and attain turbulent-flow mass transfer conditions.
c) Polymeric substrates or metabolites and mycelial growths often produce very viscous reaction mixtures which are generally pseudoplastic non-Newtonian. Again, these conditions tend to limit desirably high flow dynamics in biochemical reactors.
d) Many multicellular microbial growths, especially fungal ones, generally form relatively large cell aggregates such as mycelia, clumps or pellets, as compared to catalyst particles. Intra-particle diffusional resistances are often pronounced in these systems, e.g. leading to anaerobiosis.

e) Biochemical reactors frequently require critically close control of solute concentrations, pH, temperature, and local pressures in order to avoid damage or destruction of live or labile components which are essential to the process.
f) Very low concentrations of reactants and/or products are normally involved in biochemical reactors so that the concentration driving forces for mass transfer are often severely limited.
g) Microbial growth rates are substantially lower than chemical reaction rates so that relatively large reactor volumes and residence times are required.

As an illustration of some of the problems imposed by the above constraints, we note that an adequate oxygen supply rate to growing cells is often critical in aerobic processes. Because of its low solubility in water, gaseous oxygen, usually in the form of air, must be supplied continuously to the medium in such a way that the oxygen absorption rate at least equals the oxygen consumption rate of the cells. Even temporary depletion of dissolved oxygen could mean irreversible cell damage. In this respect, it is worth noting that the same microbial species may show large variations in its oxygen requirements, depending on the oxygen concentration to which it has been adapted[167)].

Previous studies in which the oxygen supply to a submerged growing microbial culture was stopped have shown a linear decrease in oxygen concentration with time over a large concentration range[46, 159)]. Below a certain oxygen concentration, called the "critical oxygen tension", the decrease follows a hyperbolic pattern compatible with Michaelis-Menten kinetics. The reason for the linear decrease is not clear; it may be caused by the very low saturation parameters (K_m-values) for oxygen reduction (10^{-6} to 10^{-8} M).

Often, deviations from the linear and hyperbolic oxygen concentration decrease patterns are found. As will be seen in Sect. 4, the rate controlling step in a microbial process may shift from the oxygen supply rate into the bulk liquid to the demand rate inside the cell if cell aggregates are formed which are larger than a few hundredths of millimeters. This may cause different K_m-values for the reaction with oxygen for cells at different radial locations in the cell clumps. Usually, this is seen from an increased value of the critical oxygen tension or the total absence of a linear part of the oxygen concentration decrease curve, showing the dependence on the concentration driving force at the cell surface. In some cases a constant O_2-concentration gave optimal results, viz, for tryphtophan[126)] and L-glutamic acid production[163)].

Another reason for air sparging and mechanical mixing in a fermentor is to remove carbon dioxide and other possible toxic gaseous metabolic by-products which are produced in the broth. Thus, gas-liquid mass transfer can be important in reactor design and indeed is the basis for the so-called $k_L a$ criterion often used in the design of activated-sludge waste treatment facilities.

Similar considerations may apply to other reactants or product species in biochemical reactors.

Part I. Basic Concepts

2 Mass Transfer Pathways

2.1 Rate-Controlling Steps

Figure 2 schematically describes a biochemical reactor subsystem involving two or more phases. An important example of this representation is an aerobic process in which a microbe utilizes oxygen (supplied by air bubbles which also desorb toxic carbon dioxide) and other dissolved nutrients (sugars, etc.) to grow and produce soluble extracellular metabolites. Eight resistances in the mass transfer pathways for the nutrient supply and utilization and for metabolite excretion and removal are possible at the following locations:

(1) in a gas film, (2) at the gas-liquid interface, (3) in a liquid film at the gas-liquid interface, (4) in the bulk liquid, (5) in a liquid film surrounding the solid, (6) at the liquid-solid interface, (7) in the solid phase containing the cells, (8) at the sites of the biochemical reactions.

It should be noted that all the pathways except the last one are purely physical.

Figure 2 can depict a wide range of other practical situations. The continuous phase may be liquid or gas, the latter including special cases such as "solid-state" (e.g. composting, trickle-bed reactors, and "Koji") processes while the dispersed phase may be one or more of the following phases: solid (e.g. microbial cells, immobilized enzyme particles, solid substrates), liquid (e.g. insoluble or slightly soluble substrates) or gas (e.g. air, carbon dioxide, methane).

In addition to the physical constraints considered above, the mass transfer pathways may be complicated by the biochemical reaction steps which occur concurrently inside the cellular or enzymic materials. For example, Fig. 3 illustrates various feedback mechanisms which may be generated by the biochemical changes in the reactor. First, we will examine mechanisms of the physical mass transfer (Sect. 3), then the coupled physical and biochemical reaction rates (Sect. 4). Fortunately, the uncoupled physical mass transfer steps occur in series so that

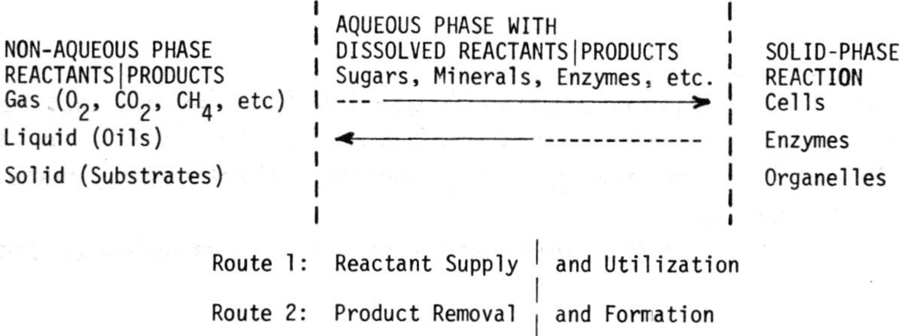

Fig. 2 Generalization of biochemical reactor conditions illustrating the importance of aqueous phase-mass transfer steps

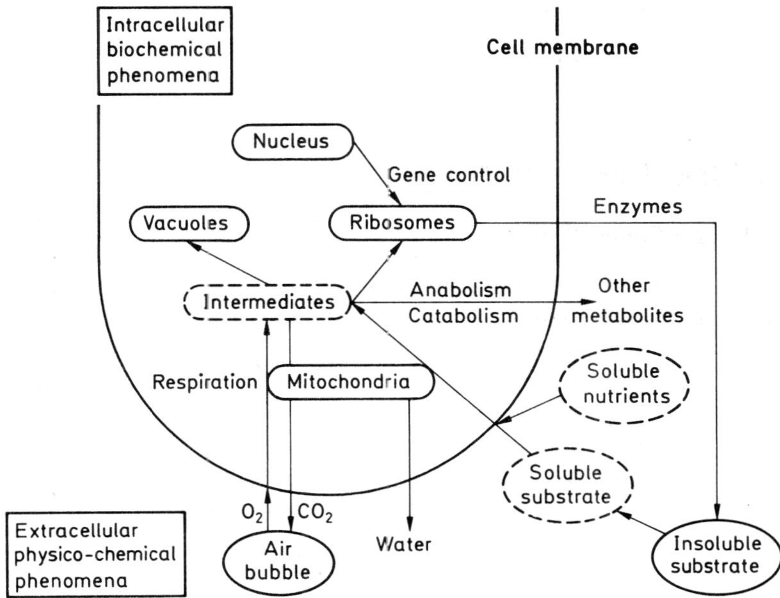

Fig. 3 Generalized interrelationships between intracellular biochemical and extracellular physical pathways in a microbial reactor

a rate-controlling step exists. In addition, this rate-controlling mass transfer step often becomes highly important since the overall rate of the biochemical reactions, which are enzymatically catalyzed, is usually relatively high.

In biochemical reactors, because of the relatively high mass diffusivities in the gases compared to the liquid media (about 10^5 fold) and their low solubilities, an aqueous liquid-phase resistance invariably controls the overall physical mass transfer rate. For example, in Fig. 2 one of the following four liquid-phase resistance is rate-controlling:

a) A combined liquid phase resistance near and at a gas-liquid interface: this resistance is often rate-controlling in aerobic reactors because of the relatively low solubility of oxygen in aqueous solutions and the retardation effects of adsorbed materials (e.g. surfactants) and electrolytes at the interface.
b) A liquid-phase resistance in the aqueous bulk medium separating the dispersed phases: this resistance is often insignificant because of the good liquid mixing promoted in practical reactor systems.
c) A liquid-phase resistance near and at the solid-liquid interface: this resistance can be significant because of the low density differences between the continuous aqueous medium and some dispersed phase (e.g. microbes, gel-entrapped enzymes, liquid drops, polysaccharides).
d) A liquid-phase resistance inside a dispersed "solid" phase: this resistance can be significant in cell flocs, mold pellets, immobilized enzyme carriers, insoluble substrate particles.

Thus, in practice, there are two basic situations of liquid-phase mass transfer which are important in biochemical reactors: external particle mass transfer and intra-particle mass transfer. The possible complex interaction of these two fundamental types of mass transfer steps in a biochemical reactor is illustrated in Fig. 3. We will examine the first situation in Sect. 3 and the second situation in Sect. 4, after reviewing some basic concepts of mass transfer theory.

2.2 Definition of Transfer Coefficients

A mass transfer coefficient can be conveniently defined by a simple mass balance for a given reactant or product species in the biochemical reactor. For example, considering the oxygen solute of the air bubbles passing through the fermentor subsystem illustrated in Fig. 3, we obtain

$$\text{Oxygen transfer rate} = k_L A(C_s - C) \tag{1}$$

where C is the local dissolved oxygen concentration in the bulk liquid at any time t, C_s the oxygen concentration in the liquid at the gas-liquid interface at infinite time (equivalent to the saturation concentration), A the interfacial area, and k_L the liquid-phase mass transfer coefficient.

Depending on the type of flow pattern inside the reactor, Eq. (1) can be incorporated in an overall oxygen balance in the liquid phase, and thus oxygen supply rates can be readily evaluated in the laboratory. For a dispersed system, integration of Eq. (1) gives:

a) For a well mixed batch liquid process:

$$-\ln(1 - E) = k_L a t\,, \tag{2}$$

where a is the interfacial area per unit volume of dispersion, t the aeration time and E the fractional approach to equilibrium:

$$E = \frac{C - C_O}{C_s - C_O}\,, \tag{3}$$

in which C_O, C and C_s = the initial, instantaneous and saturation liquid-phase oxygen concentration.

b) For a well mixed continuous-flow liquid process (chemostat):

$$\frac{E}{1 - E} = \frac{k_L a}{D}\,, \tag{4}$$

where D is the dilution rate and

$$E = \frac{C' - C'_O}{C_s - C'_O}\,, \tag{5}$$

in which C' is the constant steady-state liquid phase oxygen concentration and C'_O the oxygen concentration in the inlet medium.

According to Eq. (1), the mass transfer rate is dependent on the mass transfer coefficient, the interfacial area in the dispersion and the concentration driving force. Interfacial area is controlled by factors discussed in Sect. 5. The concentration driving force will generally follow Henry's law and may be increased by pressure as in the ICI-type reactor[57)]. We will now consider the effects of the processing conditions on the mass transfer coefficient. An aerobic system will serve as a generalized example.

Heat effects may be expected from actively growing cells. For example, the amount of heat produced by aerobically growing yeast cells is estimated to be about 1.4×11^{-15} kcal s^{-1} per cell. At the cell surface, this means a heat flux of 3×10^{-15} kcal m^{-2} s^{-1}. Methods of evaluating heat transfer rates between the dispersed cells and continuous liquid phase in media will be given in Sect. 3, by analogy to the mass transfer phenomena.

2.3 Effect of Diffusion

Fick's laws of diffusion form the bases for the current theoretical approaches to mass transfer. To varying degress, all approaches postulate the existence of fluid films at the phase boundary for interfacial transfer. For steady-state unidirectional diffusion, Fick's law takes the form

$$\text{Mass flux of component } A \text{ in } B = J_A = -D_L \frac{dC_A}{dx}, \tag{6}$$

where dC_A/dx = concentration gradient over a diffusional path of length x. Values of the diffusion coefficient D_L for binary liquid systems usually fall in the range 0.5 to 2.0×10^{-5} cm^2 s^{-1} for non-viscous liquids. D_L may be estimated in these systems by the Stokes-Einstein equation

$$D_L = \frac{kT}{6\pi r_o \mu}, \tag{7}$$

where r_o is the solute A sphere radius and μ the viscosity of the solvent B. Sherwood, Pigford and Wilke[152)] present considerable data on D_L for a variety of systems. For oxygen in water, D_L has the value of 2.10×10^{-5} cm^2 s^{-1} at 25 °C.

In high viscosity media such as polysaccharide gum or fungal broths, deviations from the Stokes-Einstein equation for low molecular weight solutes such as oxygen have been reported. Values of D_L do not decrease with increasing viscosity as would be expected from the Stokes-Einstein equation. Gainer et al.[51, 52)] provide data on D_L in polymer solutions and O_2 and CO_2 diffusion into albumin and globulin solutions. Values of D_L are found to be only slightly lower than those for solute molecules diffusing into water.

The solution to Eq. (6) for a stagnant medium (as in a cellular mass) or an external film at a particle interface in a liquid of constant concentration (achieved by convection currents) gives[91)]:

$$k_L = \frac{D_L}{X}. \tag{8}$$

This expression was obtained by Lewis and Whitman[91)], according to their "film" theory, for a more realistic situation involving a laminar falling film of thickness X. In this case of steady-state diffusion, it is seen that the mass transfer coefficient is proportional to D_L.

The assumption of a stagnant laminar-flow film next to the boundary in which the mass transfer resistance is highest is not appropriate under many practical flow conditions which require the application of Fick's law for unsteady-state diffusion given below:

$$\frac{\partial C_A}{\partial t} = -D_L \frac{\partial^2 C_A}{\partial x^2} . \tag{9}$$

To solve this equation, simplifying assumptions must be made, especially with regard to the liquid film behavior. Higbie[66)] solved this equation according to his "penetration theory" for diffusion into a "semi-infinite medium" representing a liquid film around a spherical fluid particle. Using appropriate boundary conditions, he deduced that

$$k_L = 2\sqrt{\frac{D_L}{\pi t}} , \tag{10}$$

where t (exposure time) is defined as d_B/U_B, the time required for the fluid particle to travel one equivalent diameter, which is only an approximation but appears to be a good one for the real situation. Danckwerts[37)] subsequently questioned the hypothesis of a constant exposure time and postulated a random continuous renewal of surface elements at the interface according to his "surface renewal" as a more realistic situation. He introduced a statistical parameter s (the rate of surface renewal) and found that

$$k_L \propto \sqrt{D_L s} , \tag{11}$$

which is in agreement with Higbie's conclusion on the dependence of k_L on $\sqrt{D_L}$.

For rigid non-slip interfaces, the classical boundary layer theory can be used to evaluate mass transfer coefficients. By relating the momentum with the concentration boundary layers it can be shown that[50)]

$$k_L \propto D_L^{2/3} . \tag{12}$$

Thus, according to current theories the effect of the molecular diffusivity on the mass transfer coefficient is to the power of one-half, two-thirds or unity, depending on the hydrodynamic conditions.

2.4 Effect of Interfacial Phenomena

If we consider a fluid particle (gas or liquid), moving relative to a continuous liquid phase, there are two possible extremes of interfacial movement as classified below.

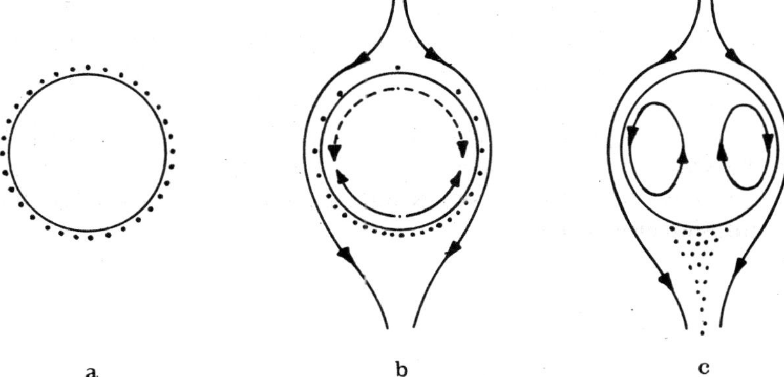

Fig. 4 Surfactant effects on bubble/drop surface-flow at **(a)** zero **(b)** low and **(c)** high relative particle velocities

(For convenience, we will consider the simplest geometry, a sphere; as seen later (p. 17) this is a good approximation of real cases.)

a) There is no internal circulation within the particle. These particles behave essentially as if they are solid with rigid surfaces. We will refer to these as particles with rigid surfaces.

b) There is a fully developed internal circulation within the particle due to an interfacial velocity. The particle behaves as a part of an inviscid continuous phase with only a density difference. We will refer to these as particles with mobile surfaces.

Examples of velocity profiles for both kinds of these particles are illustrated in Fig. 4. As pointed out above, for moving particles with rigid surfaces, k_L is proportional to $D_L^{2/3}$ as predicted by the boundary layer theory, and to $D_L^{1/2}$ for moving spheres with mobile surfaces as predicted by the potential-flow theory.

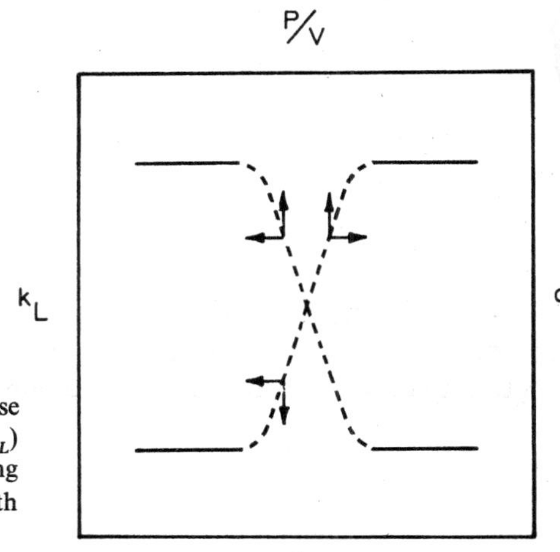

Fig. 5 Typical variations showing (i) the increase of bubble or drop mass transfer coefficient (k_L) with increasing particle size (d_B) but decreasing power input (P/V) (ii) the interfacial area (a) with rising power input and decreasing particle size

This concept has been useful in explaining many drop and bubble phenomena. For example, it has been found that trace amounts of surface-active materials can hinder the development of internal circulation by means of a differential surface pressure (see Fig. 4). Small bubbles rising slowly are apt to behave like particles with rigid surfaces. This phenomenon can lead to a decrease in k_L as the age of a bubble increases[28]. Larger bubbles, rising more quickly, may sweep their front surface free of trace impurities and therefore escape the contaminating effect of surfactants as illustrated in Fig. 4. These effects lead to significant variations of k_L with changing bubble size and agitation power as illustrated in Fig. 5.

In practice, clean bubble systems are probably rarely achieved and it is fairly safe to base a design on contaminated rigid interface behavior as discussed later (Sect. 7). However, it has been estimated by an industrial manufacturer that even a 1% increase in mass transfer rate in certain antibiotic processes could mean an increase in profitability of the order of million dollars per year!

3 External Particle Mass Transfer

3.1 Relevant Operating Variables

Because of the complex hydrodynamics usually found in the multiphase system in a biochemical reactor, a useful approach to its mass transfer problems is dimensional analysis.

For the relatively simple cases where theoretical analyses from fundamental principles are possible, the solutions can still be conveniently expressed in terms of these dimensionless groups, as shown on p. 13.

For external mass transfer, the following dimensionless groups are relevant:

$$Sh \text{ (Sherwood number)} = \frac{\text{total mass transfer}}{\text{diffusive mass transfer}} = \frac{k_L d}{D_L}$$

$$Sc \text{ (Schmidt number)} = \frac{\text{momentum diffusivity}}{\text{mass diffusivity}} = \frac{\mu}{\varrho D_L}$$

$$Gr \text{ (Grashof number)} = \frac{\text{gravitational forces}}{\text{viscous forces}} = \frac{d^3 \varrho g \Delta \varrho}{\mu^2}$$

$$Re \text{ (Reynolds number)} = \frac{\text{inertia forces}}{\text{viscous forces}} = \frac{du \varrho}{\mu} .$$

The first three analogous groups for heat transfer which can be used in later discussions include:

$$Nu \text{ (Nusselt number)} = \frac{hD}{k} \text{, analogous to } Sh \text{ ,}$$

$$Pr \text{ (Prandtl number)} = \frac{C_p \mu}{k} \text{, analogous to } Sc \text{ ,}$$

$$Gr_H \text{ (Grashof number)} = \frac{d^3 g \beta \Delta T}{\mu^2} \text{, analogous to } Gr.$$

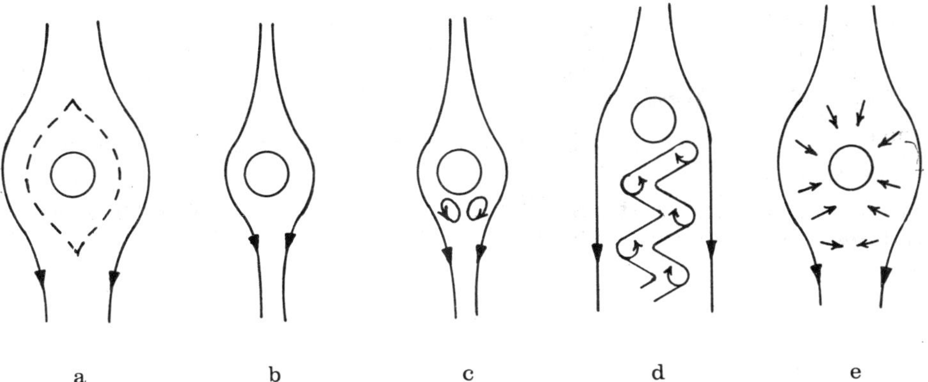

Fig. 6 Possible conditions of the momentum boundary layer around a submerged solid sphere with increasing relative velocity: **a** envelope of pseudo-stagnant fluid, **b** streamline flow, **c** flow separation and vortex formation, **d** vortex shedding, **e** localized turbulent eddy formations

In the following summary of correlations for k_L, different expressions for *Sh* are given for different flow regimes, as characterized usually by the *Re*-number. This can partly be understood by the increasing influence of the momentum boundary layer. Fig. 6 illustrates the increasing complexity of external flow conditions in which a particle may find itself in a biochemical reactor as agitation intensity varies. Whenever possible, references to theoretically derived expressions are also given.

3.2 Particles in Stagnant Environments

For non-moving submerged particles (with rigid or mobile surface) in a stagnant medium, mass transfer occurs only by radial diffusion $Re = Gr = 0$, whence it can be shown that[50]:

$$Sh = Nu = 2\,. \tag{13}$$

As the lower limit for *Sh*, we will see that this value usually vanishes for bubble mass transfer, but it may become significant when applied to small light particles, e.g. microbial cells. Pseudo-stagnant liquid environments can exist in viscous reactions and/or with well dispersed single cells as illustrated in case a of Fig. 6.

When $Re = 0$, but $Gr > 0$, it has been shown by Levich that[90]:

$$Sh = 1.1(Gr_\delta Sc)^{1/3} \tag{14}$$

which occurs when the following boundary layer thickness *Re*-number relationship is obeyed, as is the case with certain packed-bed systems.

$$\frac{U_B \delta_M \varrho}{\mu} > 0.4\,\frac{Gr_\delta^{\;0.5}}{Sc^{0.17}}\,. \tag{15}$$

In practice, this type of analysis can also be applied to "free-suspension" gas bubble systems (see p. 14).

3.3. Moving Particles with Rigid Surfaces

A range of these cases can occur in packed-bed, trickle-bed or free-rise or free-fall dispersed-phase reactor systems. For creeping flow, $Re < 1$, (e.g. certain packed-bed immobilized-enzyme reactors) the theory developed by Levich[90] shows that:

$$Sh = 0.99\ Re^{1/3}\ Sc^{1/3} = 0.99\ Pe^{1/3}\ . \tag{16}$$

Several similar theoretical expressions have been reported (e.g.[49]):

$$Sh = 1.01\ Pe^{1/3}\ . \tag{17}$$

In the range $10 < Re < 10^4$, (e.g. certain trickle-bed reactors)

$$Sh = 0.95\ Re^{1/2}\ Sc^{1/3}\ . \tag{18}$$

Various empirical and semitheoretical correlations have been reviewed recently by Rowe et al.[142]. An overall approximation is given as:

$$Sh = 2 + 0.73\ Re^{1/2}\ Sc^{1/3} \tag{19}$$

where the factor 2 accounts for radial diffusion.

For cases in which flow action just balances gravitational forces ("free-suspension"), Re can be expressed in terms of a bulk Gr, and the mass transfer coefficient is given by the following correlation developed by Calderbank and Moo-Young[28]:

$$Sh = 2 + 0.31(Sc\ Gr)^{1/3}\ . \tag{20}$$

Ignoring the radial diffusion contribution,

$$k_L\ Sc^{2/3} = 0.31 \left(\frac{\Delta\varrho\ \mu g}{\varrho^2}\right)^{1/3}, \tag{21}$$

where the strong effects of continuous-phase viscosity (e.g. polysaccharide production) and low particle density (e.g. microbial cells) on mass transfer are revealed. Thus, if oxygen demand at the cell interface is the limiting mass transfer step in a process, the performance of the reactor may be outside the control of the operator in terms of aeration and agitation. It is to be noted that the particle velocity need not be known in these cases for design purposes. Also, the correlation applies to small bubbles (<2.5 mm) in aqueous solutions, because they behave as particles with rigid interfaces. In addition, the level of agitation power input in these dispersions is necessarily low.

At high agitation intensities, turbulence is expected to affect the mass transfer rates at solid particle surfaces. However, in these cases, the actual particle velocity is unknown and conventional Reynolds numbers cannot be deduced. In this case, the concept of local isotropic turbulence may be applied. According to this concept, large "primary eddies" emerge, e.g. by impeller action, as waves in a turbulent fluid

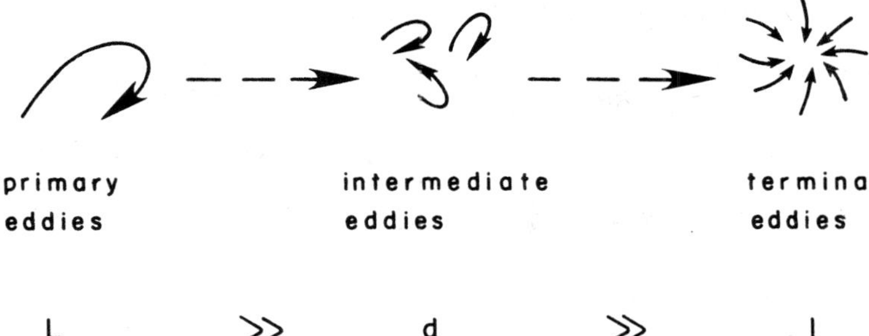

Fig. 7 Energy transfer from large primary non-isotropic eddies down to small terminal isotropic eddies according to the concept of local isotropic turbulence

field. The scale of these primary eddies, L, is of the order of magnitude of the impeller diameter. Primary eddies are non-isotropic, resulting in a net, unidirectional velocity. The primary eddies are unstable and break up into smaller eddies of intermediate size, d, which may or may not be isotropic. Eventually, these "intermediate eddies" break down into very small "terminal eddies" which have completely lost their unidirectional nature, therefore being isotropic. By these terminal eddies, most of the energy dissipation takes place. The cascade of energy transfer is illustrated in Fig. 7. The size of such terminal eddies is given by Kolmogoroff as[86]:

$$l = \frac{\mu^{3/4}}{\varrho^{1/2}} / (P/V)^{1/4} . \tag{22}$$

For local isotropic turbulence to prevail it has been estimated that the ratio L/l should be $\geqq 10^3$, where L may be approximated by the impeller diameter for mechanically stirred systems.

In order to define an appropriate Re-number characteristic of local isotropic turbulence, a velocity expression can be derived according to Batchelor[10] as:

$$\sqrt{\overline{u_d^2}} \propto (P/V)^{1/3} (d/\varrho)^{1/3} , \tag{23}$$

where $\sqrt{\overline{u_d^2}}$ is the root mean square fluctuating velocity component in which d refers to the length scale over which it operates. Thus, an isotropic turbulence Re-number, Re_e, for the particle of diameter d can be given as

$$Re_e = \frac{d^{4/3} \varrho^{2/3} (P/V)^{1/3}}{\mu} . \tag{24}$$

Calderbank and Moo-Young[28] developed a correlation for rigid-surface particle mass transfer in biochemical reactors in terms of the energy input to the system as follows:

$$Sh = 0.13\, Re_e^{3/4}\, Sc^{1/3} , \tag{25}$$

where k_L is seen to be dependent on $(P/V)^{1/4}$, a dependence which may be masked by the effect of power on interfacial area (see Sect. 7).

3.4 Moving Particles with Mobile Surfaces

Mobile-surface fluid particles show a behavior which is less sphere-like than that of rigid-surface fluid particles. By viscous interaction with the continuous phase, oscillating shape variations of liquid drops and gas bubbles occur. For $Re > 1$, mobile-surface fluid particles in free-rising or falling conditions move in a wobbling or spiral-like manner which has a marked influence on mass transfer rates. As pointed out earlier (p. 11), $Sh \propto Sc^{1/2}$ in the case of mobile interfaces, indicating a more pronounced influence of the velocity boundary layer than in the case of rigid interfaces where $Sh \propto Sc^{1/3}$.

As before, we can arrive at different correlations for different bulk flow regions. These are summarized as follows:

For creeping flow ($Re < 1$), Hadamard[58)] showed that

$$Sh = 0.65\left(\frac{\mu}{\mu + \mu_d}\right)^{1/2} Pe^{1/2}. \tag{26}$$

For gas-liquid dispersions, this reduces to:

$$Sh = 0.65\, Pe^{1/2}. \tag{27}$$

For the lower intermediate Re-numbers ($10 < Re < 100$):

$$Sh = 0.65\, Pe^{1/2}(1 + Re/2)^{1/2}. \tag{28}$$

For upper intermediate Re-numbers, $100 < Re < 1000$,

$$Sh = 1.13\left[1 - \frac{2 + 3\dfrac{\mu_d}{\mu}}{1 + (\varrho_d\mu_d/\varrho\mu)^{1/2}} \cdot \frac{1.45}{Re^{1/2}}\right]^{1/2} Pe^{1/2}. \tag{29}$$

For gas-liquid dispersions, this reduces to:

$$Sh = 1.13\left(1 - \frac{2.9}{Re^{1/2}}\right)^{1/2} Pe^{1/2}. \tag{30}$$

For higher Re-numbers, Higbie[66)] introduced an equation which takes the form:

$$Sh = 1.13\, Pe^{1/2}. \tag{31}$$

Calderbank and Lochiel[26)] derived an equation for Sh in the case of spheroidal bubbles having an eccentricity E_1:

$$Sh = [(2/3(1 + K_1)]^{1/2}\left[\frac{2.26 E_1^{1/3}(E_1^2 - 1)^{1/2}}{E_1(E_1^2 - 1)^{1/2} + \ln[E_1 + (E_1^2 - 1)^{1/2}]}\right] \times \left(\frac{ud_e}{D}\right)^{1/2}, \tag{32}$$

in which: $K_1 = -\left(\dfrac{eE_1^2 - E_1 \sin^{-1} e}{e - E_1 \sin^{-1} e}\right)$

and, $e = (1 - 1/E_1^2)^{1/2}$

thus revealing the small (and usually negligible) effects of bubble eccentricity on mass transfer coefficients for practical situations with non-viscous fluids.

In thick viscous liquids ($\mu > 70$ cp) large spherical-cap bubbles are frequently encountered and for mass transfer:

$$Sh = 1.79 \frac{(3E_1^2 + 4)^{2/3}}{E_1^2 + 4} Pe^{1/2} , \tag{33}$$

in which E_1 is the ratio of bubble width to bubble height. For spherical-cap bubbles, E_1 seems to be fairly constant at 3.5 so that

$$Sh = 1.31\, Pe^{1/2} . \tag{34}$$

As for small bubbles, large bubbles ($d_B > 2.5$ mm) in non-viscous media appear to be in a state which approximates free suspension where gravitationally induced flow is responsible for the mass transfer; for these cases, the mass transfer coefficient is given by Calderbank and Moo-Young as[28]:

$$Sh = 0.42\, Sc^{1/2}\, Gr^{1/3} . \tag{35}$$

As before, the absolute bubble velocity need not be known to evaluate *Sh*. It should be noted that the exponent of the *Sc*-number has changed from $^1/_3$ to $^1/_2$ which is consistent with the transition from rigid to mobile interface behavior.

As discussed in Sect. 5.4, high agitation intensities which promote local isotropic turbulence will lead to particle disruption rather than increased mass transfer coefficients in fluid-fluid dispersions.

3.5 Interacting Particles

In swarms of bubbles where the gas hold-up is high, the relative proximity of the bubbles alters the fluid streamlines around the bubbles, thus affecting the mass transfer coefficient k_L. Similar effects may be observed in enzyme systems using e.g. glass beads as immobilizing carriers. Gal-Or and coworkers[53, 54] have extensively investigated the effects of gas hold-up on bubble motion and mass transfer of gas dispersions in Newtonian liquids under creeping flow conditions. This work is an extension of Happel's[61, 62] "free surface cell model" describing creeping flow of a fluid over an assemblage of solid spheres.

The effects of surfactants may change the behavior of the bubbles; with increasing surfactant concentration, rigid-sphere behavior may be observed as discussed earlier (see p. 12). For strong internal circulation (mobile interface behavior), Gal-Or introduces the correlation:

$$Sh = 0.895 \left(\beta \frac{1 - \varphi^{5/3}}{Y - \varphi^{1/3} W} \right)^{1/2} Pe_{sw}^{1/2} , \tag{36}$$

where $W = 3 + 2\beta + 2\varphi^{5/3} (1 - \beta)$

$Y = 2 + 2\beta + \varphi^{5/3} (3 - 2\beta)$

$$\beta = \frac{\mu}{\mu_d + \gamma} ,$$

where γ is an "interfacial viscosity" due to adsorbed surfactant impurities. In the region of intermediate internal circulation

$$0.849\left(\beta\frac{1-\varphi^{5/3}}{Y}\right)Sh^{-2}+4.96\left\{\frac{2}{15}\left[\frac{3}{2}-2\beta\frac{(1-\varphi^{5/3})}{Y}\right]+\varphi^{5/3}\left[\beta\frac{(1-\varphi^{5/3})}{Y}-\frac{1}{2}\right]\right\}Sh^{-3}=Pe_{sw}^{-1} \tag{37}$$

and in the absence of internal circulation (solid-sphere behavior)

$$Sh=1.306\left(\frac{1-\varphi^{5/3}}{Y-\varphi^{1/3}w}\right)Pe_{sw}^{1/3}. \tag{38}$$

Thus, for both circulating and non-circulating bubbles, there is an increase in the Sherwood number at large Peclet numbers with rising gas hold-up ϕ. Generally, the Sherwood numbers of bubble swarms with strong internal circulation are always higher (by up to one order of magnitude) than those for non-circulating bubbles. In all cases, the Sherwood numbers of bubbles swarms ($\varphi > 0.1$) are higher than those for single bubbles. These results are summarized on Fig. 8.

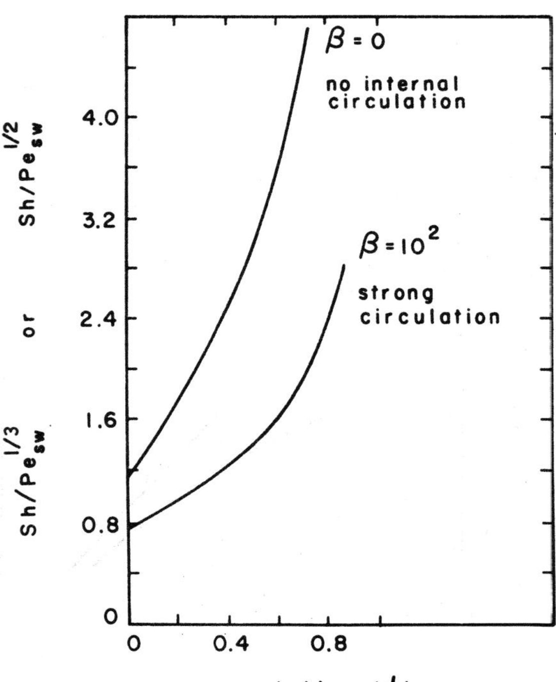

Fig. 8 Effect of gas hold-up on mass transfer coefficients as predicted from the free-cell model[53])

3.6 Non-Newtonian Flow Effects

When the liquid phase exhibits non-Newtonian behavior, the mass transfer coefficient k_L will change due to alterations in the fluid velocity profile around the submerged particles. There are only few data available on these effects but more information on the changes in the drag coefficient when spheres move into non-Newtonian fluids. The trends for both mass transfer and drag coefficient are analogous. As for Newtonian fluids, two types of interfacial behavior need to be considered.

3.6.1 Mobile-Surface Particles

For power-law fluids, Hirose and Moo-Young[67] have obtained a correction factor for k_L for single bubbles based on small pseudoplastic deviations from Newtonian behavior ($0.7 < n < 1.0$). These authors also provide some data on drag coefficients as functions of the power-law index n. Bhavaraju, Mashelkar and Blanch[12] examined both power-law and Bingham plastic fluids with mobile interfaces, using perturbation analysis, and provided the following corrections for the enhancement of mass transfer

$$Sh = 0.65\{1 - 3.24m\}^{1/2}\, Pe^{1/2}\,, \tag{39}$$

where $m = \dfrac{n-1}{2}$

and for Bingham plastic fluids

$$Sh = 0.65\{1 + 0.5\varepsilon\}^{1/2}\, Pe^{1/2}\,, \tag{40}$$

where the Bingham number (2ε) is given by

$$\varepsilon = \tau_0 R/u\mu_0\,. \tag{41}$$

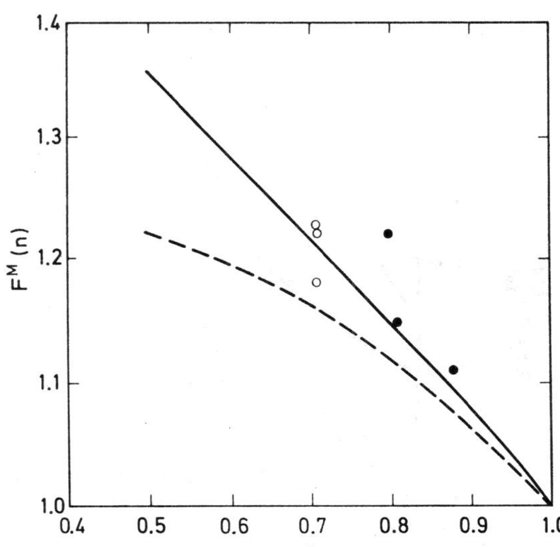

Fig. 9 Effect of pseudoplasticity on the mass transfer coefficient according to a correction factor, $Sh = 0.65F^M(n)\,Pe^{1/2}$. The solid line represents theoretical predictions of Bhavaraju et al.[12] while the dotted line describes theoretical predictions and data points of Hirose and Moo-Young[67]

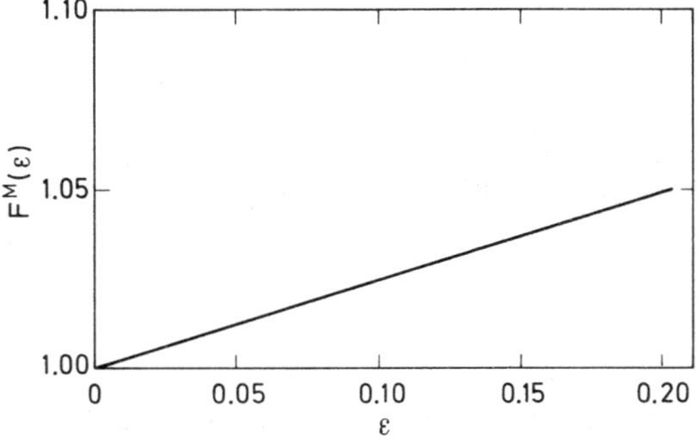

Fig. 10 Effect of the Bingham number on the mass transfer coefficient according to $Sh = 0.65F^M(\varepsilon)\ Pe^{1/2}$

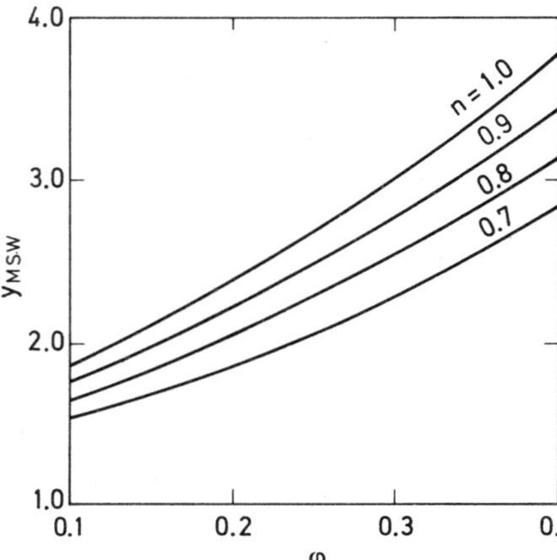

Fig. 11 Correction factor $y_{MSW} = f(n, \varphi)$ (Eq. (42)) for a swarm of bubbles with mobile interfaces as a function of gas hold-up and power-law index
$Sh = 0.65f(n, \varphi)\ Pe_{SW}^{1/2}$

In extending this work on mass transfer from single bubbles to bubble swarms, these authors found a deterioration in k_L for power-law fluids

$$Sh = 0.65f(\varphi, n)\ Pe_{SW}^{1/2}\ , \tag{42}$$

where $f(\varphi, n)$ depends on the hold-up and the power-law index. These results are shown in Figs. 9, 10 and 11.

The effect of viscoelasticity on the mass transfer coefficient has been analyzed by Moo-Young and Hirose[112)].

3.6.2 Rigid-Surface Particles

For solids and very small bubbles (less than 2 mm in diameter) or bubbles contaminated with surface active material, rigid interfacial behavior may be observed. Drag coefficients for rigid-surface single bubbles in power-law fluid can be obtained from the results of Tomita[169)], Wasserman and Slattery[173)], and Nakano and Tien[123)]. The mass transfer coefficients are obtained from the results of Wellek and Huang[174)], and Acharya et al.[1)]. The corresponding cases for solid particles have been analyzed by Moo-Young and Hirose[112)], who showed that an additional effect of "interfacial slip" from additives can occur in practice.

In the case of bubbles swarms with immobile interfaces, results can be obtained from solid-sphere calculations. Mohan and Raghuraman[106] have found upper and lower bounds on the drag force for motion in power-law fluids, and expressions for the stream function. Using these results Bhavaraju et al.[13,14] have obtained values of the mass transfer coefficients for varying gas hold-ups and power-law indices.

3.7 Effect of Bulk Mixing Patterns

In addition to the determination of the mass transfer coefficient, k_L, and the interfacial area, a, the development of gas- and liquid-phase mass balance equations for the species transferred depends on the flow behavior of both gas and liquid phases.

In low viscosity liquids it is reasonably well established that in small stirred-tanks the liquid phase can be considered to be "perfectly mixed"[99,176]. Under these conditions, the gas phase has also generally been assumed to be well mixed in tanks operating above a critical impeller speed[60]. In large tanks, however, the situation is less clear, and care must be taken to establish the behavior of both phases. In cases where the degree of gas absorption is high, the assumptions of well mixed or plug-flow of the gas phase may predict gas absorption rates which differ by a significant order of magnitude. It may thus be necessary to model both the gas and liquid phase behavior.

Russell et al.[32,151] present design equations for simple models of gas and liquid flows. For the case of well mixed gas and liquid phases, the coupled sets of mass balances are

a) gas

$$Q_1 y_0 - Q_2 y - K_G a P V_L \left(y - C \frac{H}{P} \right) = \frac{\mathrm{d}}{\mathrm{d}t} \left(\frac{P V_G}{RT} y \right), \tag{43}$$

b) liquid

$$F(C_0 - C) + K_G a P V_L \left(y - C \frac{H}{P} \right) - r V_L = \frac{\mathrm{d}}{\mathrm{d}t} (V_L C). \tag{44}$$

For the case of plug flow of the gas and a well mixed liquid phase:

a) gas

$$- K_G a P V_G \left(y - C \frac{H}{P} \right) - v_B \frac{\partial}{\partial_z} \left(\frac{P V_G}{RT} y \right) = \frac{\partial}{\partial t} \left(\frac{P V_G}{RT} y \right), \tag{45}$$

b) liquid

$$F(C_0 - C) + K_G a P V_L \left(\bar{y} - C \frac{H}{P} \right) - r V_L = \frac{\mathrm{d}}{\mathrm{d}t} (V_L C). \tag{46}$$

where the mean mole fraction $\bar{y}$ is given by

$$\bar{y} = \frac{1}{H_T} \int_0^{H_T} y \, \mathrm{d}z. \tag{47}$$

It is not always clear from literature values of $k_L a$ what assumptions concerning the flow behavior of either phase have been made, and whether the appropriate mass balance equations have been developed. If RTD information is available, overall dispersion coefficients may be incorporated into the mass balances for both phases. In particular, when $k_L a$ values have been determined by techniques such as sulfite oxidation, the behavior of the gas phase is critical, and care is needed in deciding whether the literature data reported are meaningful.

4 Intraparticle Mass Transfer

4.1 General Concepts

In some biochemical systems the limiting mass transfer step shifts from the gas-liquid or solid-liquid interfaces (Sect. 3) to the interior of solid particles. The most important classes of systems where this situation is recognized to occur is in solid-substrate matrices and compacted cell aggregates (such as microbial flocs, cellular tissues and immobilized whole cells as in trickle-bed reactors, and mold pellets) and in immobilized enzymes (gel-entrapped or supported in solid matrices). In the former, diffusion of oxygen (or other nutrients) through the particle limits the metabolic rates of the cells (Sect. 4.2) while in the latter substrate, reactant or product diffusion into or out of the enzyme carrier often limits the rate of reaction at the active sites (Sect. 4.3).

Approximating the particle to a sphere, a generalized mass balance for the above scenarios is considered.

Under steady-state conditions, the nutrient diffusion rate into a volume element will be equal to the nutrient consumption rate in this volume.

$$\frac{4}{3}\pi r^3 Q = 4\pi r^2 D_r \left.\frac{dC}{dr}\right|_r , \qquad (48)$$

where Q = specific nutrient consumption rate of the particle

D_r = nutrient diffusivity in the particle

$\left.\frac{dC}{dr}\right|_r$ = nutrient concentration gradient at radius r.

Provided Q is independent of the dissolved nutrient concentration, integration yields the equation for the nutrient concentration profile in a particle:

$$C_r = C_R - \frac{QR^2}{6D_r}\left(1 - \frac{r^2}{R^2}\right), \qquad (49)$$

where C_r, C_R = nutrient concentration at radius r and R, respectively. (This assumption may not be very realistic and will be relaxed in the next section.)

When the nutrient concentration in the center of the particle falls below a critical nutrient concentration C_T, nutrient deficiency will occur. The apparent critical nutrient concentration C_A can then be calculated from

$$R = \left(\frac{6D_r}{Q}\right)^{1/2} (C_A - C_T)^{1/2} . \qquad (50)$$

This equation indicates that for large particles (e.g. mold pellets) it is difficult to obtain adequate nutrient supply in the interior of the particle although the dissolved nutrient (e.g. oxygen) concentration in the medium may be high. The effect of the dissolved nutrient concentration on the nutrient uptake rate of the particle can be calculated as follows:

$$C_R = \frac{QR^2}{6D_r}\left(1 - \frac{r_p^2}{R^2}\right), \tag{51}$$

where r_p = radius at which $C = 0$.
For limiting nutrient conditions, the fraction of the particle that has adequate nutrient will be:

$$\frac{\frac{4}{3}\pi R^3 - \frac{4}{3}\pi r_p^3}{\frac{4}{3}\pi R^3} = 1 - \left(\frac{r_p}{R}\right)^3. \tag{52}$$

Assuming that 1) the microbe will grow but that the particle size remains constant (due to constant shear in the liquid medium), 2) the cell concentration within the particle is constant (where applicable), 3) Q is a known function of C (e.g. $Q = K \cdot C$ where K = constant), it can be shown that the mass transfer rate for the nutrient is given by

$$\varphi_m = 4\pi R^2 \left.\frac{dC}{dr}\right|_{r=R} = 4RD_rC_s\left(-1 + R\sqrt{\frac{K}{D_r}}\coth R\sqrt{\frac{K}{D_r}}\right). \tag{53A}$$

This equation can be simplified for the following cases:

a) $R\sqrt{\frac{K}{D_r}} < 0{,}3$

$$\text{then } \varphi_m = \frac{4}{3}\pi R^3 K C_s \tag{53B}$$

and the concentration within the particle is equal at all positions.

b) $R\sqrt{\frac{K}{D_r}} > 2$ (fast reaction),

$$\text{then } \varphi_m = 4\pi R D_r C_s\left(-1 + R\sqrt{\frac{K}{D_r}}\right), \tag{53C}$$

c) $R\sqrt{\frac{K}{D_r}} > 20$ (very fast reaction),

$$\text{then } \varphi_m = 4\pi R^2 C_s\sqrt{KD_r}\,. \tag{53D}$$

It is seen that in cases (b) and (c) mass transfer resistance within the particle becomes important.

4.2 Oxygen Transfer in Mold Pellets

Marshall and Alexander[98)] discovered that for several pellet-forming fungi a "cube root" growth curve fits their data significantly better than the "standard" exponential growth model. Pirt[134)] suggested that this was probably due to the effects of intra-particle diffusion: a nutrient was not diffusing into the particle fast enough to maintain unrestricted growth of the mold pellets. It was soon realized that oxygen was this limiting nutrient.

Phillips[133)] has measured oxygen diffusion in pellets of *Penicillium chrysogenum* by first assuming that diffusion is the mechanism that supplies oxygen to the interior of the microbial pellet and that the mass transfer resistance outside the pellet is comparatively small. Yano et al.[179)] and Kobayashi et al.[85)] did the same with *Aspergillus niger* pellets. Taking into account the effect of intra-particle diffusion, Kobayashi and Suzuki[82)] were able to characterize the kinetic behavior of the enzyme galactosidase within mold pellets of *Mortierella vinacea*.

Following Aiba and Kobayashi[3)] an oxygen balance on a spherical shell of a mold pellet yields

$$\frac{\partial C}{\partial t} = D_r\left(\frac{\partial^2 C}{\partial r^2} + \frac{\partial C}{\partial r}\right) - \varrho_m Q \tag{54}$$

where C = local oxygen concentration
r = radius within the sphere
D_r = molecular diffusivity of oxygen within the pellet
Q = specific respiration rate at C
ϱ_m = density of mycelia (assumed constant with r).

Noting that the respiration reaction in pellets is given by the Michaelis-Menton equation then[85)]:

$$Q = Q'_{\max}\frac{C}{K_m + C} = \bar{Q}\left(\frac{K_m + \bar{C}}{\bar{C}}\right)\left(\frac{C}{K_m + C}\right), \tag{55}$$

where $Q'_{\max} = \bar{Q}\left(\frac{K_m + \bar{C}}{\bar{C}}\right)$ = maximum value of specific respiration rate at C

$\bar{Q}$ = specific respiration at $C = \bar{C}$
$\bar{C}$ = dissolved oxygen concentration in the bulk media
= dissolved oxygen concentration at the pellet surface $r = R$ (*assumed*)
K_m = Michaelis constant.

At steady state, substitution of Eq. (55) into Eq. (54) gives

$$D_r\left(\frac{\partial^2 C}{\partial_r 2} + \frac{2}{r}\frac{\partial C}{\partial r}\right) = \varrho_m Q_{\max}\left(\frac{K_m + \bar{C}}{\bar{C}}\right)\left(\frac{C}{K_m + C}\right). \tag{56}$$

Eq. (56) cannot be solved analytically in this form without simplifying assumptions. Yano et al.[179)] and Phillips[133)] obtained a solution by assuming the respiration rate to be independent of the dissolved oxygen level. This assumption reduces the right-hand side of Eq. (56) to $\varrho_m Q_{\max}$ and allows direct integration of the equation. This analysis has led to the concept of a "critical radius" within the pellet where the dissolved oxygen concentration falls to zero thus leading to anaerobiosis. This discontinuity however fails to satisfy the differential equation for the concentration distribution.

Yoshida et al.[184] considered the respiration rate to be uniform within the pellet and hence obtained a solution.

Aiba and Kobayashi[2] assumed that $K_m/\bar{C} = 1.0$ and then solved the equation for this special case using Gill's modification of the Runge-Kutta technique on a digital computer. Bhavaraju and Blanch[11] applied the same technique to obtain solutions for values of $K_m/\bar{C}$ ranging from 0.01 to 100.

Kobayashi et al.[85] obtained solutions by defining a general modulus (ξ) and then calculating the effectiveness factor using an approximate equation. The error in approximation is claimed to be very small and avoids the use of a computer.

Effectiveness factors are widely used for intra-particle diffusion in porous catalyst pellets. Within the context of a mold pellet it can be defined as follows:

$$E_f = \frac{\text{total reaction rate with diffusion limitations}}{\text{reaction rate if all substrate is available at } \bar{C}} .$$

When $E_f = 1$, the effect of intra-particle diffusion on the overall reaction rate is negligibly small. When E_f is less than 0.5, intra-particle diffusion becomes significant.

Kobayashi et al.[85] studied three cases:

a) Uniform respiration activity throughout the mycelial pellets,
b) respiration activity as a function of age distribution within the pellet,
c) respirative activity adaptation to the local oxygen concentration within the pellet.

In case (a), the effectiveness factor (E_f) is simply the ratio of the specific respiration rate of a pellet (Q) to the respiration rate of well dispersed filamentous mycelia ($\bar{Q}$).

Theoretical and experimental results are given in Fig. 12. It is seen that the three cases considered give similar results and it is difficult to discriminate between them by use of the limited experimental data. It should be noted however that the data of Kobayashi et al.[85] agree more closely with the theoretical solutions than those of Yano et al.[179].

While the studies referred to in this section have greatly expanded our understanding of oxygen transfer in mold pellets some of the results should be received with caution for the following reasons:

a) All of the researchers assume that the oxygen concentration at the pellet surface is equal to the oxygen concentration in the bulk liquid. This assumption has not been verified and is likely to be significantly incorrect in some cases.
b) The diffusivity of oxygen in the pellet is considered to be independent of the density of the pellet. As this density may vary, this assumption can become invalid.
c) In the experimental methods used, no precautions were taken to prevent oxygen starvation at the heart of the pellet. If this occurred, the respiration rate would be lower than that predicted by the model.
d) In the work of Kobayashi et al.[85], a density function obtained by Yano et al. for "small" pellets was assumed to be applicable to "larger" pellets. While

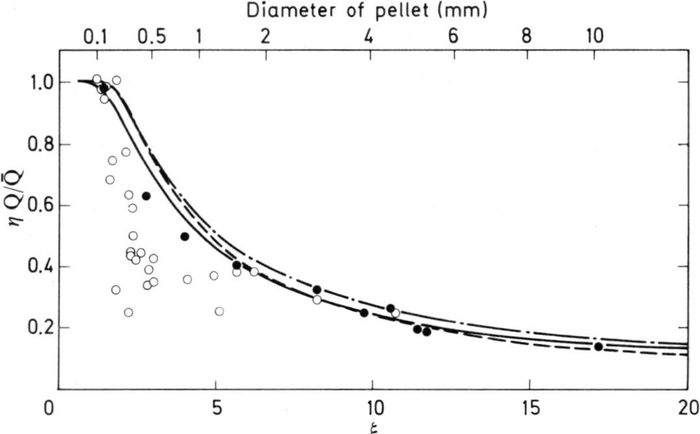

Fig. 12 Oxygen transfer in mold pellets. Comparison between theoretical curves relating η to ξ for cases a (·—·—·—), b (---------) and c (———). (○) Data of Yano et al.[179], (●) Data of Kobayashi et al.[85], C = 1.9×10^{-4} (μmol mm^{-3} O_2); $K_m = 3.0 \times 10^{-6}$ (μmol mg^{-1} min^{-2} O_2).
$\xi = R\sqrt{\varrho_m \bar{Q}/2D_r\bar{C}}$

this may be true, is has yet to be verified.

e) The studies done by Yoshida et al.[184] are highly questionable as the oxygen consumption measurements appear to have been performed in oxygen-saturated salt solutions without any nutrients. The oxygen consumption under these conditions would be much less than that during growth.

For those systems where Fig. 12 is applicable the usefulness of the diagram is clearly apparent. For a mold processes where the pellets are of identical diameter and the various constants are known (ϱ_m, $\bar{Q}$, D_r, $\bar{C}$) the ratio $Q/\bar{Q}$ is obtained directly from the diagram. The respiration rate of the culture (Q) can then be easily calculated. For a slightly more realistic system where there is a known distribution of pellet sizes which changes in a predictable manner with time, an analogous but more complicated calculation technique can be followed. Many different moments in time are examined separately. The size distribution for a given point in time is divided into many small intervals. The respiration rate for each of these intervals is obtained separately from Fig. 12 and then an overall weighted average is calculated.

After obtaining the respiration rate of the culture by this procedure, the engineer may wish to change the oxygen concentration in the bulk liquid. For instance, if intra-particle diffusion is limiting, then increasing $\bar{C}$ will reduce this effect. The bulk oxygen concentration can be increased by raising the k_L a value on the supply side (Sect. 8). The most effective and obvious way of minimizing the effect of intra-particle diffusion however is to keep the size of the pellets small; in practice, this may be achieved by judicious use of shear forces and/or surfactants in the medium[43].

4.3 Immobilized Enzyme Kinetics

Intra-particle diffusion can also have a significant effect on the kinetic behavior of enzymes immobilized on solid carriers or entrapped in gels. In their basic analysis of this problem, Moo-Young and Kobayashi[114)] made the following simplifying assumptions:

a) The enzyme membrane can be represented by a slab of width $2\,L$ and cross sectional area A.
b) There is a partition of the substrate and product between the membrane and the external solution.
c) The mass transfer resistance between the bulk solution and the membrane surface is negligible.
d) Enzymatic activity is uniform through the membrane.
e) Fick's law applies and the diffusivity of substrate and product is constant.
f) The reaction involves a single substrate.
g) The system is at steady state.

At steady state, the following mass balances on a differential section of the slab applies

$$D_s \frac{d^2S}{dl^2} - r(s) = 0\,, \tag{57A}$$

$$D_p \frac{d^2P}{dl^2} + r(s) = 0\,,$$

where S refers to the substrate and P to the product; $r(s)$ is the local rate of the enzymatic reaction.

Depending on the appropriate kinetics for the system in question, $r(s)$ may assume one of the following expressions:

Case (a): Michaelis-Menten type reaction

$$r(s) = \frac{V_m S}{K_m + S}\,. \tag{57B}$$

Case (b): Substrate inhibition (non-competitive type)

$$r(s) = \frac{V_m S}{(S + K_m)\left(1 + \frac{S}{K_i}\right)}\,. \tag{57C}$$

Case (c): Product inhibition (competitive type)

$$r(s) = \frac{V_m S}{(S) + (K_m)\left(1 + \frac{P}{K_i}\right)}\,. \tag{57D}$$

Case (d): Product inhibition (non-competitive type)

$$r(s) = \frac{V_m S}{(S + K_m)\left(1 + \frac{P}{K_i}\right)}\,. \tag{57E}$$

Moo-Young and Kobayashi derived a generalized, dimensionless form of Eq. (57A):

$$\frac{d^2y}{dz^2} = 2h_0^2 \frac{y}{(\beta_1 + \beta_2 y + \beta_3 y^2)} . \tag{57F}$$

Depending on how the parameters are assigned, this equation will reduce to one of the four cases given above. For example, for case (b)

$$\beta_1 = \alpha_1 = \frac{K_m}{S_i} ; \quad \beta_3 = \alpha_2 = \frac{S_i}{K_i} ; \quad \beta_2 = 1 + \alpha_1\alpha_2 ;$$

$$y = \frac{S}{S_i} ; \qquad z = \frac{l}{L} ; \qquad h_0 = \sqrt{\frac{V_m}{2D_s S_i}} L . \tag{57G}$$

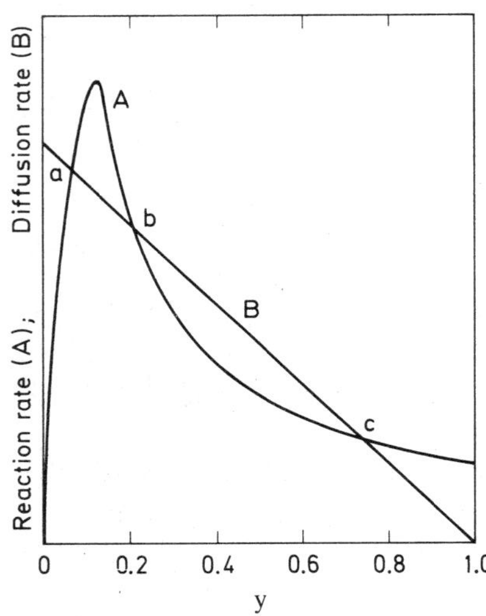

Fig. 13 Reaction rate and diffusion rate profiles for case (b)[114)]

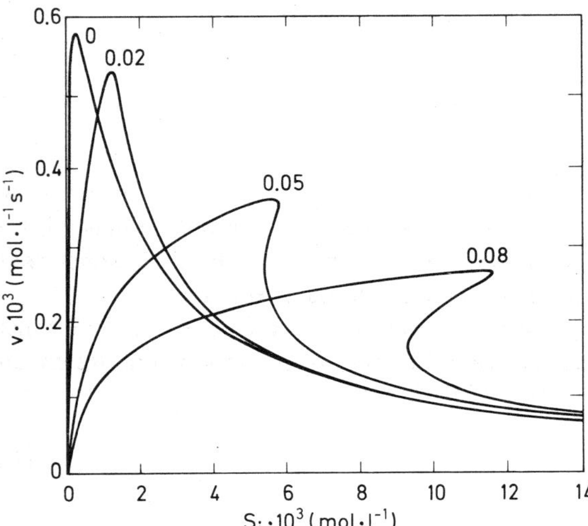

Fig. 14 Overall reaction rate as a function of substrate concentration for various widths of an artificial membrane for case (b)[114)]
$V_m = 10^{-3}$ (mol · l^{-1} s^{-1}), $K_m = 10^{-4}$ (mol · l^{-1}), $K_i = 10^{-3}$ (mol · l^{-1}), $D_s = 10^{-6}$ (cm^2 s^{-1}), parameter L (mm)

A general modulus and effectiveness factor can be defined in a similar manner to that for mold pellets. Eq. (57F) is then solved with the appropriate boundary conditions by means of a computer. Some results for case (b), which occurs frequently in practice, are given in Fig. 13 and 14. In Fig. 13 the reaction rate and the diffusion rate are plotted against a dimensionless concentration at a point within the membrane. The three points of intersection (a, b, and c) between the curve and the straight line represent possible steady states for the system. However, point (b) is unstable because a slight shift in concentration will direct the reaction to point (a) or (c).

Fig. 14 shows the overall reaction rate as a function of substrate concentration for various widths of artifical membrane (case b). When the membrane is thin, the instability problem does not occur. This problem becomes more and more severe with increasing membrane thickness.

While these results are very interesting it should be remembered that they are purely mathematical without direct experimental results for support. Also the seven assumptions listed in p. 27 limit the applicability of these results. (Some of the criticisms listed in Sect. 4.2 for oxygen diffusion in mold pellets are also applicable here.)

For immobilized-enzyme packed-bed reactors (the type usually used in practice), the effect of bulk mixing on inter-particle mass transfer should not be overlooked. Around each bound enzyme particle, there is an unmixed liquid film through which nutrients must diffuse from the bulk liquid. Depending on the degree of mixing in the bulk liquid, this inter-particle resistance can be important. Then, as has already been discussed, there is also a resistance to mass transfer within the enzyme itself (i.e. intra-particle).

Kobayashi and Moo-Young[84)] examined this situation with immobilized invertase on ion-exchange resin beads. Their experimental apparatus approximated a plug-flow reactor of the packed-bed type. They found that resistance to inter-particle diffusion becomes apparent at low flow rates but virtually disappears as the flow rate increases. They also detected that intraparticle diffusion becomes important in large particles.

4.4 Enzymatic Degradation of Insoluble Substrates

When the substrate in a biochemical reactor is a water-insoluble material (e.g. cellulose), the effects of intra-particle mass transfer may also be significant. In such systems, extracellular enzymes can break specific molecular bonds of the substrate, eventually producing water-soluble "substrate fragments" which may then be consumed by micro-organisms in the bulk liquid medium.

If the substrate is sufficiently porous, the enzyme can diffuse into it and hence degradation can proceed inside the substrate. The water-soluble substrate fragments however must also diffuse out of the solid matrix through the same pores into the bulk solution where they are still subject to enzymatic attack. The reaction can, of course, proceed at the exterior of the substrate surface and, indeed, for substrates of low porosity this is where much of the degradation takes place.

The analysis of this situation, although somewhat more complicated, is similar to that described in Sect. 4.2 and 4.3. Once again, utilization of the effectiveness factor and general modulus is convenient in solving the differential equations.

Suga et al.[160)] provided theoretical and experimental results concerning, the enzymatic breakdown of cross-linked dextran ("Sephadex") caused by a dextranase from *Penicillium funiculosum.* They found that substrate degradation proceeded at a higher rate for large radii (Fig. 15); consequently, the substrate concentration in the bulk liquid is also higher. The reaction principally occured within the particles when the pore radius is small ($\leqq 0.32 \times 10^{-6}$ cm). As the particle radius becomes large for a constant (small) pore radius, the rate of degradation becomes slower and slower, indicating the effect of intra-particle diffusion. However, the effect of the particle radius is not as significant as with larger pore radii (0.56×10^{-6} cm).

One of the factors not included in their analysis was the variation of diffusivity related to the molecular weight distributions of released substrate fragments.

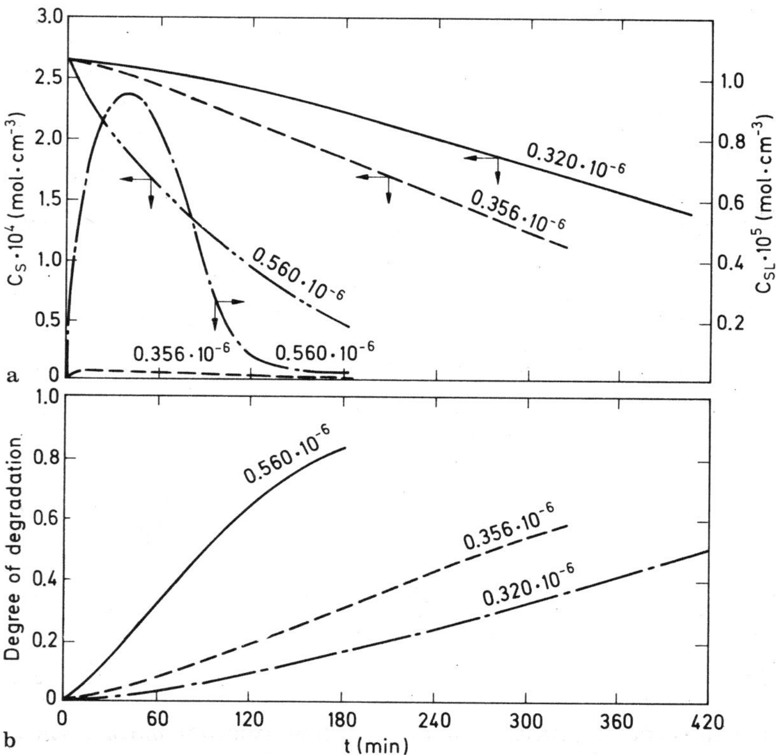

Fig. 15a Illustration of the effect of pore radius on the change in substrate concentration in the particle C_S and in solution C_{SL}: $V_m = 0.100 \times 10^{-7}$ (mol cm^{-3} s^{-1}), $K_m = 0.952 \times 10^{-6}$ (mol cm^{-3}), $R = 0.01$ (cm). Parameter: pore radius, γ (cm).
b Illustration of the effect of pore radius on the degree of degradation[160)]

5 Physical Properties of the Process Materials

5.1 Rheological Properties

The rheological properties of the materials being processed in biochemical reactors will influence the power consumption and the heat and mass transfer rates. These properties are particularly important in some antibiotic processes and systems involving semi-solid media because of the very viscous and frequently non-Newtonian behavior of the materials.

Two main classes of fluid behavior have been encountered in biochemical reactors:

1) Purely viscous fluids where
shear stress = f (shear rate)

$$\tau = f(\dot{\gamma}) , \tag{58}$$

a) Newtonian fluids

$$\frac{\tau}{\dot{\gamma}} = \mu \, (= \text{constant}) , \tag{59}$$

b) non-Newtonian fluids

$$\frac{\tau}{\dot{\gamma}} = \mu_a \neq \text{constant} , \tag{60}$$

2) viscoelastic fluids

$$\gamma = f(\tau, \text{extent of deformation}) .$$

Fortunately, the second class occurs infrequently; a rare example is bread dough.

There are three sub-classes of time-independent non-Newtonian fluids (the non-Newtonians being most frequently found in bioreactors). These include:

(a) pseudoplastics, (b) dilatants, (c) viscoplastics (e.g., Bingham plastics).

The apparent viscosity (μ_a) decreases with increasing shear rate of pseudoplastics and Binghams but increases for dilatants. Blanch and Bhavaraju[18)] have reviewed most of the literature available on the rheological behavior of fermentation broths. Most non-Newtonian broths follow the power-law model

$$\tau = K(\dot{\gamma})^n , \tag{61}$$

where K is the "consistency coefficient" and n the "flow behavior index". Thus, for pseudoplastics, $n < 1$, and for Newtonians $n = 1$, while for dilatants, $n > 1$. This behavior is illustrated in Fig. 16. Power-law models are very useful from an engineering standpoint, especially when compared to the non-linear and unquantified multi-parametric equations which have been developed from molecular considerations[156)]. However, they fail to predict the Newtonian behavior frequently observed at very high and very low shear rates and the equations are not dimensionally sound.

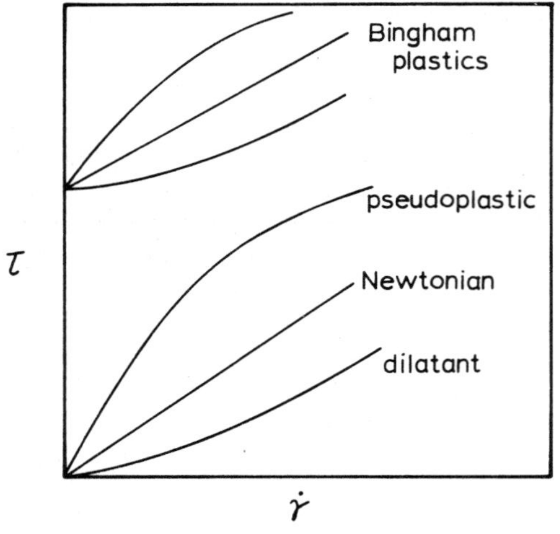

Fig. 16 General shear behavior of rheologically time-independent fluid classes

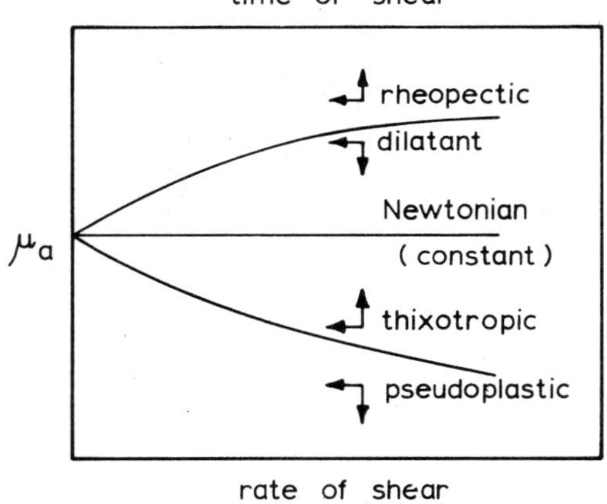

Fig. 17 Generalized variations of the apparent viscosity ($\mu = \tau/\dot{\gamma}$) of non-Newtonian fluids according to generic classifications

There are two types of time-dependent non-Newtonians:

a) thixotropic fluids which show a reversible decrease in shear stress with time at constant rate of shear and fixed temperature,

b) rheopectic or antithixotropic fluids which display an opposite effect and occur rarely.

These two effects can occur in the case of pseudoplasticity and dilatancy respectively with shear rate and time having analogous effects on μ_a (Fig. 17). In process operations, these time dependencies are only important in start-ups and systems perturbations.

Viscoelastic fluids exhibit both viscous and elastic properties. In an elastic solid, the stress corresponding to a given strain is independent of time whereas for viscoelastic substances, the stress will gradually dissipate. In contrast to purely viscous liquids, viscoelastic fluids flow when subjected to stress but part of their deformation is gradually recovered upon removal of the stress. Viscoelasticity can be modelled by a combination of Newtonian viscosity and Hookean elasticity. In this way, mechanical models have been devised[5].

The rheological parameters for power-law non-Newtonians (Eq. (61), p. 31) can be evaluated from viscometric measurements on variable shear-rate viscometers. For Couette viscometers, which are probably most commonly used, it should be noted that the values of shear rates supplied by the instrument manufacturers are almost invariably for Newtonian fluids. These values must be multiplied by a correction factor for application to non-Newtonians as shown by Calderbank and Moo-Young[28,29]:

$$C_R = 1 + \frac{S^2 - 1}{2S^2}\left(\frac{1}{n} - 1\right)\left(1 - \frac{2}{3}\ln S + \frac{1}{3}\alpha - \frac{1}{45}\alpha^3 + \frac{2}{945}\alpha^5 - \frac{1}{4725}\alpha^7 \ldots\right), \tag{62}$$

where $\alpha = \left(\frac{1}{n} - 1\right)\ln S$.

For these viscometers, the operation with relatively large volumes of liquid, when container wall-effects are insignificant, yields a correction factor equal to the reciprocal of n.

In applying equations relating power consumption to the impeller Reynolds number, or in bubble-columns where the bubble size depends on liquid properties, an apparent viscosity is usually used. The apparent viscosity in the immediate vicinity of an impeller is given by[29]:

$$\mu_a = K\dot{\gamma}^{n-1} = \frac{K}{(BN)^{1-n}}\left(\frac{3n+1}{4n}\right)^n, \tag{63}$$

where B is a geometric parameter ($= 11$) for the usual bioreactor stirred-tank conditions with $n < 1$ and $(T/D) > 1.5$. Other relationships have been reported by Skelland[156] and Metzner[102].

5.2 Basic Dispersion Properties

It is clear that since the maximum value of the concentration-driving force for mass transfer is limited (due to its low solubility), the oxygen transfer rate from the gas bubble to the medium is largely determined by k_L (which is dependent on bubble diameter) and the interfacial area a. The main variables which influence a are the bubble size (d_B), the terminal velocity of the bubble (U_B) and the hold-up (φ).

Dispersions of bubbles in reaction mixtures are subjected to shear (e.g. by mechanical agitation) which may be involved when high interfacial areas and good mixing are required. In aerobic reactors, the sparger design does not generally determine the bubble size which depends on the eventual bulk level of turbulence in the continuous phase. For example, small bubbles which are formed from tiny orifices such as sintered glass may coalesce to form large bubbles eventually in the medium if there are inadequate mixing and/or ineffective surfactants present.

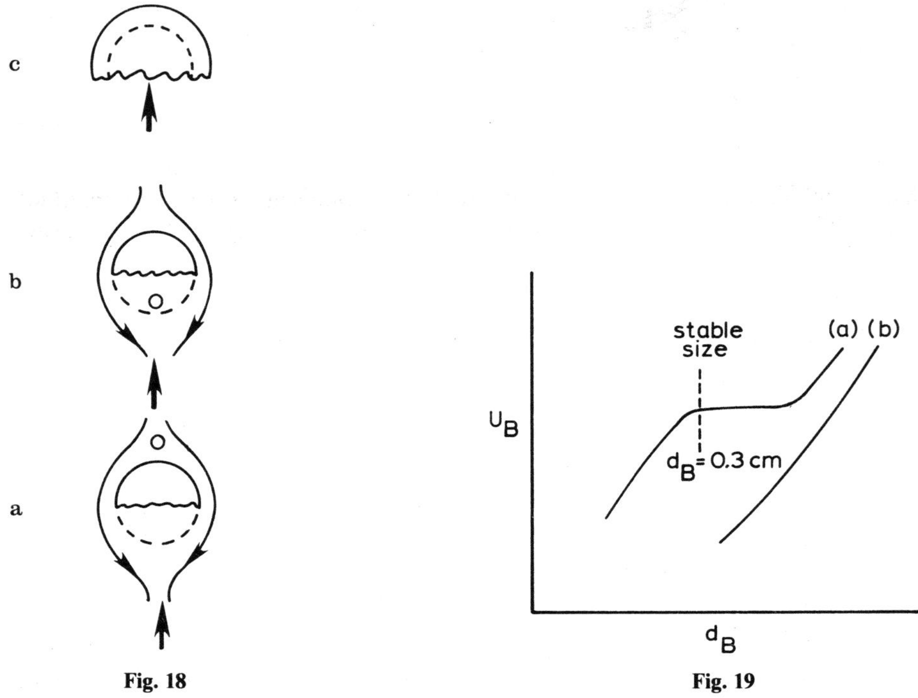

Fig. 18

Fig. 19

Fig. 18 Process involved in self-accelerating bubble coalescence: **a** the bubble is transported in the wake behind the large bubble, **b** the bubble is accelerated in the hemispherical cap bubble, **c** the smaller bubble is captured, increasing the overall bubble volume

Fig. 19 Bubble rising velocity at various bubble sizes for: **a** non-viscous, **b** viscous liquids. The plateau region on curve **a** stabilizes bubble size. Absence of a plateau on curve **b** results in continual coalescence as the bubble rises through fluid

In viscous liquids (> 35 cp.) large free-rising bubbles will coalesce if their surfaces are brought within a distance of about one bubble diameter of each other, since they carry behind them a wake of dimension approximating their own size[124]. Thus, in these types of media, rapid coalescence may take place due to the fact that a bubble in the wake of a preceding one increases its rise velocity, and collision occurs. This self-accelerating coalescence mechanism leads to the formation of fast rising spherical cap bubbles, causing deterioration of the gas dispersion. This phenomenon is illustrated in Fig. 18. Large spherical cap bubbles are stable in viscous solutions because of their low Reynolds numbers, but cannot exist in less viscous liquids where the liquid *Re*-numbers would be such as to render them unstable.

Typical patterns of rise velocities for bubbles in aqueous and viscous liquids are shown in Fig. 19. It should be noted that these aqueous solutions show a unique plateau regime in rise velocity which partly explains the bubble size stability in these systems. The rise velocity depends on the bubble volume and its interfacial mobility. Small, rigid interface bubbles follow Stokes' equation

$$U_B = \frac{g\varrho}{18\mu} d_B^2 , \tag{64}$$

which is valid for $Re < 1$, e.g. creeping flow conditions. When the interface becomes mobile, a jump in the rise velocity can be observed when the fluid is elastic. Here, the Hadamard-Rybczynski equation predicts

$$U_B = \frac{g\varrho}{16\mu} d_B^2 . \tag{65}$$

At higher bubble Reynolds numbers, the rise velocity can be obtained from the Mendelson wave analogy

$$U_B = \sqrt{\frac{2\sigma}{d_B} + \frac{g d_B}{2}} . \tag{66 A}$$

When the gravity stresses are higher than the surface tension stresses

$$U_B = \sqrt{\frac{g d_B}{2}} , \tag{66 B}$$

which agrees well with the Taylor-Davies prediction stating that

$$U_B = 1.02 \sqrt{\frac{g d_B}{2}} . \tag{67}$$

For a discussion of the effects of imposed mixing forces, see Sect. 5.4.

5.3 Gas Flow Effects on Bubble Swarms

There is a fairly extensive literature available on the behavior of gas-sparged systems, with height-to-diameter ratios ranging from unity to twenty or more[148)]. Mass transfer coefficients in these systems may depend on system geometry, sparger design and gas-flow rate. Less information is available on the influence of viscosity on bubble behavior. Due to the density difference between the gas and liquid phases, the rate of mass transfer is primarily determined by the force of gravity. In a mechanically agitated tank, turbulence forces may exceed those of gravity and would determine mass transfer rates. In gas-liquid dispersions, however, it is generally difficult to exceed gravitational forces, as agitators operate poorly under these conditions.

Provided the bubble size can be determined from the operational conditions in the vessel, its rise velocity and the superficial gas velocity will essentially determine k_L and a. In a sparged tank, there are several factors which may influence the gas bubble size. The first of these is the bubble size generated at the sparger, which is a function of the gas flow rate and sparger diameter. The bubbles so generated may either coalesce or break up, these being functions of the bulk liquid properties away from the orifice. Thus, two regions in any gas sparged tank need to be considered: an orifice region and a bulk liquid region where break-up and coalescence may occur.

5.3.1 Bubble Sizes Generated at an Orifice

In liquids with viscosities of the order of 1 *cp*, Miller[105] and Sideman et al.[154] provide equations for the determination of the bubble size as a function of gas flow rate. The orifice diameter (d_0) only influences the bubble size strongly at very low rates, where the bubble size is found by equating surface tension and buoyancy forces

$$d_B = \left[\frac{6\sigma d_0}{g(\varrho_L - \varrho_G)}\right]^{1/3}. \tag{68}$$

The gas rates for which this equation is valid are too small to be of practical interest. At moderately high gas rates in aqueous systems, Davidson[41] correlates bubble diameter with gas flow rate, Q, using orifices ranging in size from 0.1 to 1.0 cm:

$$d_B = 0.19\, d_0^{0.48}\, Re_0^{0.32}, \tag{69}$$

where

$$Re_0 = \frac{4Q\varrho_G}{\pi\, d_0 \mu_G},$$

while Leibson et al.[88] propose

$$d_B = 0.18 d_0^{1/2}\, Re_0^{1/3} \quad \text{for} \quad Re_0 < 2000. \tag{70}$$

Davidson and Schuler[40] and Kumar and Kuloor[87] provide models for bubble growth and detachment which include the effects of liquid viscosity. Explicit relationships between bubble size and gas flow rate and gas and liquid properties have however not been obtained, except at very low gas rates (0.25 to 2.5 $cm^3\, s^{-1}$)

$$V_B = \left(\frac{4\pi}{3}\right)^{1/3} \left(\frac{15\mu_L Q}{2\varrho_L g}\right)^{3/4}. \tag{71}$$

Bhavaraju, Russell and Blanch[14] have examined the formation of bubbles at an orifice in liquids with apparent viscosities ranging from 1 to 1,000 cp. Beyond the transition gas flow rate for which Eq. (68) is valid, the bubble size is given by

$$\frac{d_B}{d_0} = 3.23\, Re_{0L}^{-0.1}\, Fr_0^{0.21}, \tag{72}$$

where the Reynolds number is based on liquid rather than gas properties

$$Re_{0L} = \frac{4\varrho_L Q}{\pi \mu_L d_0}, \qquad Fr_0 = \frac{Q^2}{d_0^5 g}. \tag{73}$$

This correlation agrees well with those of Leibson et al. and Davidson, but extends the viscosity range covered. Pseudoplastic power-law fluids were examined, and the gas flow rates covered ranged up to velocity of sound through the orifice. This equation is recommended for design purposes.

At higher gas flow rates, there is a transition from the formation of discrete gas bubbles to jetting. Leibson et al.[88] relate the onset of jetting to the orifice Reynolds numbers (Re_0, based on gas properties) greater than 2000. For $Re_0 > 10{,}000$ they propose a weak dependence of the equilibrium bubble size (far from the orifice) on the Reynolds number

$$d_{Be} = 0.71\ Re_0^{-0.05}\ (d_{Be} \text{ in cm})\,. \tag{74}$$

This equilibrium size is approximately 0.45 cm being determined by break-up and coalescence processes. The jetting criterion ($Re_0 > 2000$) was determined only for systems of low viscosities, and does not apply to more viscous liquids. The break-up process appears to be related to liquid-phase velocities and is essentially independent of gas properties.

5.3.2. Bubble Size Far from the Orifice

In the region of the tank away from the orifice, the bubble size may vary, depending on the liquid properties and the liquid motions generated by the rising gas stream.

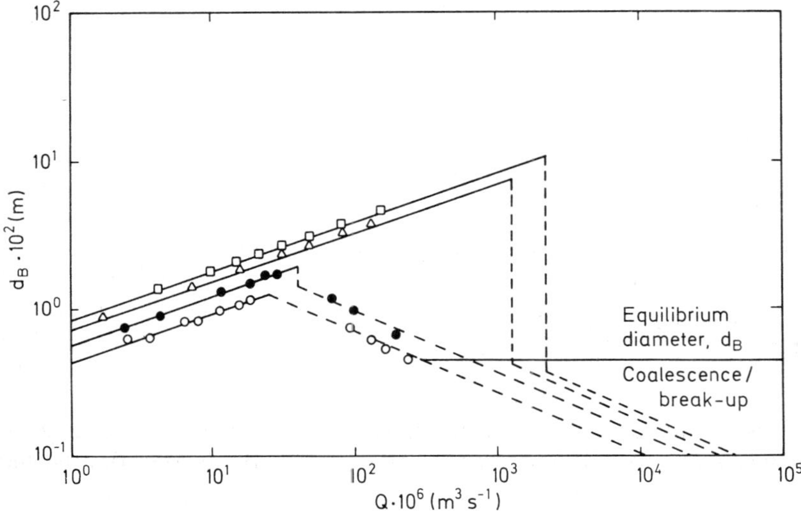

Fig. 20 Bubble size far from the orifice in a bubble column. Bubble break-up is absent in 0.15% and 0.20% carbopol solutions, as liquid circulation is laminar. ○ water, ● 0.10% carbopol, △ 0.15% carbopol, □ 0.20% carbopol[14]

If the power input from the gas phase is insufficient to generate turbulence in the liquid phase, the bubble size in the tank will be that of bubbles formed at the orifice, and may increase with liquid height in the tank due to bubble coalescence. Once the liquid is in turbulent motion, however, bubble break-up will also occur, and an equilibrium between coalescence and break-up will determine the mean bubble size. These effects are illustrated in Fig. 20[14)].

In a gas sparged tank an overall liquid circulation develops due to the density-driven flow of the gas phase. The power dissipated by the rising gas stream (essentially the change in the Gibb's free energy) causes a circulating liquid motion. An energy balance on the gas and liquid phases can be used to determine the liquid velocity. This procedure is detailed by Bhavaraju et al.[14)]. The factors which determine bubble coalescence and break-up are described in the following sections.

5.4 Bubble Coalescence and Break-up

Two extreme cases of bubble motion which may lead to bubble coalescence can be considered. These are bubbles rising through a relatively quiescent liquid phase and bubbles interacting in a turbulent liquid phase, where the liquid velocities in the bubble wake are insignificant when compared to the velocities in the bulk of the liquid. Both situations may be encountered in fermentation practice although the situation with a turbulent liquid phase is more common.

Coalescence of bubbles rising in a line takes place in several stages.

a) the approach of the following bubbles to the vortex region of the leading bubble,
b) the trailing bubble moves in the vortex of the leading bubble until the bubbles are separated only by a thin interface,
c) final thinning and rupture of the film between bubbles.

The first two stages have been examined by Narayanan et al.[124)] using aqueous glycerin solutions with viscosities up to 900 cp. The coalescence process was observed using high-speed photography. Five classes of bubble wakes were classified as a function of the bubble Reynolds number.

In viscous solutions, bubbles are predicted to coalesce at the orifice. Narayanan et al. provide estimates of the distance from the orifice for the interfacial area to be reduced by 25% and 50% as a function of the frequency of bubble formation and liquid viscosity.

The thinning of the thin film between bubbles has been examined in a series of experimental and theoretical papers by Marrucci[96, 97, 125)]. Coalescence of bubbles sitting on two neighboring nozzles has been studied several electrolyte solutions. The thinning and rupture of the film occurs in two stages. The initial film decreases in thickness due to a pressure difference between the liquid of the film and the liquid outside the border of the film. A quasi-equilibrium film thickness then results. The concentration of the surface active material within the film is different from that outside the film and, at the border of the film, a diffusion process starts. The difference in surface tension between the film and the bulk tends to be destroyed, and the film has to stretch more to keep the balance of forces properly satisfied. This diffusion-controlled mechanism at the border of the film is the major cause of further thinning of the film down to rupture.

The predictions of this theory agree well with coalescence times reported for various electrolyte solutions (where rates of coalescence are reduced as the electrolyte concentration and surface tension increase).

Considerably less information is available on coalescence of bubbles in highly agitated liquids where bubble collisions may not lead to coalescence, as the fluid may carry the bubbles apart before coalescence can occur. In electrolyte solutions, coalescence is greatly reduced under these circumstances, and higher gas hold-ups have been reported by a number of authors.

Bubble break-up is caused by the dynamic pressure forces exerted on the bubble by the turbulent liquid field. The ratio of the dynamic pressure to surface tension forces is given by the Weber number

$$We = \frac{\tau d_B}{\sigma}. \tag{75}$$

The dynamic pressure forces can be found from the isotropic turbulence theory

$$\tau \propto \varrho \left(\frac{P}{V} \frac{d_B}{\varrho}\right)^{2/3}, \tag{76}$$

so that at equilibrium the Weber number is constant and a maximum stable bubble size can be predicted from the above equations as

$$d_B \propto \frac{\sigma^{0.6}}{\left(\frac{P}{V}\right)^{0.4} \varrho^{0.2}}. \tag{77}$$

Similar relationships are available for liquid drops. In gas-sparged vessels the power per unit volume can be found from an energy balance on the gas phase[14, 72].

$$\frac{P}{V} = \frac{Q}{V}\left(\frac{P_2}{P_1 - P_2}\right) \ln \frac{P_1}{P_2} \cdot \varrho g H + \eta \frac{V_0^2}{2} \cdot \frac{1}{V}. \tag{78}$$

The first term represents the work done by the expanding gas phase (P_1 and P_2 are the pressures at the bottom and top of the tank, H is the liquid height, and Q the gas flow rate under orifice conditions). The second term accounts for the kinetic energy of the gas, where V_0 is the velocity of the gas in the orifice, and relates to the gas velocity just above the orifice through the efficiency η (generally 0.06). The kinetic energy of the gas leaving the vessel is generally negligible, and the kinetic energy term is generally small for most values of Q. Hence, the equation reduces to

$$\frac{P}{V} = \frac{Q}{V} \varrho g H \left(\frac{P_2}{P_1 - P_2}\right) \ln \frac{P_1}{P_2}. \tag{79}$$

The resulting mean bubble size can be found by inserting this value of P/V into

$$d_{Be} = 0.7 \frac{\sigma^{0.6}}{\left(\frac{P}{V}\right)^{0.4} \varrho_L^{0.2}} \left(\frac{\mu_a}{\mu_G}\right)^{0.1} \tag{80}$$

$(d_{Be}$ in $m)$.

With increasing gas flow, the liquid circulated by the power input from the gas phase will become turbulent. Prior to that point, the break-up of bubbles will occur to a limited extent if very large bubbles, which are formed at the orifice, create sufficient bulk flow. With the onset of turbulence, bubbles larger than the size predicted by Eq. (72) will break up. At very high gas rates, an equilibrium between break-up and coalescence will exist, and the equilibrium bubble size is in general larger than that predicted by Eq. (80). The actual bubble size can be found by comparing the values of d_B obtained from Eq. (80) and the equilibrium size predicted by

$$d_{Be} = 0.71\, Re_0^{-0.05}\, (d_{Be} \text{ in cm})\,. \tag{81}$$

The larger bubble size is the appropriate one. The above equation needs to be modified to account for the effects of ionic strength, surfactants and viscosity on bubble coalescence. With decreasing rates of coalescence, a smaller equilibrium bubble size is obtained. Marrucci[96] provides guidelines for the estimation of ionic strength effects on coalescence.

Part II. Equipment Performance

6 Gas-Liquid Contactors without Mechanical Agitation

6.1 Bubble Columns

Pneumatically agitated gas-liquid reactors may show wide variations in height-to-diameter ratios. In the production of baker's yeast, a tank-type configuration with a ratio of 3 to 1 is commonly used in industry. Tower-type systems may have height-to-diameter ratios of 6 to 1 or more. As would be expected, the behavior of both gas and liquid phases may be quite different in these cases. In general, the gas phase rises through the liquid phase in plug-flow, under the action of gravity, in both types of system. However, in examining the literature on methods for estimating mass transfer coefficients, Shaftlein and Russell[151)] have observed that "it is often difficult to assess experimental work in the area because investigators are careless about driving forces and the effect of any liquid-phase reaction. It may be necessary to use a different model for the experimental situation employed by an investigator and to recalculate k_La if one is interested in a particular study".

We shall examine correlations for the mass transfer coefficient. In general, correlations for k_L take the form

$$\frac{k_L}{\mu} = a_1 (LU_L/\nu)^{a_2} (\nu/D)^{a_3} , \tag{82}$$

where a_1, a_2, a_3 are constants, U_L is a liquid velocity, L a length parameter and ν the kinematic viscosity.

The most useful correlation for k_L appears to be that of Hughmark[72)]

$$Sh = 2 + C_1 \left[Re^{0.484} \, Sc^{0.339} \left(\frac{d_B g^{1/3}}{D^{2/3}} \right)^{0.072} \right]^{C_2} , \tag{83}$$

where the values for C_1 and C_2 are as follows:

	C_1	C_2
Single bubbles	0.061	1.61
Bubble swarms	0.0187	1.61

The velocity for the bubble swarm case is the slip velocity between bubbles and the liquid.

Further approaches to estimate k_L and experimental techniques are reviewed by Shaftlein and Russell[151], with comments relating to the effects of chemical reaction on the physical absorption coefficient.

The penetration theory may also be applied to obtain a useful estimate of k_L. The exposure time can be taken as that time for the bubble to rise through its own diameter,

$$t_e = \frac{d_B}{U_B} \tag{84}$$

and thus

$$k_L = \left(\frac{4 D_L U_B}{\pi d_B}\right)^{1/2} . \tag{85}$$

The use of this equation requires an iteration procedure to obtain the rise velocity from the correlations available, calculating the bubble Reynolds number to check that the appropriate rise velocity was obtained. The bubble size may be obtained as a function of liquid properties and gas flow rate (see Sect. 5.3 and 5.4).

Once the bubble size is ascertained, the determination of the interfacial area is straightforward, provided estimates of the gas holdup are available. The interfacial area is given by the following equation:

$$a = \frac{6\varphi}{d_B} . \tag{86}$$

The gas hold-up φ has been shown by several authors to depend linearly on the superficial gas velocity, up to 3 cm s^{-1} [181, 171, 71]. At higher gas rates, a decrease is observed. Akita and Yoshida[4] correlate hold-ups up to velocities of 30 cm s^{-1} with the empirical expression

$$\frac{\varphi}{(1-\varphi)^{1/4}} = 0.2 \left(\frac{g D_T \varrho}{\sigma}\right)^{1/8} \left(\frac{g D_T^3 \varrho^2}{\mu^2}\right)^{1/12} \left(\frac{V_s}{\sqrt{g D_T}}\right) . \tag{87}$$

Hughmark[72] also provides a correlation for estimating the hold-up in bubble columns. Shaftlein and Russell[151] have reviewed much of the literature on bubble sizes and hold-ups in bubble columns, tanks and sieve plates.

In tower systems containing a draft tube, the liquid circulation pattern can be expected to alter the hold-up. Chakravarty et al.[30] provide expressions for the hold-up in the inner and outer annuli while Hatch[65] proposes

$$\varphi = \frac{V_s}{1.065\, V_s + u\gamma} , \tag{88}$$

where γ = ratio of gas flow rate to gas plus liquid flow rate
$u\gamma = 32$ cm s^{-1}, $\gamma < 0.43$,
$u\gamma = 257(\gamma - 0.43) + 32$, $\gamma > 0.43$,

$\gamma = 0.43$ signifies the outset of liquid circulation within the draft tube and outer annulus.

A variety of correlations for $k_L a$ have been reported for bubble columns. Sideman et al.[154] review the literature up to 1965. The dependency of $k_L a$ on the superficial gas velocity has been reported ranging from 0.5 to 1.6. Yoshida and Akita[181] correlate the mass transfer coefficients in 7-, 14-, 30- and 60-cm diameter bubble columns as follows:

$$\frac{k_L a D^2}{D_T} = 0.6 \left(\frac{\mu}{\varrho D_L}\right)^{1/2} \left(\frac{g D_T \varrho}{\sigma}\right)^{0.62} \left(\frac{g D_T^3 \varrho^2}{\mu}\right)^{0.31} \varphi^{1.1} . \tag{89}$$

The hold-up φ can be obtained from Eq. (87) where the superficial gas velocity enters into the correlation.

A recent review of bubble columns by Schügerl et al.[148] examines single and multistage columns and a variety of liquid phases. A correlation for $k_L a$ is proposed for porous and perforated plate spargers, with continuous phases including methanol, water and ethanol (cgs units)

$$k_L a = 0.0023 (V_s / d_B)^{1.58} . \tag{90}$$

Many of the available correlations for $k_L a$ have been obtained using small-scale equipment, and have not taken cognizance of the underlying liquid hydrodynamics. Thus, their use on large-scale equipment is suspect. Bhavaraju et al.[14] propose a design procedure based on the difference in bubble size (and hence interfacial area) close to the orifice and in the liquid bulk. As described earlier, provided the liquid is turbulent, the equilibrium bubble size in the bulk will be independent of the size at formation. The height of the region around the orifice, where the bubble formation process occurs, is a function of sparger geometry and gas flow rate. In laboratory-scale equipment, the authors show that this height may be a significant fraction of the total liquid height (up to 30%). In plant-scale equipment, however, this generally represents less than 5% of the total liquid height.

Thus, correlations developed on small-scale apparatus need to be reviewed in the light of the varying interfacial area with column height. Similarly, when porous plates are used as gas spargers, bubbles smaller than the equilibrium bubbles may be generated, resulting in a decrease in a. In general, great care needs to be taken in applying correlations for $k_L a$ which are not based on an understanding of the basic fluid and bubble motions occurring in the equipment in which the correlations were developed.

For non-viscous aqueous media, the mass transfer in bubble-column types of contactors are often correlated as $k_L a = f(u_G)$. The various function relationships are summarized in Table 1.

6.2 Devices with Stationary Internals

Several laboratory-scale devices which include internal elements to enhance mass transfer rates have appeared. These include draft-tubes, multiple sieve plates staged along the length of the column, and static mixing elements.

Table 1. Summary of mass transfer correlations for bubble columns, as $k_L a = \alpha U_G^{\beta}$

Liquid	Sparger type	D (m) $\times 10^2$	L (m) $\times 10^2$	L/D	U_G (m/s) $\times 10^2$	Flow model	α	β	Ref.
Sulfite solution (0.1–0.4 N)	Porous plate	6.3 15.2	96 192	15.2 12.6	0.2–2.8	Well mixed	—	1.0	[148]
Sulfite solution (0.3—1.0 N)	Multi-orifice	10 15	153	15.3 10.2	2–20	Well mixed	—	1.0	[148]
Sulfite solution (0.3 N)	Single orifice	7.7–60	90–350	11.7 5.8	3.0–22.0	Well mixed	0.42	0.90	[181]
Water Sulfite solution (0.3 N)	Single orifice	15.2	400	26.3	3.0–22.0	Well mixed	0.24	0.90	[4, 182]
Water	Multi-orifice	20	723	36.2	0.2–9.0	Axial dispersion	0.73	0.96	[38]
0.7 N Na_2SO_4 NaCl (0.17 N) Na_2SO_4 (0.225 N)	Multi-orifice	20	723	36.2	0.2–9.0	Axial dispersion	0.75	0.89	

Table 2. Summary of mass transfer correlations for airlift type contactors, as $(k_L a_D) = \alpha U_G^{\beta}$

Liquid	Sparger type	L_r (m) $(L_d) \times 10^2$	D_r (m) $(D_d) \times 10^2$	L/D	U_G (m/s) $\times 10^2$	Flow model	α	β	Ref.
Yeast	Multi-orifice	295 (276)	15 (5)	19.7 (55.2)	1.4–4.5	Well mixed	0.6	1.09	[92]
		112 (112)	7.5 (2.5)		1.4–7.0		0.9	1.23	
Water	Multi-orifice	165 (147)	5.5 (4.5)	14.9 (32.7)	2–10	Axial dispersion	1.09	1.06	[115]
NaCl (0.15 N)	Multi-orifice	165 (147)	5.5 (4.5)	(32.7)	2–10		2.07	1.29	
Water	Single orifice (special design)	115.5–146 (114–145)	3–4 (1.3) IOD = 0.79 (3) IOD = 1.06		0.57–2.74 (1.01–3.78)	Plug flow			[115]
NaCl (0.15 N)		(114–145)	IOD = 1.06 IOD = 0.79 IOD = 1.06						

L_r, D_r: Height and diameter of the riser; L_d, D_d: Height and diameter of the draft-tuke. IOD: Internal orifice diameter

Many reports are available on draft-tube columns where liquid is circulated due to a density difference between the inner core and the surrounding annular space. Perry and Chilton[132)] provide design relationships for determining liquid circulation rates and gas hold-ups. The downcoming liquid in the annular space entrains air bubbles; thus, the hold-up in the central core and annular region will be different. Several reports on small-scale airlift columns as bioreactors have appeared. Chakravarty et al.[30)] have examined gas hold-ups at various positions in a 10 cm diameter column. Hatch[65)] has investigated hold-up and oxygen transfer rates. A rectangular airlift has been reported[55)], and airlifts with external recirculation[77) 107)] have been proposed. Table 2 summarizes the gas-liquid mass transfer correlations developed for experimental size airlift contactors according to the expression $k_L a = f(u_G)$.

Industrially, pilot plant-scale airlift devices have been examined by Kanazawa[77)], ICI[57)] and a 50 m^3 airlift is employed by Gulf[35)]. A British Petroleum SCP venture used an airlift design in a 16,000 metric ton a^{-1} plant[19)] at Lavera, France.

Static mixing elements have been incorporated into airlift devices with the objective of providing additional turbulence and hence greater mass transfer capabilities. Static mixers are becoming increasingly more common in oxidation ponds for biological waste-water treatment. Here, fine bubbles may be produced as the gas-liquid mixture rises through the mixing element. These are usually 45—60 cm in diameter and placed over sparger pipes. A fairly intense liquid circulation can be developed by such mixers, due to entrainment by the gas-liquid jet rising from the mixing element[78)].

Hsu et al.[71)] have reported data on a small (75 mm) column, operating as a bubble column and containing sieve trays and Koch static mixing elements. The mass transfer coefficient increases with superficial gas velocity, and the sieve plate and Koch mixing elements give improved mass transfer rates.

Kitai et al.[80)] examined sieve tray systems, using the sulfite oxidation technique, and reported $k_L a$ values of the same order of magnitude as mechanically agitated tanks. Falch and Gaden[45)] performed similar determinations of $k_L a$ in a multistage tower equipped with mechanical agitation. Little effect of the superficial gas velocity has been found, but the quantity $k_L a$ depends on the level of power supplied through mechanical agitation.

Using a Koch static mixer column (75 mm diameter) with activated sludge and synthetic waste in the liquid phase, $k_L a$ has been found to depend linearly on the superficial gas velocity[70)].

6.3 Special Tubular Devices

Tubular reactors and gas-liquid tubular contactors are commonly employed in the chemical process industry but have found little application yet, apart from the laboratory use, as biochemical reactors in the pharmaceutical industry. In waste-water treatment, however, a tubular external loop system is commercially available for biological oxidation using oxygen. Many aeration basins are designed so that liquid moves in plug-flow with cross-current aeration. The mathematical description of these processes is analogous to that for the tubular system.

Tubular devices are common as immobilized enzyme reactors but in general only liquid and solid phases are involved which will not be treated here. Two types of behavior of tubular systems need to be distinguished. Some systems reported in the literature show plug-flow behavior of the liquid phase with respect to gas-liquid mass transfer but are well mixed over the time scale for biological reactions. Others are tubular relative to the biological reaction. The time scale for mass transfer is generally of the order of seconds whereas it will increase to hours for biological reactors (immobilized enzyme systems may be of the order of minutes).

In aerobic systems, tubular reactors pose the problem of ensuring sufficient oxygen supply. In cocurrent gas-liquid flows, a liquid velocity of 30—60 cm s^{-1} is required to achieve bubble flow and thus sufficient interfacial area for mass transfer. This in general necessitates small diameter pipes of sufficient length to achieve the desired product concentration in the liquid phase. An alternative approach is to operate with a recycle of the liquid, as detailed by Russell et al.[146)], resulting in well mixed liquid-phase behavior with respect to the biological reaction. Devices such as draft-tube reactors with large height-to-diameter ratios, which may show some plug-flow characteristics with respect to gas-liquid mass transfer, have often (improperly) been referred to as tubular. They are essentially well mixed with regard to reaction and have been described earlier (see Sect. 6.2).

Tubular systems offer some potential advantages over conventional stirred tanks[146)]. They are essentially simple devices, in which the flow patterns of liquid and gas phases are well characterized, and thus can be scaled-up with greater confidence. In a pipeline, reactor material can be transported while reaction takes place. This concept has been exploited in sewage treatment processes[145, 81)] based on multiple aeration points in a pipeline reactor. Tubular systems may have potential use in algal cultivation, having a large surface area-to-volume ratio for maximum exposure to light. There are no dead spaces in tubular systems.

Design procedures have been worked out for tubular reactors in which the gas and liquid flows are cocurrent. Ziegler et al.[186)] describe a tubular loop system in which the liquid phase is recirculated by means of a pump. Design procedures have also been developed for both the tubular-loop configuration and the true plug-flow tubular system. Data on oxygen transfer coefficients have been reported in which k_La is correlated with the superficial gas valocity, V_s, and the power per unit volume supplied:

$$k_La = 335(P_g/V)^{0.4}\, V_s^{0.4}\,, \tag{91}$$

(k_La in h^{-1}, P_g/V in W/l, V_s in m s^{-1}).

Moser[122)] has examined small tubular devices in which agitation was provided by a second cylinder placed within the tube, the liquid volume being confined to the annular region. Agitation is provided by rotation of the inner cylinder, and gas flow can be made either co- or countercurrent to the liquid flow. Back-mixing is a function of the rate of rotation of the inner cylinder. Oxygen transfer rates in the range of 50—250 mmol l^{-1} h^{-1} have been reported to increase linearly with rising rotational speed of the inner cylinder.

Moo-Young et al.[108)] have worked out two novel approaches to tubular reactors. One design is based on a horizontal tube which contains an internal wall-scraper. This scraper partially segregates the liquid in the tube into moving compartments. Aeration is effected by orifices at the bottom of the tube, thus being cross-flow to the liquid. This design has the advantage of scraping the walls of the tube and thus keeping wall growth to a minimum. Studies on the residence time distribution indicate a high degree of plug-flow, approximating 6—10 stirred tanks in series for even relatively short 6 m pipes.

The second design proposed by these authors is a pneumatically scraped plug-flow fermenter. This device approximates a series of stirred-tanks. It consists of a vertical array of vessels interconnected by single orifices. Thus, the medium is partially segregated, but air bubbles move from one chamber to the next via the orifices. This device also displays a behavior approaching 6—10 stirred-tanks in series. The true plug-flow behavior of both devices, with respect to biological reactions, was confirmed by two cases — a lipase producing yeast growth and a cellulase producing fungal growth. Both activations are subject to catabolite repression and hence better performance could be expected in a plug-flow system.

7 Gas-Liquid Contactors with Mechanical Agitation

7.1 Non-Viscous Systems

Gas-liquid contacting phenomena are important in aerobic reactors, the most commonly used types of industrial reactors (e.g. most antibiotic productions; activated sludge treatment). The previous section dealt with gas-liquid contacting without mechanical agitation in such devices as bubble columns and airlift towers. To obtain better gas-liquid contacting, mechanical agitation is often required.

The supply side of the overall mass transfer of oxygen from the air bubbles to the cells (and not the cell demand side) is considered in this section. The discussion is confined to non-viscous aqueous media in fully baffled sparged stirred tanks with submerged impellers. Viscous liquids are treated in Sect. 7.2. Similarly, aeration by surface impellers (as used, for example, in some waste water treatment facilities), which has recently been reviewed by Zlokarnik[189], will not be covered. Since surface aeration by vortex formation is only used in some small bench-scale reactions and is of little practical interest, it also will not be considered.

In Sect. 3, basic correlations for the individual mass transfer coefficient, k_L, were described for design purposes but the overall mass transfer coefficient, k_La, is ultimately required. Evaluation of the interfacial area, a, is therefore necessary. For particulates such as cells, insoluble substrates, or immobilized enzymes, the interfacial area can be determined from direct analyses, e.g. by microscropic examination. For gas bubbles and liquid drops, a can be evaluated from semi-empirical correlations developed by Calderbank and Moo-Young[23]. By applying the theory of local isotropic turbulence, (discussed in Sect. 3 and 5), they found that the force balances according to Eqs. (75) and (76), when coupled with experimental data, give the following correlations:

For "coalescing" clean air — water dispersions

$$a = 0.55\left(\frac{P}{V}\right)^{0.4} V_s^{0.5}, \tag{92A}$$

$$d_B = 0.27\left(\frac{P}{V}\right)^{-0.17} V_s^{0.27} + 9 \times 10^{-4} \tag{92B}$$

and for "non-coalescing" air-electrolyte solution dispersions

$$a = 0.15 \left(\frac{P}{V}\right)^{0.7} V_s^{0.3}, \tag{93A}$$

$$d_B = 0.89 \left(\frac{P}{V}\right)^{-0.17} V_s^{0.17}. \tag{93B}$$

In Eqs. (92A)—(93B), (P/V) is in W/m^{-3} and V_s is in m s^{-1}. It is seen that there is a significant effect of electrolyte on the correlations. In general, it is found[21,46,99] that the effect of surfactants such as sodium lauryl sulfate on a and d_B is similar to that caused by electrolytes such as NaCl and Na_2SO_4. Electrolytes and surfactants inhibit bubble coalescence resulting in the formation of smaller bubbles and increased interfacial areas than those in clean water systems.

At very high gas flow rate, liquid blow-out from the vessel may occur. In addition, Eqs. (92A)—(93B) are applicable provided that the impeller is not flooded by too high a gas flow rate as determined by the equations which are applicable to standard turbine impellers[175]

$$\left(\frac{N D_I}{(\sigma \varrho)^{0.25}}\right) > 1.22 + 1.25 D_T/D_I \tag{94}$$

and that there is no gross surface aeration due to gas back-mixing at the surface of the liquid as determined by the equation[23]

$$\left(\frac{N D_I}{V_s}\right)^{0.3} \times R_e^{0.1} < 2 \times 10^4. \tag{95}$$

The efficiency of gas-liquid contacting has already been described separately in terms of the fundamentals of k_L and a; the overall correlations should therefore have universal applicability.

Several investigators have developed empirical overall correlations[73,138,143,180].

Cooper, Fernstrom and Miller[34] relate $k_L a$ directly to operating variables using the so-called "sulfite oxidation" technique. They established the following correlations for geometrically similar systems:

For vaned disc impellers

$$k_L a \propto \left(\frac{P}{V}\right)^{0.95} V_s^{0.67} \tag{96A}$$

and for paddle impellers

$$k_L a \propto \left(\frac{P}{V}\right)^{0.53} V_s^{0.67}. \tag{96B}$$

In general, these workers also found an effect of H_L/D_T on $k_L a$. For H_L/D_T ratios between 2 and 4 (when multi-impellers are used) there is a 50% increase in

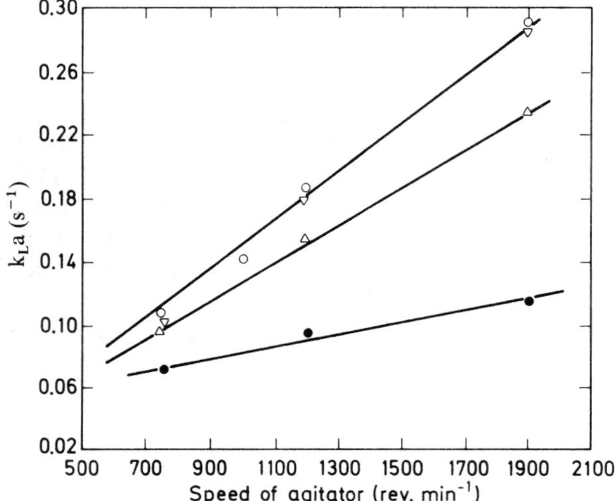

Fig. 21 Effect of different types of agitator on liquid-side mass transfer coefficient. System: CO_2 — Na_2CO_3 + $NaHCO_3$. T = 20 cm. ○: 6-straight blade disk turbine, D/T = 0.52; ▽: 6-curved blade turbine, D/T = 0.50; △: 4-curved blade turbine, D/T = 0.50; ●: 5-blade axial flow turbine, D/T = 0.50[99]

k_La over that for $H_L/D_T = 1$. Mehta and Sharma[99] have also detected that k_La decreases with increasing H_L/D_T. However, because of the unresolved quantitative effect of the chemical reaction in the sulfite oxidation technique[175, 93], correlations based on the sulfite method may not yield true values of k_La. As illustrated here, this technique is probably applicable for comparing the relative efficiency of different aeration devices.

Mehta and Sharma[99] have also reported that k_La for air-electrolyte dispersions is affected by the blade arrangement on turbine type impellers, especially at high speeds as indicated in Fig. 21. They have also found that k_La decreases by a factor of 2.5 when the impeller-to-tank diameter ratio, D_I/D_T, is reduced from 0.46 to 0.29 at the same impeller speed. As shown on p. 51, the impeller speed is probably not as good a correlating parameter of k_La as the power per unit volume for different gas-liquid contacting devices.

In recent years, more definitive studies have been carried out using physical rather than chemical reaction measurements for the evaluation of k_La. On this basis, Smith, Van't Riet, and Middleton[157] have found that, in general, the following correlations apply to a wide variety of agitator types, sizes, and D_I/D_T ratios:
For "coalescing" air-water dispersions

$$k_La = 0.01\left(\frac{P}{V}\right)^{0.475} V_s^{0.4}\,. \tag{97A}$$

For "non-coalescing" air-electrolyte dispersions

$$k_La = 0.02\left(\frac{P}{V}\right)^{0.475} V_s^{0.4}\,. \tag{97B}$$

Both relations are in *SI* units, e.g. k_La is in s^{-1}, (P/V) in $W\,m^{-3}$, V_s in $m\,s^{-1}$. The accuracy of Eqs. (97A) and (97B) is $\pm 20\%$ and $\pm 35\%$, respectively, with

95% confidence. These equations indicate that the overall k_La in "non-coalescing" systems is higher by a constant factor of two than that in "coalescing" systems under the same aeration-agitation conditions. This implies that the rate of increase in the interfacial area is higher than the rate of decrease in k_L (discussed in Sect. 3) during the transition from "mobile" to "rigid" interfacial behavior in going from an uncontaminated "coalescing" dispersion to a contaminated "non-coalescing" one. Some workers (e.g.,[99, 188]) have shown that the overall k_La is affected not only by the presence of an electrolyte in water but also by its concentration. However, we are only interested in a narrow concentration range of electrolytes (about 0.15 molar) normally used in practice.

Overall correlations for k_La may also be derived by combining the individual correlations for k_L and a developed by Calderbank and Moo-Young[28] and previously discussed for 6 flat-bladed disc turbines and standard agitator-tank configurations. The equations, found for geometrically similar systems include:

For "coalescing" clean air-water dispersions

$$k_La = 0.025\left(\frac{P}{V}\right)^{0.4} V_s^{0.5}\,. \tag{98A}$$

For "non-coalescing" air-electrolyte solution dispersions

$$k_La = 0.0018\left(\frac{P}{V}\right)^{0.7} V_s^{0.3}\,. \tag{98B}$$

Eqs. (98 A) and (98 B) suggest that the effect of the contaminants, such as electrolytes, on the overall k_La is not constant for all aeration-agitation conditions as implied by Eqs. (97A) and 97B). Eqs. (98 A) and (98 B) indicate that the relative effect of (P/V) with respect to V_s in "non-coalescing" air-electrolyte dispersions is reversed

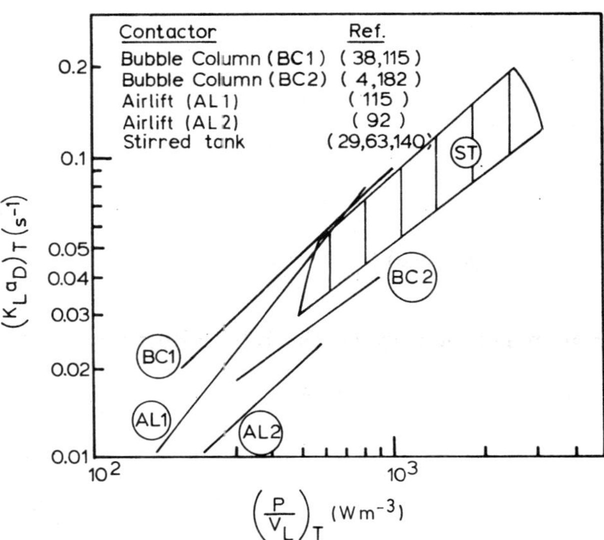

Fig. 22 Aeration efficiencies of various gas-liquid contacting devices (air-electrolyte systems)

Table 3. Summary of $k_L a$ correlations for stirred tanks (6-bladed turbine), as $k_L a = \lambda \left(\frac{P_m}{V_L}\right)^m (U_G)^n$

Liquid	Method	D_T (m)	D_I/D_T	$\frac{P_m}{V_L}$ (W/m³)	U_G (m/s)	λ	m	n	Ref.
		$\times 10^2$		$\times 10^{-2}$	$\times 10$				
Water	Physical absorption/ desorption (Winkler method)	15–50	0.333	2.6–53	0.3–1.8	0.024	0.4	0.5	32, 24, 29)
Electrolyte solutions	Physical absorption/ desorption (Winkler method)	15–50	0.333	2.6–53	0.3–1.8	0.018	0.74	0.26	32, 24, 29)
Water	Physical absorption/ desorption (oxygen probes)	15.0	0.333	0.3–180	0.1–0.5	—	0.4	0.35	140, 141)
KCl solution (0.22 N)	Physical absorption/ desorption (oxygen probes)	15.0	0.333	0.3–180	0.1–0.5	—	0.71	0.36	140, 141)
(0.1 N)	Physical absorption/ desorption (oxygen probes)	15.0	0.333	0.3–180	0.1–0.5	—	0.63	0.62	140, 141)
Water	Physical absorption/ desorption (oxygen probes)	15.0	0.333	4.4–100	0.37–1.11	0.0275	0.42	0.43	63)
Electrolyte solution Na_2SO_4 + KOH $\Gamma = 0.1$	Physical absorption/ desorption (oxygen probes)	15.0	0.333	4.4–100	0.37–1.11	0.017	0.52	0.43	63)

for the "coalescing" non-electrolyte dispersions. The apparent discrepancies between the two pairs of equations may be due to the relative imprecision of the former in attempting to include all agitator types compared to the latter which is limited to a given geometry. A graphical comparison of the equations is shown in Fig. 22. It is clear from this figure that the correlations are not mutually exclusive. Table 3 summarizes the previous correlations in the above forms of Eqs. (98A) and (98B).

The above correlations for $k_L a$ apply to the aeration of non-viscous liquids without particulates. The corresponding correlations for viscous liquids are given in Sect. 7.2 where cases are included in which enhanced viscosity may be imparted by mycelial growth and by the presence of colloidal matter or polymeric metabolites (e.g. xanthan gum). The effects of discrete particles such as single cells (e.g. bacterial cells or yeasts) are not expected to affect significantly correlations (97A) and (97B)[75] since their concentrations are sufficiently low in media so that the physico-chemical properties of the liquid is not changed appreciably.

The relative effect of power input (P/V) on $k_L a$ for gas-liquid contacting with and without mechanical agitation is illustrated in Fig. 22 for electrolyte solutions which simulate non-viscous aqueous media[115]. From the graph it is seen that for comparable power inputs, the magnitude of $k_L a$ obtained is about the same whether mixing is performed mechanically in stirred-tanks or pneumatically in bubble-column or airlift devices. However, only mechanically agitated systems are capable of attaining high values of $k_L a$ as is required or example in some antibiotic processes and in the activated sludge method of treating waste-water. Non-mechanically agitated systems would result in liquid blow-out before reaching these high aeration rates.

It should be stressed that none of the overall correlations for $k_L a$ has universal applicability. The problem is that any scale-up procedure based on equalizing $k_L a$ of both scales according to a given correlation may cause other criteria to be violated as discussed in Sect. 9. Thus, $k_L a$ may not be the only criterion that can be used rigorously for scale-up. Depending on biological demands and tolerances, other criteria may be more important. For example, an increase in the $k_L a$ value can sometimes result in a damage to the organisms in highly turbulent fermentation broth. For this reason, one correlation may prove more valuable than another for a given culture. For example, Taguchi[165] bas found that a modified Cooper correlation[34] predicts the glucamylase process better than does the Richards correlation[138]. For further discussions on scale-up, see Sect. 9.

7.2 Viscous Systems

There are several experimental approaches which have been taken in examining the effect of viscosity on the overall mass transfer coefficient $k_L a$ in stirred-tank gas-liquid contactors. Two types of viscous reactions need to be distinguished.

a) Fungal cultivations (such as penicillia, actinomycetes) where the viscosity is due to the mycelial structure, and the continuous phase is essentially aqueous,
b) Reactions (such as polysaccharide production) where the viscosity is due to polymers excreted by the organism into the continuous phase. These reactions result in an essentially homogeneous, viscous liquid phase.

The first type can be simulated by materials such as paper pulp which exhibits a macroscopic structure analogous to fungal hyphae suspended in water. The second type may be simulated by aqueous polymer solutions of known properties. A third approach is to use actual media although there are difficulties in reproducing the broth characteristics. The behavior of the two classes of broths can be expected to differ with respect to bubble behavior in the continuous phase.

Sideman et al.[154)] have proposed the following general correlation for the liquid side-mass transfer coefficient in mechanically agitated gas-liquid dispersions

$$\frac{k_L a D_I^2}{D} = A\left(\frac{\mu_a}{\varrho D}\right)\left(\frac{\mu_a V_s}{\sigma}\right)^{\beta}\left(\frac{D_I^2 N \varrho}{\mu_a}\right)^{\gamma}\left(\frac{\mu_d}{\mu_a}\right)^{\delta}. \tag{99}$$

Additional dimensionless terms could be incorporated to account for geometric variables, e.g., H/D_T, D_I/D_T. Eq. (99) can be altered to relate $k_L a$ to the specific power input, i.e. by relating P/V to impeller diameter and impeller velocity. In this modified form the relationship would be independent of tank geometry; however, P/V is itself a complex function of the impeller type, gas velocity, etc. For a specified geometry and liquid, Eq. (99) reduces to the familiar forms used in Sect. 7.1.

For systems with viscosities not far from those of aqueous systems (1 to 12 cp), Mehta and Sharma[99)] report an increase in the mass transfer coefficient with rising viscosity due to an increase in interfacial area. The higher viscosity results in a higher dispersion stability and reduced rates of bubble coalescence.

Perez and Sandall[131)] examined the absorption of CO_2 into aqueous carbopol solutions (0.25, 0.75 and 1.00%). The experimental data are correlated by the equation

$$\left(\frac{k_L a D_I^2}{D}\right) = 21.2\left(\frac{N \varrho D_I^2}{\mu_a}\right)^{1.11}\left(\frac{\mu_a}{\varrho D}\right)^{0.5}\left(\frac{D_I V_s}{\sigma}\right)^{0.45}\left(\frac{\mu_d}{\mu_a}\right)^{0.69}. \tag{100}$$

These authors have used the term $\left(\frac{D_I V_s}{\sigma}\right)$ to establish this correlation. Yagi and Yoshida[178)] and Blanch and Bhavaraju[18)] have found that this gas flow number has dimensions and should be of the form $\left(\frac{\mu_a V_s}{\sigma}\right)$. The apparent viscosity, μ_a, used by these authors is a function of the flow behavior index n and consistency K.

$$\mu_a = k(11N)^{n-1}\left(\frac{3n+1}{4n}\right)^n. \tag{101}$$

Yagi and Yoshida[178)] examined viscous solutions and viscoelastic solutions of carboxymethylcellulose and sodium polyacrylate and obtained the mass transfer coefficients by oxygen desorption. Since they observed no effect of gas viscosity, they established for viscous Newtonian fluids the following correlation

$$\left(\frac{k_L a D_I^2}{D}\right) = 0.06\left(\frac{N \varrho D_I^2}{\mu_a}\right)^{1.5}\left(\frac{D_I N^2}{g}\right)^{0.19}\left(\frac{\mu_a}{\varrho D}\right)^{0.5}\left(\frac{\mu_a V_s}{\sigma}\right)^{0.6}\left(\frac{N D_I}{V_s}\right)^{0.32}. \tag{102}$$

For viscoelastic fluids, the Deborah number (characteristic material time/process time) was included in the correlation to yield

$$\left(\frac{k_L a D_I^2}{D}\right) = 0.06\left(\frac{N\varrho D_I^2}{\mu}\right)^{1.5}\left(\frac{D_I N^2}{g}\right)^{0.19}\left(\frac{\mu_a}{\varrho D}\right)^{0.5}\left(\frac{\mu_a V_s}{\sigma}\right)^{0.6}\left(\frac{\mu_a D_I}{V_s}\right)^{0.32}$$
$$\times(1 + 2De^{1/2})^{-0.67}\,. \quad (103)$$

For purely pseudoplastic fluids (as is the case with most media), this correlation reduces to that obtained for Newtonian fluids, although this has not been examined experimentally. The non-Newtonian fluids used by Yagi and Yoshida display significant viscoelasticity. There is a discrepancy between the results of Perez and Sandall and those of Yagi and Yoshida in the dependencies of $k_L a$ on apparent viscosity μ_a and the impeller Reynolds number

$$k_L a \propto Re^{1.11}\,\mu_a^{-0.19} \quad \text{(Perez and Sandall)}\,, \quad (104A)$$

$$k_L a \propto Re^{1.5}\,\mu_a^{-1.10} \quad \text{(Yagi and Yoshida)}\,. \quad (104B)$$

Yagi and Yoshida suggest that this difference may be due to the small-scale equipment used by Perez and Sandall and to the possibility of some surface active material affecting the interfacial areas.

Loucaides and McManamey[94] examined sulfite oxidation rates in paper pulp suspensions, these simulating filamentous media. Tank and impeller geometries were both varied, vessel volumes ranging from 5 to 72 l. Analogous to the results of of Mehta and Sharma in non-viscous solutions, $k_L a$ was found to correlate well with variations in tank diameter (for constant D_I/D_T):

$$k_L a \doteq C_1\left(\frac{D_T}{H_T}\right) N D_I D_T^{-0.5} + C_2\,. \quad (105)$$

By varying the impeller blade dimensions, variations in power per unit volume were carried out at constant impeller speed. At low power per unit volume there was a linear increase in $k_L a$ which correlated with P/V having an exponent of 0.9 to 1.2. Beyond the breakpoint, the exponent relating the P/V dependence was 0.53. In both regions, $k_L a$ depended on the superficial gas velocity to the 0.3 power. These results are similar to those reported earlier by Blakebrough and Sambamurthy[17], and Hamer and Blakebrough[59]. They have been obtained by using smaller scale vessels and paper pulp suspensions.

The general area of aeration of viscous non-Newtonian media has been recently reviewed by Banks[8] and Blanch and Bhavaraju[18]. Details on rheology and fluid dynamics may be fould there.

8 Agitation Power Requirements

8.1 Relevant Operating Variables

Mixing is used to promote or enhance the mass and heat transfer rates in a biochemical reactor. When mixing is induced pneumatically or fluid circulation by pumping, it is fairly easy to evaluate the power consumption from pressure drop considerations as in Sect. 5.2. However, for mechanically induced mixing, power consumption is more difficult to evaluate from operating variables. Since 1.0 to 4.0 kW per 1000 l is normally required and the usual range of capacities in industrial units is 40,000 to 160,000 l, the power requirement is often 40 to 600 kW per unit. This is an important consideration in process economics.

Several mechanical mixer parameters have been used for design purposes; the common ones are:

a) $\frac{P}{V}$ (power-per-unit-volume) affecting mass and heat transfer and suspension efficiency;
b) N (impeller speed) which is proportional to the pumping capacity of the rotating agitator and exerts a great influence on the mixing time;
c) ND (impeller tip speed) relating to dispersion efficiency;
d) Re (impeller Reynolds number) which influences the hydrodynamics.

The power characteristics of standard mixer configurations[128] are reviewed below. The agitator shaft is positioned in the center of a vertical cylindrical tank with wall baffles which is the most common position for bioreactors. These reference geometries for radial-flow (e.g. turbines) and axial-flow (e.g. propellers) impellers provide adequate mixing in most cases. Under some circumstances, these configurations are not optimal and may even be impractical. Most published results, however, are based on these configurations. With other agitator types an analogous treatment can be applied.

8.2 Newtonian Systems

8.2.1 Ungassed Stirred-Tanks

Applying dimensional analysis, it can be shown that[154]

$$\frac{P}{D^5 N^3 \varrho} = f\left(\frac{D^2 N \varrho}{\mu}, \frac{DN^2}{g}, \text{geometric factors}\right). \tag{106}$$

The first term basically defines the inertia forces and is called the Power number (or in Europe the Euler number); the second term is the Reynolds number the physical significance of which has already been discussed; the third term is the Froude number which takes into account gravity forces. In a simplified form, the correlation for dynamic similarity is given as the Power factor:

$$\Phi = \frac{Po}{(Fr)^n} = C(Re_l)^m, \tag{107}$$

in which C is a constant depending on geometrical conditions.

The function $(Fr)^n$ is very often equal to unity, e.g. in fully baffled tanks, or in unbaffled tanks for $Re < 300$ when gravitional effects on the liquid surface are the same. The correlation for Po as a function of Re and Fr is available in a number of standard texts[156]. The exponent n is a function of the Reynolds number:

$$n = \frac{a - \log_{10} Re}{b}, \tag{108}$$

in which a is a function of D/T. The value of b is dependent on the impeller type. Values of a and b for various impellers can be found in Ref.[132].

Although the complete functional relationship in Eq. (106) is rather complex and can only be represented graphically, simple analytical expressions can be derived as follows:

a) In the turbulent flow region,

$$Po = \text{constant}$$

and

$$P \propto \varrho N^3 D^5 . \tag{109}$$

Thus, P is strongly dependent on diameter but independent of liquid viscosity. For standard impeller types with different blade ratios:

$$Po = 160 \frac{WL(D - W)}{D^3} . \tag{110}$$

b) In the laminar flow region,

$$Po \propto 1/Re_I$$

and

$$P \propto \mu N^2 D^3 . \tag{111}$$

Here, P is proportional to viscosity.
For various impeller types, a modified Power number is given as[29]:

$$Po' = Po\, f(W, L, D) . \tag{112}$$

c) In the transition from laminar to turbulent flow, the Po vs. Re_I change is gradual, covering a Re range change of about 10^3, and cannot be given a general expression.

8.2.2 Gassed Stirred-Tanks

The power required to agitate gassed liquid systems is less than that for ungassed liquids since the apparent density and viscosity of the liquid phase decrease upon gassing. For Newtonian liquids this decrease may be as much as two-thirds of the ungassed power. The reduction in drawn power is a result of the formation of gas cavities on the trailing edge of the impeller blade. The shape and number of these carities have been observed by Bruijn et al.[20] with a rotating television camera. The reduction in gassed power is generally given as a function of the ratio of the superficial gas velocity to the impeller tip speed, this defining the aeration number, Na.

$$Na = Q/ND^3 . \tag{113}$$

The reduction in power in the turbulent regime is usually expressed as the ratio of gassed to ungassed power (P_g/P), although Judat[76] questions this usage and proposes P_g vs. Na. Curves of P_g/P versus the aeration number (Na) are available for a large variety of impeller and vessel geometries as reported by Judat[76], Calderbank[22], Aiba et al.[2], Zlokarnik[188].

Oyama and Endo[130] have expressed the relationship between gassed and ungassed power as

$$\frac{P_g - P_\infty}{P - P_\infty} = \exp(-a/Na) , \tag{114}$$

where P_∞ is the power consumption at very high gas rates and a a constant. Michel and Miller[104] propose a correlation of the form

$$P_g = C[P^2 \mu D^3/Q^{0.56}]^{0.45} \tag{115}$$

which appear to fit not only Newtonian liquids but, according to Taguchi[164], non-Newtonian systems as well. It should be noted that Eq. 115 is not dimensionally sound and predicts unrealistic results for very small Q values.

8.3 Non-Newtonian Systems

8.3.1 Ungassed Stirred-Tanks

The prediction of the power consumption of ungassed non-Newtonian systems has been fairly extensively examined, and three main methods are available. Metzner and Otto[103] propose an average shear rate in the vessel which is proportional to the impeller speed. From this, an apparent viscosity can be defined and then non-Newtonian liquid related to a viscous Newtonian liquid. This approach does not depend on a model of the rheological behavior of the liquid and extends well beyond the laminar region.

Calderbank and Moo-Young[29] define a generalized impeller Reynolds number for power-law fluids,

$$Re'_I = \frac{D^2 N \varrho}{\mu_a} , \tag{116}$$

where μ_a is given in Eq. (63). This procedure allows the use of a conventional power curve formerly developed for Newtonian fluids.

Foresti and Liu[47] also provide a correlation for power-law fluids in the laminar region. All three methods are described in some detail by Skelland[156].

8.3.2 Gassed Stirred-Tanks

Bruijn et al.[20] observed the formation of gas cavities behind impeller blades with a rotating television camera and noted that the decrease in gassed power consumption with rising gas flow was due to an increasing number of large gas cavities formed behind the blades of the impeller. With viscous liquids (substantially Newtonian in nature), the authors reported a change in the shape of the gas cavities. This transition occured over the range 5 to 300 cp. Once these cavities are formed in viscous liquids, they are stable, even after the gas flow ceases. Thus, reduced power levels can be maintained at very low gas flow rates; this means that the aeration number should not have any effect on gassed power consumption. This, indeed, appears to be the case for pseudoplastic and viscoelastic fluids.

Edney and Edwards (see Refs.[8, 18]) have reported power measurements in dilute aqueous solutions of carboxymethylcellulose and polyacrylamide at various impeller speeds and gas flow rates, the gassed power being almost independent of the aeration number. Similar results were obtained by Ranade and Ulbrecht[135, 136] using viscoelastic polyacrylamide solutions and viscous Newtonian corn syrup. Ulbrecht's results also show a dependency of the Power number on viscosity in addition to that contained in the Reynolds number used to correlate the results. Taguchi and Miyamoto[167] report on the gassed power consumption of *Endomycopsis* media, observing also an independence of P_g/P_0 on Q/ND^3 and a dependence on viscosity. Fig. 23 describes a plot of P_0 vs. the Reynolds number taken from Ranade and Ulbrecht[136] and defined for the power-law fluids used in the above

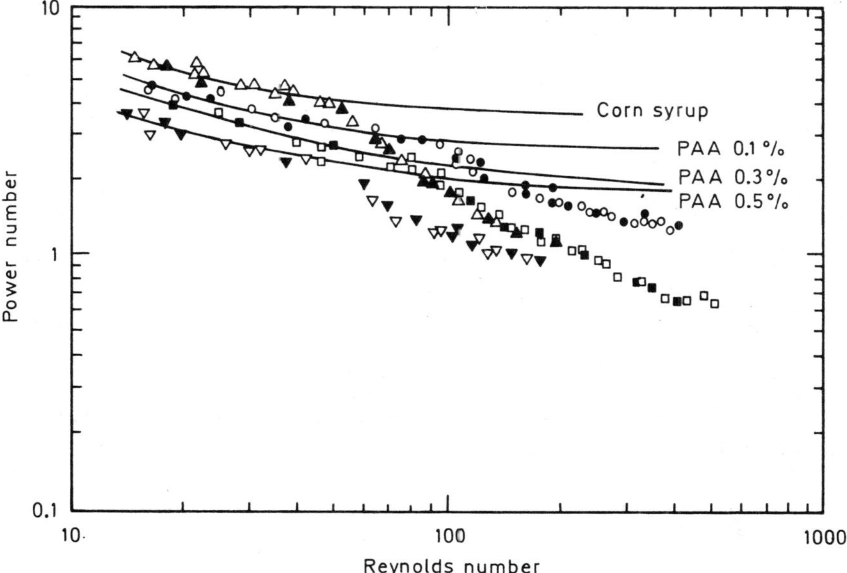

Fig. 23 Data of Ulbrecht et al.[136] of power consumption in gassed non-Newtonian solutions. Corn syrup is viscous but Newtonian. Solid lines refer to unaerated polyacrylamide solutions
○, ● 0.1% PAA; 0.5, 5.0 l/min^{-1} air, ▽, ▼ 0.5% PAA; 0.5, 5.0 l/min^{-1} air,
□, ■ 0.3% PAA; 0.5, 5.0 l/min^{-1} air, △, ▲ corn syrup; 0.5, 5.0 l/min^{-1} air

studies with an apparent viscosity of

$$\mu_a = K(BN)^{n-1} . \tag{117}$$

The value of B has been determined by several authors (Calderbank and Moo-Young[29], Metzner et al.[101]) to be about 11.5.

A problem in the experimental determination of the effect of pseudoplasticity on gassed power consumption results when the size of the bubbles formed at the sparger are of the same dimension as the impeller blade. This will be the case in laboratory-scale but not in industrial scale-equipment. Here, the impeller may spin in a "gas donut" and have the bulk of the fluid motionless. It should be noted that vibrational problems may arise if large bubbles move away from the impeller causing it to move through fluids of differing densities.

9 Scale-Up Considerations

9.1 Gas-Liquid Mass Transfer Basis

When, at the laboratory scale, the optimum process conditions are found for the growth of a specific organism or its metabolic productivity, there is a need to transfer these findings for use in larger units. There may be several criteria for optimal growth and, unfortunately, their effects on scaling problems are not all identical.

We have already implied that if k_La drops below a certain value, growth is hampered or destroyed. Thus, k_La is frequently used as a basis for scaling-up, especially in aerobic biological waste treatment systems. If we make the k_La values identical for the smaller and the greater vessel, the operating conditions are then derived.

Table 4 gives an example of scale-up at constant k_La. From this table it is seen that if it is desirable to maintain equal volumetric gas flow rates (VVM), then the linear gas velocity through the vessel will increase almost directly with the scale ratio. However, this linear velocity is also an important factor in the design of the reactor, e.g. the mixing energy required to disperse the gas stream and upper limit

Table 4. Scaling-up based on constant k_La for gas-liquid contacting in a sparged stirred-tank reactor

Property	Lab. reactor	Plant reactor		
	(80 l)	(10000 l)		
H_L/T	1	1	1	2.8
P/V	1	1	>1	1
VVM	1	1	0.2	0.1
V_s	0.1	0.5[a]	0.1	0.1
k_La	1	—	1	1

[a] indicates impractical liquid "blow-out" conditions

before liquid blow-out action begins. If the percentage of oxygen required is relatively small it may be possible to reduce the volume of gas per volume of liquid per minute on scale-up but to increase the gas absorption efficiency. Table 4 illustrates that this may be achieved by changing geometric configurations while allowing P/V to remain constant.

It appears that the use of k_La is often a reasonable design approach. An increase in k_La can sometimes have an adverse effect because of damage to organisms in highly turbulent fermentation broth and/or oxygen poisoning. Other problems such as gross coalescence are also important in non-mechanically stirred reactors.

Observations similar to those described above can be applied to the k_La criteria at the cell-liquid interface demand for oxygen.

9.2 Fluid-Flow Basis

Another common design approach is based on equal agitation power per unit volume of liquid. As with k_La, there appears to be a process minimum for P/V. In addition to its relationship to k_La, the gas dispersion efficiency is also determined by the power dissipation per unit volume of liquid. For constant P/V in turbulent flow, it is to be noted that the impeller tip speed and hence the shear increases with the cube-root of the ratio of the vessel diameters so that many flow parameters cannot be maintained constant on scale-up (see Table 5).

Scale-down considerations are also important in reactor design. Often, the real problem is to reproduce on a smaller scale the flow non-idealities which exist in a plant scale (e.g. dead space and by-passing) so that meaningful metabolic rate data can be obtained. Solomon[158)] concluded that lack of good mixing in viscous broths may prevent the attainment of the steady state in some cultivation systems. In these cases, mixing times appear to provide a better scale-up criterion.

10 Concluding Remarks

We have examined at length how mass transfer phenomena in a biochemical reactor influence the supply of reactants to living cells and other biocatalytic particulates and the removal of products and intermediates from them. Since it is these phenomena that usually control the performance of the biochemical reactor it is imperative that the design engineer be conversant with this material. However,

Table 5. Examples of incompatible flow parameters on scaling-up a geometrically similar ungassed stirred-tank reactor[128)]

Parameter	Lab. reactor	Plant reactor			
	(20 l)	(2500 l)			
P/V	1	1	25	0.2	0.0016
N	1	0.34	1	0.2	0.04
ND	1	1.7	5	1	0.2
Re	1	8.5	25	5	1

in biochemical reactors the complex hydrodynamics (upon which mass transfer depends) are difficult to characterize rigorously. Hence, empirical results and educated guesses are often an integral part of the design calculations. An attempt has been made to present a quantitative appreciation of the difficulties involved in biochemical reactor design from a mass-transfer viewpoint. With this material, reasonable and safe estimates for some of the design criteria (e.g. $k_L a$, P/V) can be made.

Information on intra-particle mass transfer is also lacking. The effect of particle density on intra-particle diffusivity and the effect of mass transfer at the solid-liquid interface have yet to be examined. In addition, experimental results showing the influence of mass transfer on reaction rates need to be extended.

It is clear that much more fundamental information is required on media rheology, cell and bubble motion, mixing non-idealities, diffusion in heterogeneous materials, and the interaction between these factors in order to put process design on a more rational basis. Armed with this information, we should be able to improve existing microbial systems and, probably more important, we should be encouraged to exploit commercially, the special capabilities of different reactor configurations (e.g. those used in pipeline, tower and loop-chemical reactors) and different media composition (e.g. "insoluble" substrates such as oil and wood, non-assimilatable additives such as certain polymers and surfactants, and variable specific nutrient concentration). Unfortunately, this prerequisite information is presently lacking.

11 List of Symbols

Roman Letters

A total interfacial area
a specific interfacial area (based on unit volume of dispersion)
B non-Newtonian mixing factor
C concentration of solute in bulk liquid
$\bar{C}$ concentration of solute in bulk media (as opposed to the interior of a particle)
C_A concentration of component A
C_O initial concentration of solute
C_p heat capacity
C_r nutrient concentration at r
C_R nutrient concentration at R
C_s saturation concentration of solute
C_T critical nutrient concentration
D dilution rate; impeller diameter; diffusivity
D_L liquid-phase diffusivity
D_p diffusivity of product in membrane
D_r intra-particle molecular diffusivity
D_s diffusivity of substrate in membrane
D_T tank or column diameter
d diameter of particle as an equi-volume sphere
d_B bubble diameter

d_{Be}	equilibrium bubble diameter
d_0	orifice diameter
E	fractional approach to equilibrium
E_1	ratio between bubble width and bubble height
E_D	eddy diffusivity
E_f	effectiveness factor
F	volumetric liquid flow rate feeding reactor
G	molar gas flow rate (subscript 1 indicates inlet and 2 outlet)
g	acceleration due to gravity
H	Henry's law coefficient
H_L	liquid height in reactor
H_T	total height of dispersion in reactor
h	heat transfer coefficient
J_A	mass flux of component A in B
K	consistency coefficient of power-law fluids
K_i	inhibition constant
K_m	Michaelis constant
K_G	overall gas phase mass transfer coefficient
k	Boltzman constant; thermal conductivity
k_L	liquid phase mass transfer coefficient
$k_L a$	volumetric mass transfer coefficient
L	impeller blade length; $^1/_2$ membrane thickness in Sect. 4.3
l	characteristic length; length of terminal eddies; distance from center of membrane in Sect. 4.3
N	speed of agitator
N_b	number of wall baffles in stirred tank
n	fluid behavior index of power-law fluids; Froude number exponent in Eq. (107)
n_B	number of blades on impeller
P	agitator power requirements for ungassed liquids; product concentration in Sect. 4.3
P_g	agitator requirements for gas-liquid dispersions
P_s	total pressure
P_1, P_2	pressure at bottom and top of tank
Q	specific nutrient consumption rate (when nutrient is oxygen — specific, respiration rate at C; volumetric gas flow rate
$\bar{Q}$	specific respiration rate in bulk media
Q'_{max}	maximum value of specific respiration rate at C (within a particle)
R	universal gas constant; outer radius of a sphere
r	radius; reaction rate per unit volume
r_o	sphere radius of solute
r_p	radius within a particle at which a dissolved nutrient becomes zero
$r(s)$	reaction rate of substrate
S	substrate concentration; ratio of cup to bob diameter
S_i	concentration of substrate at surface of membrane
s	surface renewal rate
T	temperature; tank diameter

t time
U characteristic linear velocity
U_B bubble velocity
$\bar{U}_d^2$ mean square fluctuating velocity component
U_L liquid velocity
U_O velocity of gas at orifice
U_T terminal velocity of particle
u relative particle velocity
V volume of fermentor contents
V_G volume of gas
V_L volume of liquid
V_m maximum reaction rate
V_s superficial gas velocity
VVM volume of air per unit volume of medium per minute
W width of impeller blade
W_b width of wall baffles
X film thickness at the interface
x diffusional distance
y mole fraction of component in gas phase; dimensionless concentration
$\bar{y}$ mean mole fraction defined by Eq. (48)

Greek Letters

$\dot{\gamma}$ shear rate
δ_0 diffusion boundary layer thickness (for mass transfer)
δ_M momentum boundary layer thickness
ε Bingham number
η ratio of gas velocity just above orifice to initial velocity
μ viscosity (dynamic)
μ_a apparent viscosity (dynamic)
μ_s interfacial viscosity
ν kinematic viscosity of continuous phase
ξ General modulus in Sect. 4.2 $\left(= R\sqrt{\varrho_m \bar{Q}/2D_r\bar{C}}\right)$
ϱ density of continous phase
ϱ_d density of dispersed phase
ϱ_m density of mycelia
σ interfacial tension between dispered and continuous phases
τ shear stress
φ hold-up of dispersed phase
φ_m intra-particle mass transfer rate for a nutrient
Φ fower factor in Eq. (107)
Δ difference

Subscripts

B bubble
d dispersed phase
G gas phase
I impeller

i interfacial
L liquid phase
s surface
o initial condition
∞ equilibrium conditions

Abbreviations for Dimensionless Groups

De Deborah number
Fr Froude number
Fr_o orifice Froude number
Gr Grashof number for mass transfer based on particle-environment density difference
Gr_H Grashof number for heat transfer
Gr_δ Grashof number for mass transfer based on the momentum boundary layer thickness
N_A aeration number
Nu Nusselt number
Pe Peclet number for mass transfer
Pe_{SW} Peclet number for bubble swarms
Po power number
Pr Prandtl number
Re Reynolds number for moving particles
Re' generalized Reynolds number for power-law fluids
Re_I impeller Reynolds number
Re_e isotropic turbulence Reynolds number
Re_0 orifice Reynolds number (based on gas properties)
Re_{0L} orifice Reynolds number (based on liquid properties)
Sh Sherwood number
Sc Schmidt number
We Weber number

12 Acknowledgement

The authors are grateful for the invaluable assistance of many students who carried out various literature and laboratory research during the preparation of this manuscript. Special thanks are due to Gerald Andre for checking the final draft.

13 References

1. Acharya, A., Mashelkar, R. A., Ulbrecht, J. J.: Chem. Eng. Sci. *32*, 863 (1977)
2. Aiba, S., Humphrey, A. E., Millis, N. F.: Biochemical engineering, 2nd edit. New York: Academic Press 1974
3. Aiba, S., Kobayashi, K.: Biotech. Bioeng. *13*, 583 (1971)
4. Akita, K., Yoshida, F.: Ind. Eng. Chem. Proc. Des. Develop. *13*, 84 (1974)
5. Alfrey, T., Gurney, E. F. In: Rheology (Eirich, F. R., ed.) vol. 1, chap. 11. New York: Academic Press 1956
6. Atkinson, B.: Biochemical reactors, London: Pion Press 1973
7. Bailey, J. E., Ollis, D. F.: Biochemical engineering fundamentals. New York: McGraw-Hill 1977
8. Banks, G. T.: In: Topics in enzymes and fermentation biotechnology (Wiseman, ed.), vol. 1, p. 72. New York: Wiley 1977
9. Bayzara, Z. S., Ulbrecht, J. J.: Biotech. Eng. *20*, 287 (1978)
10. Batchelor, G. K.: Proc. Camb. Phil. Soc. *47*, 359 (1951)
11. Bhavaraju, S. M., Blanch, H. W.: J. Ferment. Technol. *53*, 413 (1975)
12. Bhavaraju, S. M., Mashelkar, R. A., Blanch, H. W.: AIChE J. *24*, 1063 (1978), AIChE J. *24*, 1070 (1978)
13. Bhavaraju, S. M.: Ph. D. Thesis, University of Delaware, Newark, Delaware 1978
14. Bhavaraju, S. M., Russell, T. W. F., Blanch, H. W.: AIChE J. *24*, 454 (1978)
15. Bischoff, K. B., Himmelblau, D. M.: Ind. Eng. Chem. *57*, (12) 54 (1965)
16. Bischoff, K. B.: AIChE J. *11*, 351 (1965)
17. Blakebrough, N., Sambamurthy, K.: Biotech. Bioeng. *8*, 25 (1966)
18. Blanch, H. W., Bhavaraju, S. M.: Biotech. Bioeng. *18*, 745 (1976)
19. BP-Kirkpatrick Award: Chem. Eng. *80*, 62 (Nov. 26, 1973)
20. Bruijn, W., Riet, K., Smith, J. M.: Trans. Inst. Chem. Engrs. *52*, 88 (1974)
21. Bull, B.: Biotech. Bioeng. *13*, 529 (1971)
22. Calderbank, P. H.: In: Biochemical and biological engineering science (Blakebrough, ed.), vol. 1, p. 101. New York: Academic Press 1967
23. Calderbank, P. H.: Trans. Inst. Chem. Engrs. *36*, 443 (1958)
24. Calderbank, P. H.: ibid. *37*, 173 (1959)
25. Calderbank, P. H., Lochiel, A. C.: Chem. Eng. Sci. *19*, 471 (1964)
26. Calderbank, P. H., Lochiel, A. C.: Chem. Eng. Sci. *19*, 485 (1964)
27. Calderbank, P. H., Moo-Young, M., Bibby, R.: Chem. Eng. Sci., Chem. Reaction Engng. Suppl. *20*, 91 (1965)
28. Calderbank, P. H., Moo-Young, M.: Chem. Eng. Sci. *16*, 39 (1961)
29. Calderbank, P. H., Moo-Young, M.: Trans. Inst. Chem. Engrs. *37*, 26 (1959); *39*, 337 (1961)
30. Chakravarty, Y., et al.: Biotech. Bioeng. Symp. No. *4*, 363 (1973)
31. Charles, M.: Adv. Biochem. Eng. *8*, 1 (1978)
32. Cichy, P. T., Russell, T. W. F.: Ind. Eng. Chem. *61*, 15 (1969)
33. Cognated Brit. Pat. Appl. 23328/73, 53921/73, 42923/74, and 763/75
34. Cooper, C., Fernstrom, G., Miller, S.: Ind. Eng. Chem. *36*, 504 (1944)
35. Cooper, P. G., Silver, R. S., Royle, J. P. In: SCP II (Tannenbaum and Wang, eds.), MIT Press 1975
36. Cullen, E. J., Davison, J. F.: Chem. Eng. Sci., *6*, 49 (1956)
37. Danckwerts, P. V.: Ind. Eng. Chem. *43*, 146 (1951)
38. Deckwer, W., Burckhart, R., Zoll, G.: Chem. Eng. Sci. *29*, 2177 (1974)
39. Deindoerfer, F. H., Humphrey, A. C.: Ind. Eng. Chem. *53*, 755 (1961)
40. Davidson, J. F., Schuler, B. O. G.: Trans. Inst. Chem. Engrs. *38*, 335 (1960)
41. Davidson, L.: Ph. D. Thesis, Columbia Univ., New York 1951
42. Davies, J. T., Rideal, E. K.: Interfacial phenomena. New York: Academic Press 1963
43. Elmayergi, H., Scharer, J. M., Moo-Young, M.: Biotech. Bioeng. *15*, 845 (1973)
44. Fair, J. R., Lambright, A. J., Anderson, J. W.: Ind. Eng. Chem. Proc. Des. Dev. *1*, 33 (1962)
45. Falch, E. A., Gaden, E. L.: Biotech. Bioeng. *11*, 927 (1969)

46. Finn, R. K.: Bact. Rev. *18*, 245 (1954)
47. Foresti, R., Liu, T.: Ind. Eng. Chem. *51*, 860 (1959)
48. Le François, L., Mariller, C. G., Mejane, J. V.: Effectionnements Aux Procédés de Cultures Fongiques et de Fermentations Industrielles. Brevet d'Invention, France No. 1.102.200 (1955)
49. Friedlander, S. K.: AIChE J. *7*, 347 (1961)
50. Frössling, N.: Ger. Beitr. Geophys. *32*, 170 (1938)
51. Gainer, J. L.: Ind. Eng. Chem. Fund *9*, 381 (1970)
52. Gainer, J. L., Metzner, A. B.: AIChE. — Inst. Chem Eng. Joint Meet. (London) June 13—17 (1965)
53. Gal-Or, B., Resnick, W.: Chem. Eng. Sci. *19*, 653 (1964)
54. Gal-Or, B., Waslo, S.: Chem. Eng. Sci. *23*, 1431 (1968)
55. Gasner, L. L.: Biotech. Bioeng. *16*, 1179 (1974)
56. Goldstein, L., Levin, Y., Katchalshi, E.: Biochem. *3*, 1913 (1964)
57. Gow, P. G. et al. In: SCP II (Tannenbaum, Wang, eds.), MIT Press 1975
58. Hadamard, J.: Compt. Rend. Acad. Sci. *152*, 1735 (1911)
59. Hamer, G., Blakebrough, N.: J. Appl. Chem. *13*, 517 (1963)
60. Hanhart, J., Westerterp, K. R., Kramers, H.: Chem. Eng. Sci. *18*, 503 (1963)
61. Happel, J.: AIChE J. *4*, 197 (1958)
62. Happel, J., Brenner, H.: Low Reynolds number hydrodynamics. New York: Prentice Hall 1965
63. Hassan, I. T. M., Robinson, C. W.: AIChE J. *23*, 48 (1977)
64. Hassan, I. T. M., Robinson, C. M.: Chem. Eng. Sci. *35*, 1277 (1980)
65. Hatch, R. T.: Ph. D. Thesis, MIT, Cambridge, Mass. 1973
66. Higbie, R.: Trans. Amer. Inst. Chem. Engrs. *31*, 365 (1935)
67. Hirose, T., Moo-Young, M.: Can. J. Chem. Eng. *47*, 265 (1969)
68. Hirose, T., Ali, S., Moo-Young, M.: Proceedings of the Fifth Internat. Congr. Rheology, vol. 2, p. 232. Univ. Park Press 1970
69. Hirose, T., Moo-Young, M.: Chem. Eng. Sci. *25*, 729 (1970)
70. Hsu, K. H., Erickson, L. E., Fan, L. T.: Biotech. Bioeng. *17*, 499 (1975)
71. Hsu, K. T., Erickson, L. E., Fan, L. T.: Biotech. Bioeng. *19*, 247 (1977)
72. Hughmark, G. A.: Ind. Eng. Chem. Proc. Dev. *6*, 218 (1967)
73. Hyman, D.: Adv. Chem. Eng. *3*, 157 (1962)
74. Johnson, M. J.: J. Bact. *94*, 101 (1967)
75. Joosten, G.: Chem. Eng. Sci. *32*, 563 (1977)
76. Judat, H.: Chem. Ing. Techn. *100*, 865 (1976)
77. Kanazawa, M. In: SCP II (Tannenbaum, Wang, eds.) MIT Press 1975
78. Kenics Mixers, Techn. Inf. Bull. 1977
79. Kishinevskii, M. Kh.: J. Appl. Chem. U.S.S.R. *24*, 593 (1951)
80. Kitai, A., Goto, S., Ozaki, A.: J. Ferm. Technol. *47*, 356 (1969)
81. Koch, C. M., Zandi, I.: J. Water Pol. Control. *45*, 2537 (1973)
82. Kobayashi, H., Suzuki, M.: Biotech. Bioeng. *18*, 37 (1976)
83. Kobayashi, T., Moo-Young, M.: Biotech. Bioeng. *13*, 893 (1971)
84. Kobayashi, T., Moo-Young, M.: Biotech. Bioeng. *15*, 47 (1973)
85. Kobayashi, T., van Dedem, G., Moo-Young, M.: Biotech. Bioeng. *15*, 27 (1973)
86. Kolmogoroff, A. N.: Compt. Rend. Acad. Sci. URSS *30*, 301 (1941); *31*, 538 (1941); *32*, 16 (1941)
87. Kumar, R., Kuloor, N. R.: Adv. Chem. Eng. *8*, 755 (1970)
88. Leibson, I. et al.: AIChE J. *2*, 296 (1956)
89. Levenspiel, O.: Chemical reaction engng. New York: Wiley 1972
90. Levich, V. G.: Physicochemical hydrodynamics. New York: Prentice Hall 1962
91. Lewis, W. K., Whitman, W. G.: Ind. Eng. Chem. *17*, 1215 (1924)
92. Lin, C.: Biotech. Bioeng. *18*, 1557 (1976)
93. Linek, V.: Chem. Eng. Sci. *21*, 777 (1966)
94. Loucaides, R., McManamey, W. J.: Chem. Eng. Sci. *28*, 2165 (1973)
95. Lloyd, J. R., Moran, W. R.: J. Heat Transfer *96*, 443 (1974)
96. Marrucci, G.: Chem. Eng. Sci. *24*, 975 (1969)

97. Marrucci, G., Nicodemo, L., Acierno, D. In: Cocurrent gas-liquid flow, p. 95. New York: Plenum Press 1969
98. Marshall, K. C., Alexander, M.: J. Bacteriol. *80*, 412 (1960)
99. Mehta, V. D., Sharma, M. M.: Chem. Eng. Sci. *26*, 461 (1971)
100. Metz, B., Kossen, W. W. F.: Biotech. Bioeng. *19*, 781 (1977)
101. Metzner, A. B. et al.: AIChE J. *7*, 3 (1961)
102. Metzner, A. B.: In: Adv. chem. engng. (Drew, T. B., Hoopes, H. W., eds.), vol. 1. New York: Academic Press 1956
103. Metzner, A. B., Otto, R. E.: AIChE J. *3*, 3 (1957)
104. Michel, B. J., Miller, S. A.: AIChE J. *8*, 262 (1962)
105. Miller, D. N.: AIChE J. *20*, 445 (1974)
106. Mohan, V., Raghuraman, J.: AIChE J. *22*, 259 (1976)
107. Moo-Young, M.: Can. J. Chem. Eng. *53*, 113 (1975)
108. Moo-Young, M., Binder, A., van Dedem, G.: Biotech. Bioeng. *21*, 593 (1979)
109. Moo-Young, M., Chan, K. W.: Can. J. Chem. Eng. *49*, 187 (1971)
110. Moo-Young, M., Cross, J. V.: Can. J. Chem. Eng. *47*, 369 (1969)
111. Moo-Young, M., Fulford, G. D., Godsalve, E.: Can. J. Chem. Eng. *51*, 368 (1973)
112. Moo-Young, M., Hirose, T.: Can. J. Chem. Eng. *50*, 128 (1972)
113. Moo-Young, M., Hirose, T., Geiger, K. H.: Biotech. Bioeng. *11*, 725 (1969)
114. Moo-Young, M., Kobayashi, T.: Can. J. Chem. Eng. *50*, 162 (1972)
115. Moo-Young, M. et al.: Proc. Internat. Conf. on Two-Phase Flow, p. 1049. Dubrovnik: Hemisphere Press 1979
116. Moo-Young, M.: Process Biochem. *11*, 32 (1976)
117. Moo-Young, M., Shimizu, T., Whitworth, D.: Biotech. Bioeng. *13*, 241 (1971), *15*, 649 (1973)
118. Moo-Young, M., Shoda, M.: Ind. Eng. Chem. Proc. Des. Dev. *12*, 410 (1973)
119. Moo-Young, M., Tichar, K., Dullien, F. A. L.: AIChE J. *18*, 178 (1972)
120. Moo-Young, M., van Dedem, G.: Proc. 4th Internat. Conf. Global Impacts of Appl. Microbiology. Vol. 2, p. 945. Brazilian Soc. of Microbiol. 1976
121. Moo-Young, M., Vocadlo, J. J., Charles, M. E.: Can. J. Chem. Eng. *50*, 544 (1972)
122. Moser, A.: Biotech. Bioeng. Symp. Ser. *4*, 399 (1973)
123. Nakano, Y., Tien, C.: AIChE J. *14*, 145 (1968)
124. Narayanan, S., Goossens, L. H., Kossen, N. W. F.: Chem. Eng. Sci. *29*, 2071 (1974)
125. Nicodemo, L., Marrucci, G., Acierno, D.: Ing. Chimico Italiano *8*, 1 (1972)
126. Niitsu, H., Fufita, M., Tervi, G.: J. Ferm. Technol. *47*, 194 (1969)
127. Okazaki, M., Moo-Young, M.: Biotech. Bioeng. *20*, 637 (1978)
128. Oldshue, J. Y.: Biotech. Bioeng. *13*, 3 (1966)
129. Onda, K. et al.: Chem. Eng. (Japan) *32*, 801 (1968)
130. Oyama, Y., Endo, K.: Chem. Eng. (Japan) *19*, 2 (1955)
131. Perez, J. F., Sandall, O. C.: AIChE J. *20*, 770 (1974)
132. Perry, R. H., Chilton, C. H.: Chemical Engineers Handbook, 5th edit., New York: McGraw-Hill 1973
133. Phillips, H. H.: Biotech. Bioeng. *8*, 456 (1966)
134. Pirt, S. J.: Proc. Royal Soc. London, B *166*, 369 (1966)
135. Ranade, V. R., Ulbrecht, J. J.: AIChE J. *24*, 796 (1978)
136. Ranade, V. R., Ulbrecht, J. J.: Paper at 2nd Europ. Conf. on Mixing (F6), March 1977
137. Reith, T.: Third C.H.I.S.A. Congr., Czechoslovakia, Sept. 1969
138. Richards, J. W.: Prog. Ind. Microbiol. *3*, 141 (1961)
139. Righelato, R. C. In: The filamentous fungi. Smith, J. E., Berry, D. R. (eds.), vol. 1, p. 79, London: Edward Arnold 1975
140. Robinson, C. W., Wilke, C. R.: Biotech. Bioeng. *15*, 755 (1973)
141. Robinson, C. W., Wilke, C. R.: AIChE J. *20*, 285 (1974)
142. Rowe, P. N., Claxton, K. T., Lewis, J. B.: Trans. Inst. Chem. Engrs. *43*, 14 (1965)
143. Rushton, J., Chem. Eng. Prog. *53*, 319 (1956)
144. Rushton, J. H., Costich, E. W., Everett, H. H.: Chem. Eng. Prog. *46*, 467 (1950)
145. Russell, T. W. F.: Can. J. Chem. Eng. *50*, 179 (1972)
146. Russell, T. W. F., Dunn, I. J., Blanch, H. W.: Biotech. Bioeng. *16*, 1261 (1974)
147. Satterfield, C. N., Mass transfer in heterogeneous catalysis, MIT Press 1970

148. Schügerl, K., Oels, U., Lücke, J.: Adv. Biochem. Eng. *7*, 1 (1977)
149. Schügerl, K. et al.: Adv. Biochem. Eng. *8*, 63 (1978)
150. Schultz, J., Gaden, E.: Ind. Eng. Chem. *48*, 2209 (1956)
151. Shaftlein, R. W., Russell, T. W. F.: Ind. Eng. Chem *60* (5), 13 (1968)
152. Sherwood, T. K., Pigford, R. L., Wilke, C. R.: Mass transfer. New York: McGraw-Hill 1975
153. Sideman, S.: Ind. Eng. Chem. *58*, 32 (1966)
154. Sideman, S., Hortacsu, Ö., Fulton, J. W.: Ind. Eng. Chem. *58*, 32 (1966)
155. Sigurdson, S. P., Robinson, C. W. In: Develop. Ind. Microbiol. Underkofler, L. A. (ed.), vol. 18, p. 529. Washington, D.C.: Amer. Inst. Biological Sci. 1977
156. Skelland, A. H. P.: Non-newtonian flow and heat transfer. New York: Wiley 1967
157. Smith, J., Van't Riet, K., Middleton, J.: Sec. Europ. Conf. on Mixing, Prepr. April 1977
158. Solomon, G. L. et al.: Continuous cultivation of microorganisms, p. 345. New York: Academic Press 1962
159. Strohm, P., Dale, R. F., Peppler, H. S.: Appl. Microbiol. *7*, 235 (1959)
160. Suga, K., van Dedem, G., Moo-Young, M.: Biotech. Bioeng. *17*, 185 (1975)
161. Suga, K., van Dedem, G., Moo-Young, M.: Biotech. Bioeng. *17*, 433 (1975)
162. Sumino, Y., Kanzaki, T., Fukuda, H.: J. Ferm. Technol. *46* (12), 1040 (1968)
163. Sumino, Y., Kanzaki, T., Tervi, G.: J. Ferm. Technol. *46*, 1040 (1968)
164. Taguchi, H.: Adv. Biochem. Eng. *1*, 1 (1971)
165. Taguchi, H.: J. Ferm. Technol. *46*, 823 (1968)
166. Taguchi, H. et al.: J. Ferm. Technol. *46*, 823 (1968)
167. Taguchi, H., Miyamoto, S.: Biotech. Bioeng. *8*, 43 (1966)
168. Thiele, E. W.: Ind. Eng. Chem. *31*, 916 (1936)
169. Tomita, Y.: Brit. Soc. Mech. Engrs. *2*, 469 (1959)
170. Toor, H. L., Marchello, J. M.: AIChE J. *4*, 97 (1958)
171. Towell, G. D., Strand, C. P., Ackerman, G. H.: AIChE-Inst. Chem. Eng. Symp. Ser. (London) *10*, 10 (1965)
172. Treybal, R. E.: Mass-Transfer Operations New York: McGraw-Hill 1968
173. Wasserman, M. L., Slattery, J. C.: AIChE J. *10*, 383 (1964)
174. Wellek, R. M., Huang, C. C.: Ind. Eng. Chem. Fund. *9*, 480 (1970)
175. Westerterp, K.: Chem. Eng. Sci. *18*, 157 (1963)
176. Westerterp, K., R. van Dierendonck, L. L., DeKraa, J. A.: Chem. Eng. Sci. *18*, 157 (1963)
177. Wimpenny, J. W. T.: Process Biochem. *4*, 19 (1969)
178. Yagi, H., Yoshida, F.: Ind. Eng. Chem. Proc. Des. Develop. *14*, 488 (1975)
179. Yano, T., Kodama, T., Yamada, K.: Agr. Biol. Chem. *25*, 580 (1961)
180. Yoshida, F.: Ind. Eng. Chem. *52*, 435 (1960)
181. Yoshida, F., Akita, K.: AIChE J. *11*, 9 (1965)
182. Yoshida, F., Akita, K.: Ind. Eng. Chem. Proc. Des. Develop. *12*, 76 (1973)
183. Yoshida, F., Miura, Y.: Ind. Eng. Chem. Proc. Des. Develop. *2*, 263 (1963)
184. Yoshida, F. et al.: J. Ferm. Technol. *45*, 1119 (1967)
185. Yoshida, T., Taguchi, H., Teramoto, S.: J. Ferment. Technol. *46*, 119 (1968)
186. Ziegler, H. et al.: Biotechn. Bioeng. *19*, 507 (1977)
187. Zieminski, S. A., Caron, M. M., Blackmore, R. B.: Ind. Eng. Chem. *6*, 233 (1967)
188. Zlokarnik, M.: Adv. Biochem. Eng. *8*, 133 (1978)
189. Zlokarnik, M.: Adv. Biochem. Eng. *11*, 157 (1979)

Oxygen Transfer Into Highly Viscous Media

K. Schügerl
Institut für Technische Chemie, Universität Hannover,
Callinstr. 3, 3000 Hannover 1, Federal Republik of Germany

Highly viscous media often occur in biotechnology, fermentation and food processing industries as well as in chemical and pharmaceutical industries. The present paper considers the behavior of aerated highly viscous media employing stirred tank reactors, sparged single and multistage tower reactors.

Measuring methods to determine rheological properties, hydrodynamical behavior, power input, and mass transfer as well as behavior of single bubbles and bubble swarm are reviewed.

Hydrodynamical properties, power inputs, oxygen transfer rates, volumetric mass transfer coefficients and heat transfer coefficients are considered as a function of the mean operating parameters employing stirred tank reactors with different impellers. Also sparged single and multistage tower reactors are treated, especially hydrodynamical properties, oxygen transfer rates and volumetric mass transfer coefficients. Relationships are given to lay out stirred tank and tower reactors with highly viscous media.

1 Introduction

Rheological properties of culture media strongly influence bioreactor performance, especially if aerobic microorganisms are employed. Low-viscosity media are fairly easy to handle, the mixing rate and oxygen transfer rate, OTR, in them are usually high enough to ensure an unproblematic bioreactor operation, as long as microorganism specific growth rate is not extremely high.

A medium viscosity increase can be caused by

— high substrate concentration (especially at the start of a batch culture),
— secretion of highly viscous products (e.g. pullulane, xanthane)[1] (especially at the end of batch culture),
— microorganism morphology[2–5] (especially at the end of batch cell cultivations),
— substrates and cell morphology interaction (e.g. *Penicillium chrysogenum*),
— products and cell morphology interaction (e.g. *Pullularia pullulans*).

If substrates or products are responsible for the high viscosity, the medium usually has Newtonian character. High viscosity due to microorganisms is very often coupled with non-Newtonian behaviour.

The unfavorable influence of high viscosity on mixing and oxygen transfer rates can be attributed to several processes. They are briefly considered in the following.

a) Since the molecular transfer parameters (diffusivity, heat conductivity, viscosity) are insufficient to maintain adequate momentum, mass, and heat transfer rates, which are necessary for effective industrial production, these parameters are intensified by turbulence. By means of the turbulent exchange parameters (turbulent diffusivity and viscosity), which maintain a much higher transport process intensity than the molecular transport parameters are able to do, the production intensity can be increased by many orders of magnitude. The transition from laminar to turbulent flow occurs at a particular (critical) Reynolds number $Re_c = u_c \cdot d/\nu$ if Newtonian liquids are employed, where d and u are the characteristic length and flow velocity, respectively. The higher the kinematic viscosity, ν, the higher the particular (critical) velocity, u_c, at which flow transition occurs. Enforcing high velocity needs high power input, which yields a high energy dissipation rate and produces large amounts of heat. To avoid very high power input and heat development in highly viscous media, lower Reynolds numbers are employed than in low viscosity liquids. This leads to lower transport intensities (mass and heat transport rates), higher

mixing times, Θ, and lower dynamic turbulence pressure, τ_T, than those attained in low viscosity media.
For non-Newtonian media, the same statement holds; only the definition of the critical Reynolds number is more difficult.

b) At very low gas flow rates single bubbles are formed at the gas distributor. With increasing viscosity the primary bubble size or diameter, d_p, becomes larger[6]. This reduces the gas/liquid interfacial area.

c) The primary bubbles are usually destroyed by coalescence and/or redispersion processes.
If the dynamic turbulence pressure, τ_T, is low in the system, the bubbles grow by coalescence until they attain the (stable) dynamic equilibrium bubble size, which is governed by τ_T[7]. Since with increasing viscosity, τ_T diminishes, the dynamic equilibrium bubble size or diameter, d_e, becomes larger. How rapidly this maximum size is attained depends on the coalescence rate. The coalescence rate is also enlarged by increasing viscosity[8-10].
If the dynamic turbulence pressure is large in the system, the primary bubbles are dispersed. The final bubble size is determined by the local d_e, which is controlled by the local τ_T. Again, with increasing viscosity, τ_T, is diminished, d_e enlarged, and the gas/liquid interfacial area reduced.

d) As long as small spherical bubbles are present, the specific geometric gas/liquid interfacial area, with regard to the liquid volume, can be calculated by the following relationship:

$$a = \frac{6E_G}{d_S(1-E_G)}, \tag{1}$$

where E_G is the relative gas hold-up and d_S the actual Sauter bubble diameter

$$d_S = \frac{\sum f_i d_i^3}{\sum f_i d_i^2}, \tag{2}$$

f_i is the bubble frequency with diameter d_i.
Large bubbles have no spherical shape (rotation ellipsoids, spherical caps, slugs); therefore, Eq. (1) cannot be applied. However, by suitable definition of d_S, a relationship similar to Eq. (1) can be developed, i.e. Eq. (1) can be used for qualitative considerations.
For low τ_T, $d_p \leqq d_S \leqq d_e$, and for high τ_T, $d_S \simeq d_e \leqq d_p$.
The specific interfacial area, a, can be enlarged by increasing τ_T (e.g. agitator speed) and thus reducing $d_e = d_S$, or by increasing E_G (e.g. the gas flow rate).
However, in highly viscous media, the agitator speed and gas flow rate are limited. By increasing the agitator speed, the stirrer efficiency diminishes and the heat production increases. By raising the gas flow rate, large slugs are formed and/or the stirrer is flooded, which sharply reduces its efficiency. These limits of stirrer rotation speed, N, and of gas flow rate, Q_G, also exist in low-viscous liquids, but they have much higher values than those in highly viscous media.

e) Small bubbles are formed during bubble dispersion[10-11] which have long residence times in the system. Because of their low oxygen content, they are quickly exhausted and do not contribute to the OTR. Furthermore, they influence the dynamics of the two-phase fluid system in such a way that the apparent viscosity of the system decreases, resulting in an increase of the large bubble ascending velocity. Thus, the large bubbles leave the system with nearly initial oxygen content. With rising viscosity, the bubble fraction increases which breaks up and/or coalesces into very small and very large bubbles.

f) The oxygen transfer rate, *OTR*, is given by Eq. (3):

$$\text{OTR} = k_L a\,(O_L^* - O_L)\,, \tag{3}$$

where $k_L a$ = volumetric mass transfer coefficient,
k_L = gas/liquid mass transfer coefficient,
O_F^* = dissolved oxygen saturation concentration in the medium, assumed to prevail at the interface,
O_F = dissolved oxygen concentration in the medium bulk.

According to the film theory (cf.[12])

$$k_L = \frac{D_m}{\delta}\,, \tag{4a}$$

or to the penetration theory of Higbie

$$k_L = 2\sqrt{\frac{D_m}{\pi t_c}}\,, \tag{4b}$$

where D_m is the molecular diffusivity of oxygen in the medium and δ the film thickness.

Since D_m usually diminishes and δ as well as the contact time, t_c, increase with rising viscosity, k_L is reduced considerably, if the viscosity of the medium is increased.

From this short consideration, one can recognize which parameters are responsible for the unfavorable viscosity effect.

The aim of the present survey is, first to consider these parameters separately and second to analyze their interrelationships with regard to the calculation of the OTR in highly viscous media based on recent literature data considering model media.

2 Materials and Methods

2.1 Employed Media

Generally, glycerol, glucose and PEG solutions are employed as Newtonian model media and different polymer solutions as non-Newtonian model media.
The Newtonian media are characterized by their viscosity, pseudo-plastic media by

their fluid consistency index, k, and flow behavior index, n, viscoelastic media by their n and k value as well as by their relaxation time.

In the literature CMC (carboxymethylcellulose) solutions are most frequently used as pseudoplastic and PAA (polyacrylamide) solutions as viscoelastic media. The rheological properties of these media depend not only on their concentrations but also on their molecular weight. Therefore, the properties of these model media are described in that chapter of this review article where they are used.

Several trade names are employed in this article. These are compiled in the list of symbols.

2.2 Methods for Measuring the Rheological Behavior

The flow equation (i.e. the velocity gradient, dv/dx, as a function of the stress τ) (cf.[13]):

$$\frac{dv}{dx} = f(\tau) \tag{5}$$

gives the most general description of the rheological properties of fluids as long as viscoelastic behavior is not present or very slight. This flow equation can be calculated from the experimentally measured shear diagrams (shear rate versus shearing stress); it should be noted, however, that this calculation is not always possible. In contrast to the shear diagram, the flow equation is independent of experimental conditions (e.g. the type of viscosimeter) used for the determination of the viscosity.

There are many methods available to estimate the rheological behavior of fluids, but there are only a few that furnish true fluidity values. These include the capillary, the falling sphere, the Couette, and the torsional pendulum methods. Until now, the evaluation of the flow equation from the shear diagram was only possible for the capillary and the Couette methods (cf.[13]). The capillary viscosimeter cannot be employed for cultivation broths because of the adverse wall effects arising in the capillary. The Couette viscosimeter can only be used if some important conditions are fulfilled (see below). As for the falling sphere and torsional pendulum viscosimeters, Eq. (5) cannot be calculated from the shear diagram (only partial solutions are known, cf. Jain[14]). In addition to this, in falling sphere viscometers the wall effects and the disturbances which occur on the upward facing surface of the sphere are too large and cannot be neglected. Therefore, only the Couette viscometer can be used for the estimation of the non-Newtonian behavior of cultivation broths. The Newtonian behavior can be determined by employing either the Couette or the torsion pendulum viscometer. For a better comparison between the rheological properties of the model media and those of the cultivation broths, a modification of the Couette viscometer consisting of concentric cylinders with a rotating inner cylinder has been applied. The inner diameter of the outer cylinder, R_a, must be significantly larger than the outer diameter of the inner cylinder, R_i, to avoid falsification due to wall effects. Furthermore, measurements must be made using different cylinder lengths to eliminate the end effects. If the inner cylinder rotates with the angular velocity Ω and the outer cylinder is in a fixed position, then the angular velocity of the fluid at distance r from the axis of rotation will be ω, at $r = R_i$, $\omega = \Omega$, and at $r = R_a$, $\omega = 0$.

The velocity gradient at distance r is

$$\frac{dv}{dx} = -\frac{d\omega}{dr}, \tag{6}$$

while the shear stress is

$$\tau = \frac{M_i}{2\pi r^2 L}, \tag{7}$$

where M_i is the torque exerted on the inner cylinder and L the length of the inner cylinder. From Eqs. (6) and (7) it follows that

$$d\omega = \frac{1}{2} f(\tau) \frac{d\tau}{\tau}. \tag{8}$$

Integration of Eq. (8) with $s^2 = R_i^2/R_a^2 = \tau_i/\tau_a$ gives

$$\Omega = \frac{1}{2} \int_{s^2\tau_i}^{\tau_i} \frac{f(\tau)}{\tau} d\tau. \tag{9}$$

The relationship between Ω and τ is experimentally determined to obtain the shear diagram; the relationship $dv/dx = f(\tau)$ (flow equation) is to be calculated from Eq. (9). The evaluation of the flow equation from the shear diagram can be carried out using the methods of Mooney[15], Krieger[16-18], Pawlowski[19], Schulz-Grunow[20], Tillmann[21] or by a combination of the methods of Pawlowski and Weymann[22].

On comparing a non-Newtonian with a Newtonian liquid, one can define the apparent shear rate prevailing at the inner cylinder

$$\left(\frac{dv}{dx}\right)_{i\,app} = \frac{2\Omega}{1 - s^2} \tag{10a}$$

and an apparent viscosity for the Couette flow:

$$\eta_{app}(\tau_i) = \frac{\tau_i}{\left(\frac{dv}{dx}\right)_{i\,app}(\tau_i)}. \tag{10b}$$

This yields the apparent flow curve

$$\left(\frac{dv}{dx}\right)_{i\,app}(\tau_i) = \frac{1}{1 - s^2} \int_{s^2\tau_i}^{\tau_i} \frac{\left(\frac{dv}{dx}\right)(\tau)}{\tau} \tag{10c}$$

or

$$\frac{1}{\eta_{app}(\tau_i)} = \frac{1}{(1 - s^2)\,\tau_i} \int_{s^2\tau_i}^{\tau_i} \frac{1}{\eta(\tau)}\, d\tau\,. \tag{10d}$$

In the case of Ostwald-de Waele liquids with the simple power law

$$\tau = K\left(\frac{dv}{dx}\right)^n$$

and with Eqs. (*d*) and (*c*) we obtain

$$\frac{\eta(\tau_i)}{\eta_{app}(\tau_i)} = n\,\frac{1 - s^{2/n}}{1 - s^2}\,, \tag{11}$$

i.e. the relationship between $\eta(\tau)$ and $\eta_{ap}(\tau_i)$ is only a function of the viscosimeter geometry, s, and of n[176].

For the characterization of viscoelastic media special rheometers are employed in which the liquid is stressed in a concentric space between a sphere and a hollow sphere by rotation of one of the spheres[23-25]. By analysis of the torque and wall pressure characteristics it is possible to determine the rheological parameters. The estimation of the rheological behavior of the liquids permits to define three functions which are characteristic of the properties of the liquid:
A shear stress function, τ, and two independent functions of the normal stress differences, σ_I, and σ_{II}, where

σ = normal stress difference between the flow direction and directions indifferent to the flow and

σ_{II} = normal stress difference between the shear direction and directions indifferent to the shear.

Using these three functions, the following material constants can be defined[25]:

$$\eta_0 = \lim_{D\to 0} \frac{\tau}{D} \quad \text{initial viscosity,} \tag{12a}$$

$$\left.\begin{aligned} t_{01} &= \frac{1}{2}\lim_{D\to 0}\frac{\sigma_I + \sigma_{II}}{D} \\ t_{02} &= \frac{1}{2}\lim_{D\to 0}\frac{\sigma_I - \sigma_{II}}{D} \end{aligned}\right\} \text{initial relaxation times.} \tag{12b}$$

By applying the rheological function of state:

$$\bar{\sigma} = -\alpha\bar{I} + \alpha_1\bar{D} + \alpha_2\bar{D}^2 + \alpha_3\bar{\bar{D}} + \alpha_4\bar{D}^3 + \ldots \tag{13}$$

where $\bar{\sigma}$ stress tensor
I unit tensor
$\bar{D}$ deformation velocity tensor
$\bar{\bar{D}}$ first derivative of $\bar{D}$
α_i coefficients.

The relations between η_0, t_{01}, t_{02} and α_i are given by:

$$\eta_0 = \frac{1}{2}\alpha_1 , \quad (13a)$$

$$t_{01} = \frac{\alpha_2}{\alpha_1} , \quad (13b)$$

$$t_{02} = -\frac{\alpha_3}{\alpha_1} . \quad (13c)$$

According to Tanner the following power functions can be applied[26]:

$$\tau = K(D)^n , \quad (14a)$$

$$\sigma_I = h_1(D)^{\lambda 1} , \quad (14b)$$

$$\sigma_{II} = -h_2(D)^{\lambda 2} . \quad (14c)$$

2.3 Methods for the Determination of Gas Solubilities in Media

After degassing of the medium the oxygen pressure is measured in the closed system above the liquid until a constant value is attained[27-31]. The effect of elctrolytes on the solubility of gases in aqueous solutions is usually described by the Sechenov equation:

$$\log (C_0/C) = K_S C_{el} . \quad (15)$$

Here C_0 is the gas solubility in pure water and C the solubility in the solution with a molar concentration of electrolyte, C_{el}. The salting out constant K_S can be calculated according to the model of Krevelen and Hoftijzer from the ionic strength. This model was modified by Danckwerts[12]:

$$\log (C_0/C) = \Sigma h_i I_i \quad (16)$$

where I_i is the ionic strength attributed to salt i and $h_i = h_G$ (gas), h_+ (cation), h_- (anion) which are empirical constants for species i. They are tabulated in[12].

According to Deckwer[28] C can be calculated by

$$\log (C_0/C) = \Sigma H_i I_i . \quad (17)$$

The constants H_i are compiled in[28,29]. When considering organic compounds (glucose, saccharose, glycerol) the following model is recommended[29]:

$$\log (C/C_0) = a + b(H_i I_i) . \quad (18)$$

The constants a and b are listed in[29].

For aqueous polymer (PEG) solutions, the following relationship was found:

$$\log (\alpha_0/\alpha) = b' C_{Poly} \quad (19)$$

where α_0 and α are the Bunsen constants $\left(\frac{\text{mol gas under standard conditions}}{\text{mol liquid}}\right)$ in pure water and in aqueous solution, respectively, and C_{Poly} is the molar concentration of the polymer (gl^{-1})

$$b' = 2.2 \times 10^{-4}\ lg^{-1} \text{ for PEG 20000 to 200}^{30)} .$$

2.4 Methods for the Determination of the Diffusivity of Dissolved Gases in Media

Diffusivities are determined by means of the gas absorption rate under defined fluid dynamic conditions (laminar jet, laminar film (on cylinder), laminar film (on sphere), etc. (e.g.[12, 32–35])).

Using the model of Hayduk and Chang[33], the diffusivity can be calculated by

$$D_m \eta^A = K_1 , \tag{20}$$

where D_m is the gas diffusivity in the medium (cm^2s^{-1}) and η the dynamic viscosity of the medium ($mPa \cdot s$); A (—) and K_1 ($cm^2\ s^{-1}$) are empirical constants. For example, for CO_2 in polyvinyl alcohol (PVA 100000 and 49000) and PEG 20000 and 10000 the following constants are valid[34]:

$$K_1 = 1.925 \times 10^{-5}\ cm^2s^{-1} ,$$
$$A = \frac{7.233}{M_p^{0,45}} ,$$

where M_p is the molecular weight of the polymer.

2.5 Methods for Measuring the Interfacial Properties

The most common characterization of the liquid surface is by its surface tension, σ.. However, in biological media as well as in polymer solutions, σ is time-dependent. It takes a long time (about 1000 min) to attain the equilibrium surface tension[36, 37]. The surface tension as a function of time can be measured by an automatic tensiometer which uses the method of Lecomte de Noüy. The readings have to be corrected according to Harkins and Jordan[38]. If measurements are carried out without lamellae-tear off, the ring method is equivalent to the static slab method, i.e. it can also be applied to systems with interfacial films[39]. The liquid surface hast to be renewed before the σ/t curve is registered.

However, neither the "short age" (after some seconds) nor the "long age" (after some hours) surface tension characterizes the surface satisfactorily, since during the gas dispersion and bubble coalescence only the dynamic surface behavior is effective. Surface tension measured by the methods of de Noüy, Wilhelmy etc. is more or less a static property. Applying the capillary and/or longitudinal wave method, the surface viscosity and elasticity can be determined[40, 41]. This method is based on the determination of the amplitude decay and phase shift of capillary and/or longitudinal waves which were produced by a small oscillating razor blade on the liquid surface, at some distance from the signal transmitter. However, by

Table 1. Properties of the media employed by Voigt et al.[36] and Hecht et al.[37]

Concentration	ϱ at 20 °C	η	k	n	$D \times 10^5$	σ^b
wt-%	$g \times cm^{-3}$	$mPa \times s$	$Pa \times s^n$	—	$cm^2\ s^{-1}$	$mN \times m^{-1}$
Glycerol						
50	1.127	6	—	—	0.53	69.1
70	1.183	24	—	—		66.8
90	1.237	240	—	—		63.7
97	1.255	950	—	—		62.6
CMC (carboxylmethylcellulose) (Tylose C300, Hoechst AG)						
1.0	1.000[a]	—	0.09	0.82	2.28[a]	59.2
1.4	1.001[a]	—	0.24	0.77	2.28[a]	59.7
2.0	1.003[a]	—	0.72	0.71	2.28[a]	58.7
PAA (polyacrylamide) (Separan AP30, Dow Chemical)						
0.2	1.001	—	0.11	0.63		47.3
0.5	1.004	—	0.75	0.44		44.9
1.0	1.008	—	3.00	0.38		46.3

[a] at 30 °C, adapted from Yagi[112]
[b] The surface tensions σ of CMC and PAA solutions are time-dependent[36, 37]
The values compiled here are the equilibrium surface tensions

this frequency response method only the dynamic properties of the "long age" surface can by measured. The gas/liquid interface displays viscoelastic properties even at very low polymer concentrations (1 gl^{-1} CMC and 0.1 gl^{-1} PAA solutions)[175].

2.6 Properties of the Media Employed

On the authors laboratory glycerol, CMC and PAA solutions were used as model media with Newtonian, pseudo-plastic and/or viscoelastic properties. In Table 1 the properties media are compiled. In Figure 1 the viscosity of aqueous glycerol solution as a function of its concentration is shown. In Figs. 2 and/or 3, the flow behavior index, n and fluid consistency index, K, are plotted as a function of the CMC and/or PAA concentrations.

2.7 Methods for Determining the Relative Gas Hold-Up, E_G

In bubble columns it is easy to determine E_G, if the height of the bubbling layer, H, can be measured:

$$E_G = \frac{V - V_L}{V} = \frac{H - H_L}{H}, \tag{21}$$

where V = volume of the bubbling layer
V_L = volume of the bubble free layer
H_L = height of the bubble free layer.

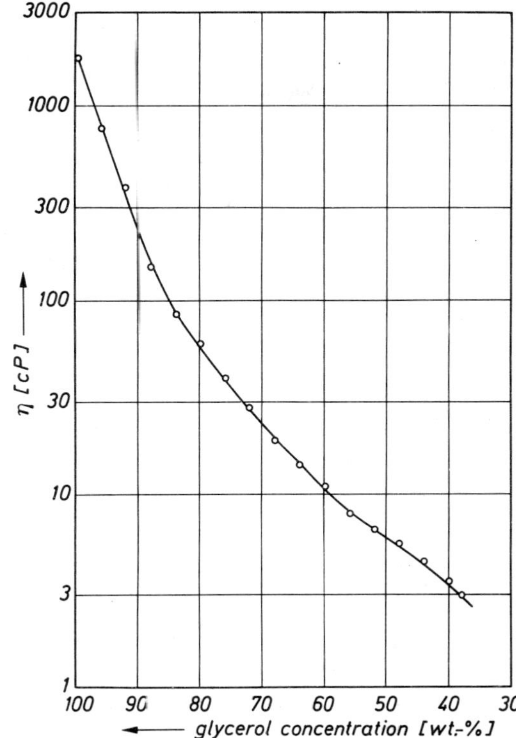

Fig. 1 Viscosity as a function of the glycerol concentration[164)]

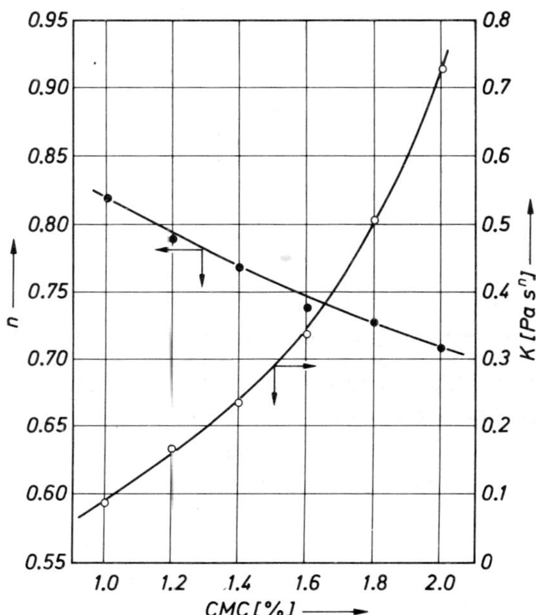

Fig. 2 K and n as a function of the CMC concentration[36)]

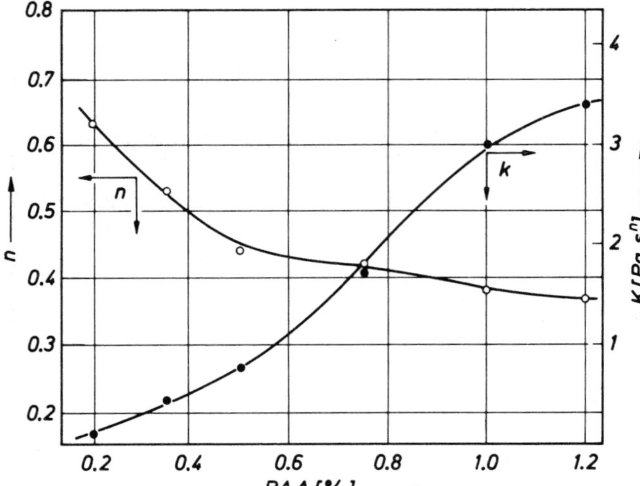

Fig. 3 *K* and *n* as a function of the PAA concentration[37)]

If foam is formed, the determination of H e.g. by the naked eye or an electrical conductivity measurement is difficult.

Also, if large slugs are present, the measurement of H is fairly inaccurate.

Sometimes, it is useful to distinguish between gas hold-up due to very "small" bubbles, E_{GK}, and due to "intermediate to large" bubbles, E_{GG}. E_{GK} is attributed to bubbles which have fairly large residence times in the system. E_{GG} is due to bubbles which quickly ascend in the two-phase system and can be calculated by

$$E_{GG} = E_G - E_{GK}\,, \tag{22}$$

where E_G is the overall relative gas hold-up under steady — state conditions.

Immediately after the gas flow has been turned on, the fraction of small bubbles becomes very low: thus, $E_G \simeq E_{GG}$.

After a steady state has been obtained, E_G is measured again. Then, the gas flow is turned off and some seconds afterward, the gas hold-up due to the "small" bubbles, E_{GK}, is measured. E_{GG} is calculated by relationship, Eq. (22). The E_{GG} values evaluated at the beginning of the measurement and at the steady state are identical[11)].

In the author's laboratory an arbitrary time delay of 15 s was employed for the determination of E_{GK}. As a result, the contribution of the "small" bubbles to the convective gas flow can be neglected. In Fig. 4, the time dependence of E_{GK}, E_{GG} and E_G are shown for glycerol solutions. In highly viscous media about 6 h are needed to attain the steady state.

In Figures 5 to 8, the steady-state values of E_{GK}, E_{GG} and E_G are shown for 50, 70, 90 and 95% glycerol solutions, using different perforated plates as gas distributors. In a 50% glycerol solution the particular relative gas hold-up due to the very small bubbles, E_{GK}, does not depend on w_{SG} (Fig. 5). When employing a 0.5 mm perforated plate, E_{GK} is negligibly small. By increasing the hole diameter, d_H, of the perforated plates, E_{GK} increases. According to this, with $d_H = 0.5$ mm,

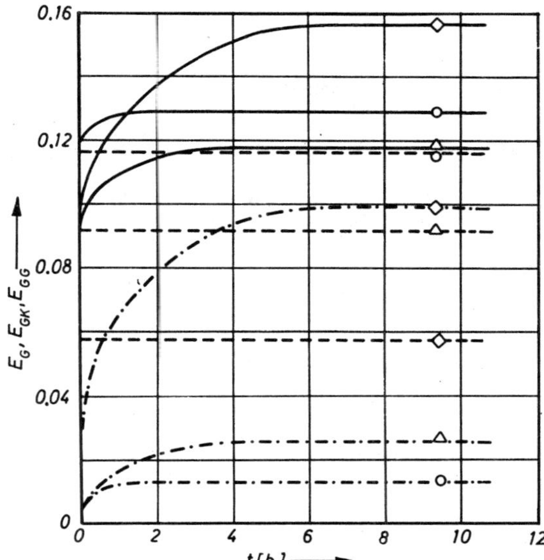

Fig. 4 Time dependence of gas hold-up E_G and of particular gas hold-up fractions E_{GK} and E_{GG} in glycerol solution[11]. Perforated plate d_H = 1.0 mm, w_{SG} = 3.8 cm s^{-1}. ——— E_G; —·—·— E_{GK}; — — — E_{GG}; ○ 50%; △ 70%; ◇ 95% glycerol solutions

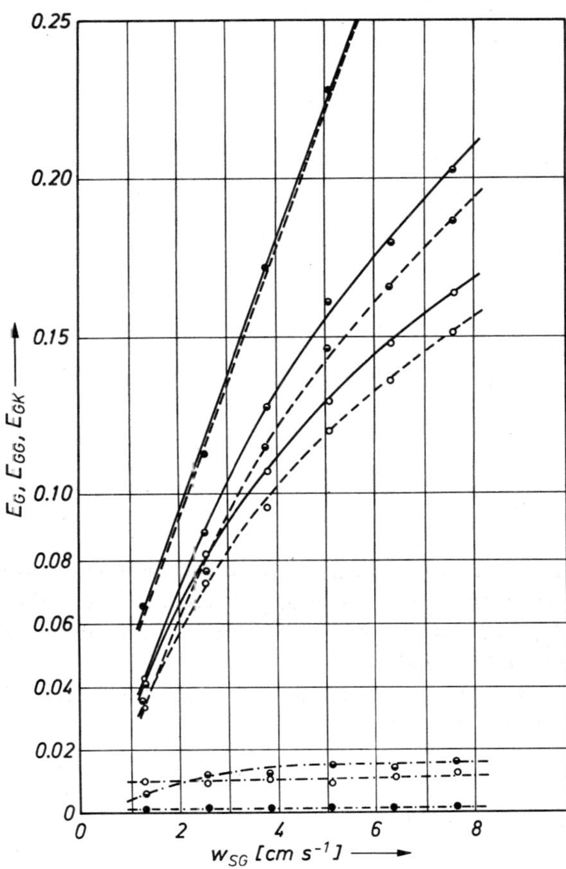

Fig. 5 Steady-state values of E_G, E_{GK} and E_{GG} in 50% glycerol solution as a function of w_{SG}. Perforated plate aerators[11]. ——— E_G; — — — E_{GG}; —·—·— E_{GK}; ● d_H = 0.5 mm; ◒ d_H = 1.0 mm; ○ d_H = 3.0 mm

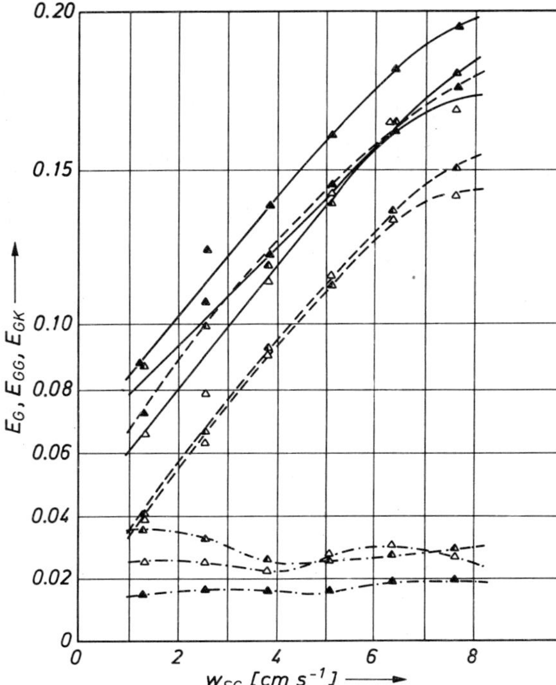

Fig. 6 Steady-state values of E_G, E_{GK} and E_{GG} in 70% glycerol solution as a function of w_{SG}. Perforated plate aerators D_c = 14 cm, H_S = 35 cm[11]. ——— E_G; – – – E_{GG}; —·—·— E_{GK}; ▲ d_H = 0.5 mm; ▲ d_H = 1.0 mm; △ d_H = 3.0 mm

$E_G \simeq E_{GG}$ and with increasing d_H:$E_G > E_{GG}$. With rising w_{SG}, E_{GG} and E_G increase. The use of 0.5 mm perforated plate produces the highest E_{GG} and E_G values.

In 70% glycerol solutions the small bubble fraction is significantly higher (Fig. 6). Again, E_{GK} is independent of w_{SG}. The application of perforated plates yields the smallest E_{GK}. By enlarging d_H, E_{GK} increases. E_{GG} and thus E_G increase with rising w_{SG}. A 0.5 mm plate again yields the highest E_{GG} and E_G values, respectively. In 90% glycerol solutions, E_{GK} has already attained a value of 5 to 6% (Fig. 7). Furthermore, for $w_{SG} < 3$ cms^{-1}, E_{GK} is larger than E_{GG}. Again E_{GK} does not depend on w_{SG} and E_{GG} is strongly influenced by w_{SG}. There is only a slight effect of d_H on E_{GG} and E_G, respectively.

In 98% glycerol solutions, E_{GK} values are high (7 to 15%) (Fig. 8). In the range $w_{SG} < 5$ cm/s, they are higher than E_{GG}. With increasing w_{SG}, E_{GK} slightly diminishes but E_{GG} markedly increases. d_H has a pronounced effect on E_{GK} and practically no effect on E_{GG}. When applying 0.5 mm perforated plates, the highest E_{GK} values are produced, in contrast to those observed in solutions of lower glycerol concentrations. The use of a 3.0 mm perforated plate yields the lowest E_{GK} values.

One can see from Figs. 5 to 8 that E_{GK} considerably increases with growing viscosity of the liquid.

The residence time of the small bubbles in the liquid is long. After turning off the gas flow, they leave the liquid at different rates, depending on their size. Figure 9 describes the variation of E_{GK} as a function of time t, after the gas flow has been turned off. E_{GK} diminishes at first quickly and later slowly due to the

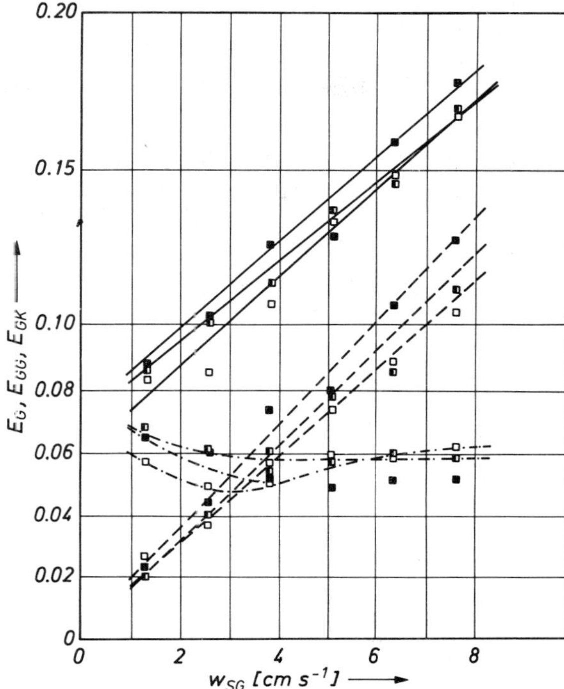

Fig. 7 Steady-state values of E_G, E_{GK} and E_{GG} in 90% glycerol solution as a function of w_{SG}. D_c = 14 cm, H_S = 35 cm, perforated plate aerators[11]. ———— E_G; — — — E_{GG}; —·—·— E_{GK}; ■ d_H = 0.5 mm; ◧ d_H = 1.0 mm; □ d_H = 3.0 mm

very small bubbles. In a 95% glycerol solution a long time is needed (ca. 24 h) to completely remove the small bubbles. Sometimes, it is appropriate to determine the gas hold-up caused by large slugs, E_{GS}, and consider only the gas-hold-up of the slug-free system, e.g. to calculate the specific interfacial area, a, according to Eq. (1). The fraction of E_G due to slugs, E_{GS}, can attain considerable values ($E_{GS}/E_G \simeq 0.5$).

2.8 Methods for Measuring the Bubble Size Distribution

Several methods of measuring bubble size distributions[42-51] are known; flash photography[41,42], electrical conductivity[43-45], electrooptical measurements[43,46-49] etc.

In highly viscous media the electrooptical method cannot be employed[48,49]. The electrical conductivity probe may be applied; however, the signal evaluation is much more difficult to perform than in low viscosity systems[51]. Also, the evaluation of photographs e.g. by means of a semiautomatic particle analyzer, is not as simple as for spherical bubbles[49]. The bubble diameter is defined as the diameter of the circle which covers the same surface area as the projection of the bubble on a photograph. It is assumed that the bubbles display axial symmetry with regard to their vertical axis. In the absence of axial symmetry, the error in the estimation of the equivalent bubble diameter is considerable. In Fig. 10 the distributions of bubble diameters are shown, which were measured in a 391 cm high bubble column at three different distances form the aerator using a 1% CMC solution. It is seen that the primary unimodal distribution becomes bimodal and trimodal with increasing

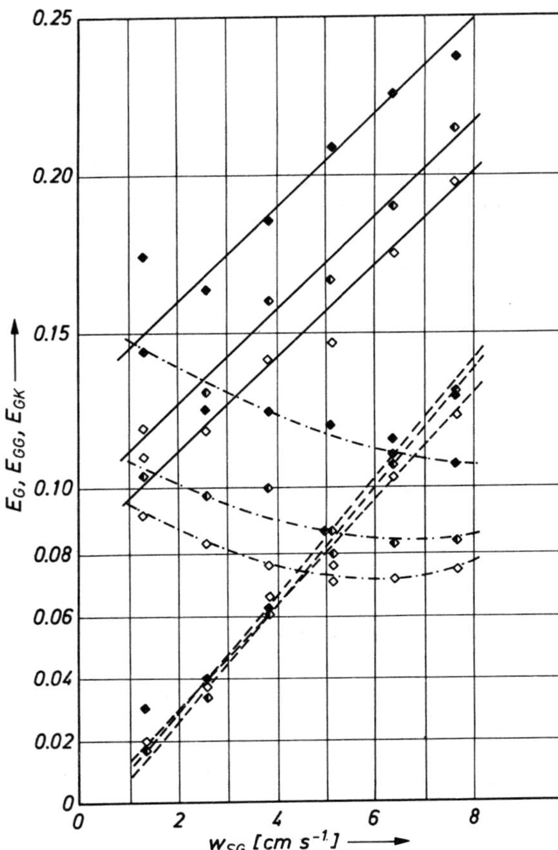

Fig. 8 Steady-state values of E_G, E_{GK} and E_{GG} in 95% glycerol solution as a function of w_{SG}. Perforated plate aerators. D_c = 14 cm, H_S = 35 cm[11]. ——— E_G; – – – E_{GG}; —·—·— E_{GK}; ◆ d_H = 0.5 mm; ◐ d_H = 1.0 mm; ◇ d_H = 3.0 mm

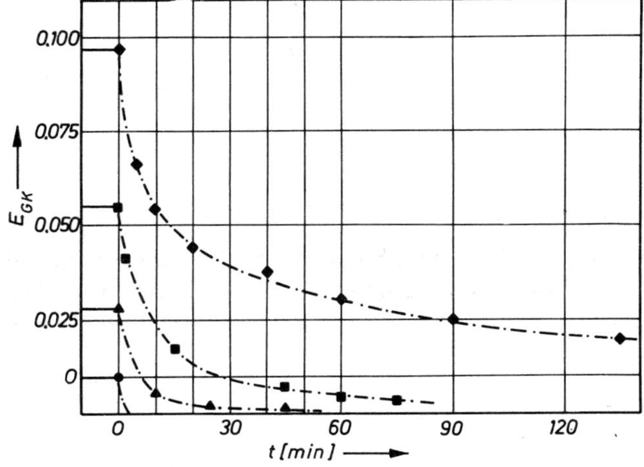

Fig. 9 Time dependence of E_{GK} at different glycerol concentrations after turning off the gas flow. D_c = 14 cm, H_S = 35 cm. Perforated plate, d_H = 3.0 mm, w_{SG} = 1.27 cm[11]. ◆ 95%; ■ 90%; ▲ 70%; ● 50% glycerol solutions

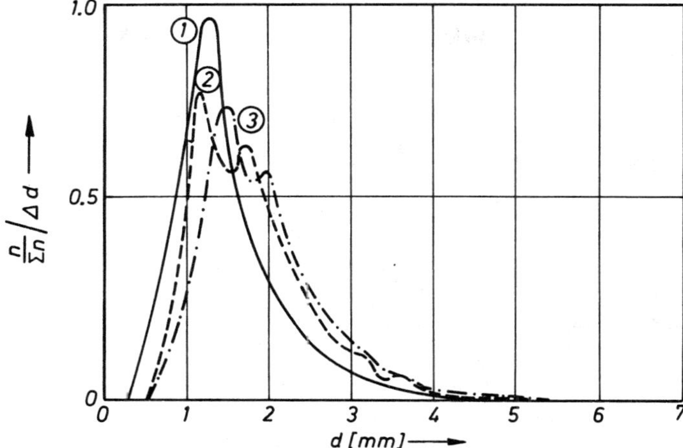

Fig. 10 Distribution of bubble diameters in a bubble column (D_c = 14 cm, H_c = 391 cm) employing 1% CMC solution. w_{SL} = 1 cm s^{-1}, w_{SG} = 1.06 cm. Longitudinal distance from the gas distributor X: ① X = 49.5 cm; ② X = 124.5 cm; ③ X = 180 cm[84]

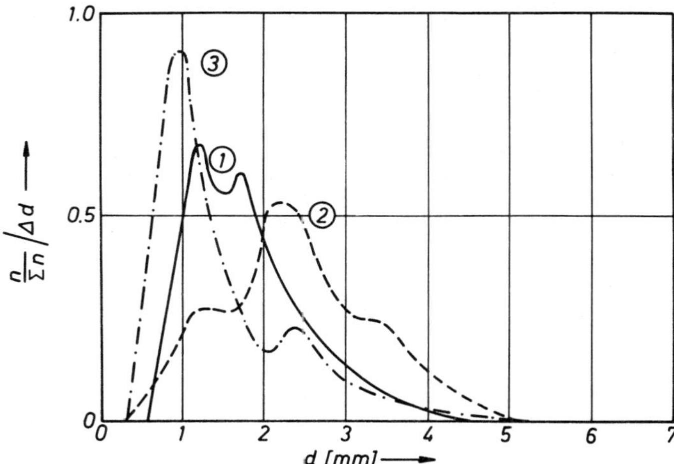

Fig. 11 Distribution of bubble diameters in a bubble column 14 cm in diameter 1.2% CMC solution. w_{SL} = 1 cm s^{-1}, X = 124,5 cm. ① w_{SG} = 1.07 cm s^{-1}; ② w_{SG} = 2.67 cm s^{-1}; ③ w_{SG} = 5.33 cm[84]

distance from the gas distributor. With rising CMC concentration and superficial gas velocity, this multimodal distribution becomes more significant (Fig. 11). At w_{SG} = 5.33 cms^{-1}, large slug-like bubbles are formed which are not plotted in Fig. 11. Fig. 11 demonstrates that with increasing w_{SG} the frequency of small and very large bubbles rises and that of the intermediate ones is reduced. The same effect has also been found for glycerol solutions[10].

2.9 Methods for Determining the Bubble-Swarm Velocity

The relative velocity of the bubble swarm, w_s, with regard to the liquid velocity is given by Eq. (23):

$$w_S = \frac{w_{SG}}{E_G} - \frac{w_{SL}}{1 - E_G}, \tag{23}$$

where w_{SG} is the superficial gas velocity and w_{SL} the superficial liquid velocity.

The influence of w_{SG} on w_S depends on the rheological character of the medium. In 50, 90 and 95% glycerol solutions, w_S is nearly constant and independent of w_{SG}[11]. In CMC solutions, w_S rises with increasing w_{SG} (Figs. 12 and 13).

w_S is strongly affected by the relative gas hold-up, E_G. In highly viscous liquids, the fraction of "small" bubbles, which have either very low or no ascending velocity, can be large. These bubbles do not provide a contribution to the convective gas transfer. Therefore, in Eq. (21) only the gas hold-up fraction, E_{GG}, due to the "intermediate-to-large" bubble should be inserted (see also 2.7).

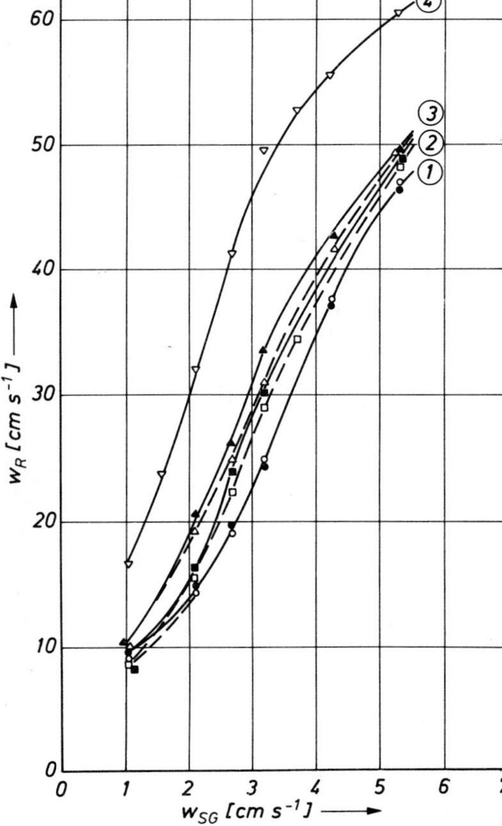

Fig. 12 Relative gas velocity (bubble-swarm velocity) as a function of w_{SG} at different CMC concentrations. $D_c = 14$ cm, $H_C = 391$ cm[84].

No.	CMC (%)	w_{SL} (cm s^{-1})	Symbol
①	1	1.0	●
		1.5	○
②	1.2	1.0	■
		1.5	□
③	1.4	1.0	▲
		1.5	△
④	1.6	1.2	▽

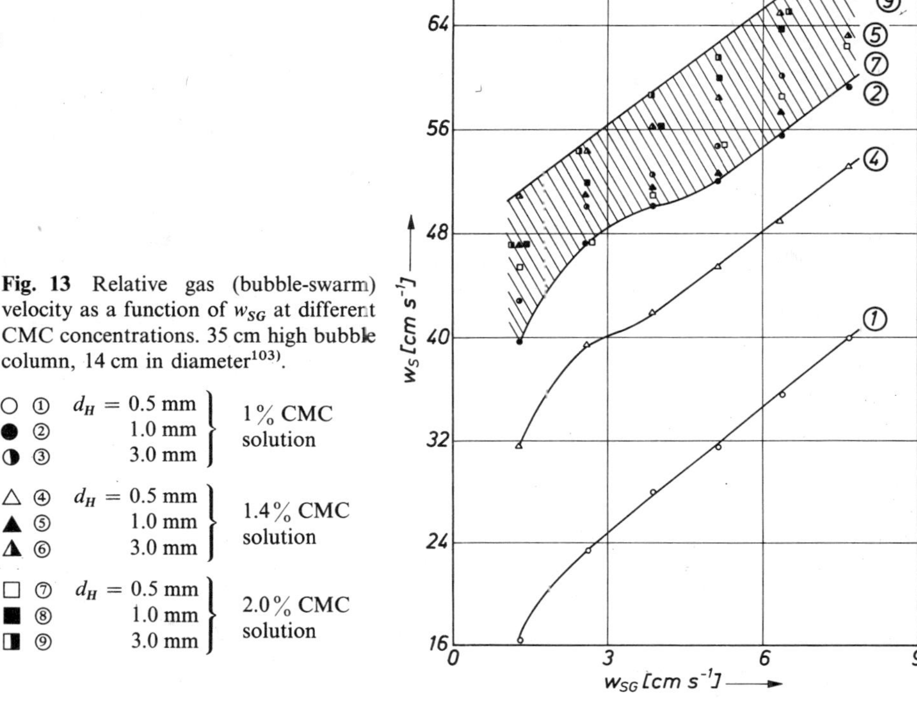

Fig. 13 Relative gas (bubble-swarm) velocity as a function of w_{SG} at different CMC concentrations. 35 cm high bubble column, 14 cm in diameter[103]).

○ ①	d_H = 0.5 mm	1% CMC solution	
● ②	1.0 mm		
◑ ③	3.0 mm		
△ ④	d_H = 0.5 mm	1.4% CMC solution	
▲ ⑤	1.0 mm		
◭ ⑥	3.0 mm		
□ ⑦	d_H = 0.5 mm	2.0% CMC solution	
■ ⑧	1.0 mm		
◨ ⑨	3.0 mm		

2.10 Methods for Measuring the Oxygen Transfer Rate, OTR, and the Volumetric Mass Transfer Coefficient, k_La

OTR and/or k_La can be determined by non-steady state or steady-state methods.

Several papers on the measurement of k_La in stirred tank reactors have been published. Most of the methods described use a non-steady-state method. The variation in concentration of oxygen dissolved in the liquid is measured by means of a polarographic oxygen electrode. Since the response signal is often influenced by the electrode transient behavior, several models have been developed with the aim of eliminating this detector effect (e.g.[52] to [55])).

Dang et al.[56]) studied the two-diffusion resistances of the oxygen transfer from the medium to the detector electrodes or resistances in series which are caused by the liquid film and the electrode membrane. In addition, the gas residence time was considered to be a first-order time delay. Dang et al. calculated k_La from the first moment of the response function. The accuracy of k_La is strongly influenced by the transient detector behavior and the gas residence time. These are determined in separate measurements, often under conditions which are not identical with those in k_La measurements. The difference in the first moments of the electrode response function and the sum of electrode and reactor response functions are small, especially at high k_La and gas hold-up values and high gas residence times. This leads to large errors in the k_La determination.

A furter non-steady-state method was developed by Joosten et al.[57]. This method is based on the different residence time behavior of gases which are readily soluble and those which are sparingly or insoluble in the liquid. This method can be employed only for very short mean gas residence times and low $k_L a$ values[64].

Steady-state methods have been applied to the determination of $k_L a$ in bubble columns (e.g.[58] to [62]) for some years. Since high bubble columns can be described only by models with distributed parameters, it is necessary to measure the position-dependence of the concentration of dissolved oxygen in the liquid. This was performed by means of several oxygen measurement points distributed along the column. The application of the steady-state technique to short bubble columns[63] or to stirred tank reactors[64] is a very simple method for such measurements if the liquid phase is so well mixed that it can be described by an ideal stirred reactor model. Measurements of the O_2 partial pressures in the liquid at the reactor inlet and outlet permit the calculation of $k_L a$ from the O_2 balance, provided that the variation in O_2 concentration in the gas phase is only slight:

$$k_L a = \frac{1}{\tau_M} \frac{\bar{p} - \bar{p}_0}{100 - \bar{p}} . \tag{24}$$

Here τ_M = mean liquid residence time in the reactor,
$\bar{p} = p/p^*$
$\bar{p}_0 = p_0/p^*$
p_0 and p = O_2 partial pressures at the inlet and outlet of the reactor
p^* = O_2 saturated liquid (set to 100 on the recorder).

Two reactors are employed; in one of them O_2 is absorbed while in the other O_2 is desorbed by N_2 and the liquid is recirculated. At first, both of the reactors are aerated until a steady state has been attained. This value is set to $p = 100$. Thereafter, both units are purged with nitrogen until a steady state has established. This value is set to $p = 0$. Finally, the reactor is aerated and the desorber purged with nitrogen until the steady state has been attaoned. This value is $\bar{p}$ (on the recorder).

Since the oxygen electrodes are calibrated under the same conditions (liquid aeration rate, stirrer speed) applied during $k_L a$ measurements, no falsification of the detector signal due to these factors occurs. If model media are employed, the frequently unknown Henry coefficient, He_{O2}, is not needed for the determination of $k_L a$.

2.11 Methods for Measuring the Mass Transfer Coefficient, k_L

The same methods, which are suited to the determination of the diffusivity of dissolved gases (see 2.4) can also be applied to the measurement of k_L. However, the usual equipment (laminar jet, laminar film (on cylinder or on sphere)) yields data which are difficult to adapt to practical systems.

To evaluate k_L values, which are more applicable in practice, stirred cell (e.g.[65–67]) or single bubbles (e.g.[68–70]) can be employed.

To have a definite gas/liquid interfacial area in the stirred cell a perforated plate (with a suitable free cross sectional area) is used. The main variable is the stirrer speed. The dissolved oxygen concentration, C_t, is measured in the liquid bulk by an oxygen electrode as a function of time, t. With known saturation concentration, C_s, and interfacial area, A, k_L can be calculated by Eq. (25):

$$k_L = \frac{-\ln\left(1 - \frac{C_t}{C_s}\right)}{At}. \tag{25}$$

The single bubble method uses a closed column, which is completely filled by the liquid. If a gas bubble is injected into this system and allowed to rise through the liquid, the pressure in the column will change as the bubble ascends and gradually dissolves. If the bubble surface area, A, is calculated by means of photographs, k_L can be determined by Eq. (26):

$$-(k_L a) = \frac{1}{S}\left\{\alpha - \frac{dP_T}{dt} + \frac{V_{bo} + \alpha(P_T - P_{TO})\left(\frac{dP_T}{dt} - \gamma u\right)}{P_T + Y - \gamma h}\right\}, \tag{26}$$

where s = solubility of the gas in the liquid ($cm^3 cm^{-3}$)
α = overall compressibility of the column and liquid therein ($cm^3\ (cm\ H_2O)^{-1}$)
P_T = pressure detected by transducer and compensated for drift (cm-H_2O)
P_{To} = P_T at $t = 0$ (bubble release)
V_{bo} = bubble volume at $t = 0$ (cm^3)
γ = specific gravity of liquid (—)
u = bubble rising velocity (cms^{-1})
Y = atmospheric pressure (cm-H_2O)
h height of the bubble above the transducer (cm).

2.12 Methods for Measuring Power Input

In an agitated tank the power input is easily measured. By measuring the torque on the shaft, M_N, and the rotational speed, N, the power, P, can be calculated by Eq. (27):

$$P = M_N 2\pi N. \tag{27}$$

In bubble columns, P can be calculated by Eq. (27):

$$P = M_G\left\{RT \ln \frac{p_{in}}{p_{out}} - gH - \frac{1}{2}(w_{G,out}^2 - w_{G,in}^2)\right\}. \tag{28}$$

Where
M_G = gas mass flow
R = gas constant
T = absolute temperature
p_{in} = gas pressure at column inlet

p_{out} = gas pressure at column outlet
H = height of the bubbling layer
$w_{G,\,out}$ = linear gas velocity at the outlet
$w_{G',\,in}$ = linear gas velocity at the inlet.

However, since the second and third terms together make up only 0.2% of the overall power input, the power input due to gas expansion dominates:

$$P = M_G\, RT \ln \frac{p_{in}}{p_{out}}\,. \tag{29}$$

In the power inputs, Eqs. (27) to (29), the energy losses due to mechanical energy (e.g. during the gas compression) are not considered.

2.13 Methods for Determining the Longitudinal Liquid Dispersion Coefficient, E

Since in stirred tank reactors the liquid is well mixed ($E \simeq \infty$), E plays only a role in tower or tube reactors. It can be determined by three different methods:

— by measuring a tracer residence time distribution and evaluating E by a non-linear fitting of the curves calculated by means of the dispersion model compared to the measured one (cf.[71]);
— by measuring the steady-state longitudinal tracer concentration profile in the reactor and employing a stationary plane source (e.g.[72]);
— by fitting the longitudinal concentration profile of dissolved oxygen calculated by means of dispersion model to the measured one. This fitting yields $k_L a$ as well as E. However, the accuracy of E is poor, since the sensitivity of the profile for E is low (e.g.[73]).

3 Single Bubble and Bubble Swarm Behavior in Highly Viscous Media

3.1 Bubble Formation, Coalescence and Break-Up

Employing a pseudo-plastic medium with rheological behavior according to Ostwald-de Waele

$$\tau = K\left(\frac{\mathrm{d}v}{\mathrm{d}x}\right)^n, \tag{30}$$

where τ = shear stress
$\frac{\mathrm{d}v}{\mathrm{d}x}$ = shear velocity
K = fluid consistency index and
n = flow behavior index

the volume of a bubble, V_B, produced by the orifice at constant flow conditions and for the orifice Reynolds numbers:

$$4000 > Re_0 = \frac{\varrho_L d_0^n u_0^{2-n}}{K} > 1000,$$

(ϱ_L = liquid density, d_0 = orifice diameter, u = orifice gas velocity), can be calculated Eq. (31)[6]:

$$V_B = \left(\frac{9X_nK}{2^n\varrho_L g}\right)^{3/(1+3n)} \left(\frac{4\pi}{3}\right)^{1/(1+3n)} \left(\frac{1+4n}{3}\right)^{3n/(1+3n)} Q_G^{3n/(1+3n)}, \tag{31}$$

where X_n is defined by $C_D = \frac{X_n}{Re^*}$, (32)

C_D = drag coefficient,

$$Re^* = \frac{\varrho_L d^n U^{2-n}}{K} = \text{bubble Reynolds number where} \tag{33}$$

d = bubble diameter, U = bubble velocity.

In the case of a Newtonian liquid with $n = 1$ and $K = \eta$, the dynamic viscosity of the liquid (Eq. (31)) reduces to:

$$V_B = \left(\frac{15\eta Q_G}{2\varrho_L g}\right)^{3/4} \left(\frac{4\pi}{3}\right)^{1/4}. \tag{34}$$

This is the well known relationship developed by Davidson and Schüler[74].

The relations were controlled by a CMC solution having $n = 0.68$ and $K = 3.04$ Pas^n[6].

For $Re_0 > 4000$, the bubble formation is very complex[75]. According to the operating conditions, bubble coalescence occurs very close to the orifice. Furthermore, groups of two bubbles are formed. Because of the great variation of the bubble formation in this region no quantitative relationship exists to calculate V_B. Depending on the conditions, the primary bubbles interact and coalesce or are redispersed. It is generally accepted that there is a critical distance at which the leading bubble begins to exert a significant influence on the following one. This interaction is due to the drag reduction of the following bubble caused by the wake of the preceding bubble. The length of the wake increases with increasing viscosity. The wake shape does not influence the coalescence rate[9]. The higher the apparent wake velocity, w_{app}, behind the preceding bubble A, the stronger the drag reduction of the following bubble B. According to Crabtree et al.[8], the ratio of w_{app} to the preceding bubble velocity, U_A, is given by:

$$\frac{w_{app}}{U_A} = \frac{6.2r_{eA}}{x}, \tag{35}$$

where $r_{eA} = \left(\frac{3V_{BA}}{4\pi}\right)^{1/3} =$

is the equilibrium radius of bubble A and x = separation distance of the bubbles A and B.

This relationship was evaluated using large bubbles (10 to 40 cm^3) in 67% sugar solutions ($\gamma = 1.97$ Stokes). When employing such large bubbles, coalescence also occurs if they were 70 cm apart. This demonstrates the wide range of the capture mechanism. The investigations of Acharya and Ulbrecht[77] indicate that the viscoelastic property of the medium increases both the collision and the coalescence times of gas bubbles. Thus, in media with strong elastic components the overall coalescence rate may be reduced considerably.

In bubble columns there is a very significant liquid velocity profile with the maximum velocity at the center[76]. Since the drag of the ascending bubbles is smallest at the column center they tend to move toward the center. Thus in the upper part of the column, the large bubbles are vertically aligned and, as a result, their coalescence probability is increased.

However, an interaction between the bubbles does not always lead to coalescence but can yield break up as well, depending on the relative position of the interacting bubbles. Coalescence takes place when more than about half of the projected area of the following bubble overlaps with that of the leading bubble at a critical distance. In contrast break up occurs when the overlapping is less than about half of the projected area of the following bubble[10].

3.2 Bubble Ascending Velocity. Behavior of Very Small and Very Large Bubbles

Since the rheological behavior of several media can be described by the power law of Ostwald-de Waele, only these media will be considered here. The following relationship holds between the single bubble velocity, U, and the bubble volume, V_B[78]:

$$U = \left[\frac{\varrho_L g}{K} \frac{2^{1+n}}{X_n} \left(\frac{4\pi}{3}\right)^{(2-n)/3}\right]^{1/n} V_B^{(1+n)/3n} . \tag{36}$$

(For the explanation of the symbols see Eq. (31)).

In Eqs. (31) and (36), X_n depends on n for non-Newtonian fluids. Since this dependence is not known, the corresponding values for Newtonian fluids, in the first approximation, are used: $X_n = 24$ for the Stokes regime (a gas bubble moves in the Stokes regime when the liquid is in creeping flow, the bubble spherical and the interface rigid), $X_n = 16$ for the Hadamard regime (a gas bubble moves in the Hadamard regime when the liquid is in creeping flow, the bubble spherical and the interface free), and $X_n = 48$ for the Levich regime (a gas bubble moves in the Levich regime when the Reynolds number is high, the bubble spherical and the interface free). The Stokes regime is easy to realize if the bubble volume is small enough. The Hadamard regime in viscous liquids may not be realized; only in pure liquids, which are free of surfactants and which exhibit low viscosities, is this regime possible. The same is true of the Levich regime. To make the Reynolds number large, the bubble must be large. However, the shape of large bubbles is severely distorted from spherical.

It is worthwhile to point out that for pseudoplastic fluids ($n < 1$) the velocity U increases with growing volume more rapidly than for Newtonian liquids, e.h. for the Stokes regime.

$$n < 1; \frac{d \log U}{d \log V_B} = \frac{1 + n}{3n} > \frac{2}{3} . \tag{37}$$

In the Hadamard regime, as Eq. (32) can be written

$$C_D = \frac{16\,F_1(n)}{Re^*}\,, \tag{38}$$

where $F_1(n)$ is a function of n. Hirose et al.[82] and Bhavaraju et al.[83] developed relationships for $F_1(n)$ assuming viscoelastic behavior. With decreasing n, $F_1(n)$ increases. This was experimentally proved by Acharya et al.[77].

Larger bubbles have a shape wich approaches the "spherical cup" bubble. For spherical cup bubbles with free interface and potential flow, Eq. (39) is valid[78, 79]:

$$U = 25.0\,V_B^{\,1/6}\,. \tag{39}$$

According to Calderbank[68] this equation gives for single bubbles slightly lower values of U in highly viscous liquids than the measured ones which were corrected for wall effects[68]. Uno and Kintner found that no wall-effect correction is necessary if $(d/D) < 0.1$[80]. Here, d is the bubble diameter and D_c the column diameter. If completely developed slug flow prevails, the Dimitrescu relationship (40) should be valid[81]:

$$U_{\text{slug}} = 0.35\,\sqrt{gD_c}\,. \tag{40}$$

Mendelson[87] recommended for large Reynolds numbers the relationship (41):

$$U = \sqrt{\frac{\sigma}{R\varrho_L} + gR}\,, \tag{41}$$

where R is the equivalent bubble radius and σ the surface tension.

The validity of Eq. (41) was proved experimentally by single bubbles in CMC, PAA and PEO solutions[77].

However, a comparison of the bubble swarm velocities, w_S, in highly viscous media with bubble-ascending velocities, U, theoretically calculated and confirmed by single-bubble measurements, indicate w_S and U do not agree[11, 84]. The bubble swarm velocities w_S, are always higher than the corresponding single bubble ascending velocities, U. Similar discrepancies were also found in low viscosity media[85]. This enhancement of the rising velocity of large cap bubbles in a uniform swarm of smaller bubbles was explained by the change in their shape. However, this explanation does not hold for highly viscous systems, since in these systems also "intermediate" bubbles show this behavior. Franz et al.[11] explained this acceleration of the bubble swarm by the presence of "small" bubbles, which considerably alter the rheological property of the system. The viscosity of the liquid-"small" bubble system is smaller than that of the pure liquid. However, this alone cannot explain these large differences. It seems that the presence of small bubbles considerably reduces the drag coefficient of "intermediate" and "large" bubbles. Also, the very large bubbles and slugs ascend at much higher velocity than the ones calculated by Eqs. (39) or (40)[84]. In 90 and 95% glycerol solutions[11], bubble swarm velocities were measured which attained 100 cms^{-1}. According to Eq. (40) the maximum ascending velocity amounts to 41 cms^{-1}. In 1 to 2% CMC solutions "very large" bubble and slug velocities were measured which again amounted to about 100 cms^{-1}, i.e. more than twice as high as predicted by theory. On the other hand, it was found that the fraction of bubble swarms, containing bubbles smaller than 1 mm does not ascend at all[11)]. It takes a long time before the liquid is free of "small" bubbles.

This means that very small bubbles do not ascend at all, "large" and "very large" bubbles ascend more quickly than the theory predicts. The "intermediate"-to-"large" bubble fractions also ascend more rapidly than single bubbles. Therefore, the author suspects that the measured or calculated single-bubble ascending velocity, U, cannot be employed to calculate the bubble swarm velocity, w_S, in highly viscous media.

3.3. Mass Transfer Coefficient

As already mentioned in 2.11 two different methods are available for the determination of the mass transfer coefficient, k_L:

1) the stirred cell and
2) the single bubble methods.

In the stirred cell method the following relationship was considered[66)]

$$Sh_N = 1.13 \times 10^{-3}\, Re_N^{1.24}\, Sc^{0.5}\,, \tag{42}$$

for glycerol solutions in the range:

$$\begin{aligned} Re_N &= 6 \times 10^2 \text{ to } 2 \times 10^4 \\ Sc &= 400 \text{ to } 54200 \\ Sh_N &= 180 \text{ to } 1040\,. \end{aligned}$$

In Eq. (42)

$Re_N = = \frac{N d_N}{\nu}$ is the stirrer Reynolds number

$Sc = \frac{\nu}{D_m}$ is the Schmidt number and

$Sh_N = \frac{k d_N}{D_m}$ is the stirrer Sherwood number, where

d_N = diameter of the propeller
N = rotation speed of the stirrer.

Now, $Sh_N = 1.19 \times 10^{-2}\, Re_N^{0.98}\, Sc^{0.5}$ (43)

for CMC solutions in the range

$$\begin{aligned} Re_N^* &= 324 \text{ to } 2603 \\ Sc &= 2510 \text{ to } 19740 \\ Sh_N &= 88 \text{ to } 307 \end{aligned}$$

where

$$Re_N^* = \frac{Nd_N}{\nu_r},$$

$$Sc^* = \frac{\nu_r}{D_m},$$

$$\nu_r = \frac{\eta_r}{\varrho_L} = \text{representative kinematic viscosity.}$$

The representative viscosity η_r was calculated according to Metzner et al.[86] employing Eq. (44)

$$\overline{\left(\frac{dv}{dx}\right)} = q^* N . \tag{44}$$

For $n < 1$, a marine type propeller, and a vessel-to-stirrer diameter ratio 2:1 q^* is equal to 10[86]. Hence, the representative viscosity becomes:

$$\eta_r = \frac{\tau}{10N}. \tag{45}$$

Based on the single-bubble method, the following relationships were recommended: (1) the Levich equation[88] (for free circulation)

$$Sh_B = 0.65\ Pe_B^{0.5} \tag{46}$$

for small spherical carbon dioxide bubbles in 90% glycerol solutions[68] in the range

$$Re_B \ll 1, Pe_B \gg 1$$

where $Pe_B = \dfrac{d_e V}{D_m}$, the bubble Peclet number

$Sh_B = \dfrac{k_L d_{eq}}{D_m}$, the bubble Sherwood number

$Re_B = \dfrac{d_{eq} U}{\nu}$, the bubble Reynolds number and

d_{eq} = diameter of the spherical bubble having the same volume as the bubble in question;

(2) the Boussinesq equation[89] (potential flow solution)

$$Sh_B = 1.13\ Pe_B^{0.5} \tag{47}$$

for 99% glycerol solutions[68] where $Re_B \gg 1$, $Pe_B \gg 1$.

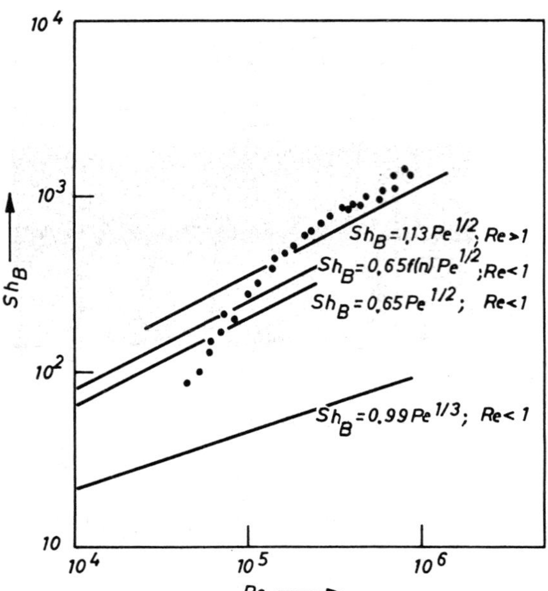

Fig. 14 Sh_B number as a function of Pe_B number for 0.5% PAA solution[90)]

The measurements of Zana and Leal[90)] indicate that for $Pe_B > 10^6$, Eq. (47) is more suitable for the calculation of Sh_B for 90% glycerol solution than Eq. (46). They also employed viscoelastic media: 0.1, 0.5 and 1.0% per weight Separan AP 30 (PAA) solutions as well as a mixture of 0.523% per weight Separan AP 30, 53.9% glycerol, and 45.6% water (to absorb CO_2).

Figure 14 shows that Sh_B is enhanced with increasing Pe_B.. This is due to the medium viscoelastic property.

The degree of increase in mass transfer rates correlates well with the power law index n. The viscoelastic values of mass transfer rates represent the additive contribution of shear-thinning and separate elastic effects. The increase in mass transfer is also greater than predicted by the power-law model of Hirose et al.[82)], which predicts an enhancement in the mass transfer rate over its Newtonian value:

$$Sh_B = 0.65 \left[1 - \frac{4n(n-1)}{2n+1} \right]^{0.5} Pe_B^{0.5} . \tag{48}$$

This also indicates that the viscoelasticity causes a significant enhancement of the mass transfer rate.

4 Stirred Tank Reactors

In mechanical stirred tank reactors the component mixing and the gas dispersion as well as the heat removal are intensified by mechanical agitators. A large number of agitator types is employed in industry. Some stirrer types, usually with vanes, i.e. turbine, paddle, MIG etc., are more frequently used in highly viscous aerated

Fig. 15 Gas trails behind the blades of turbine agitator in PAA solution[92, 93]

reactors than others. If gas dispersion is necessary, the special agitators used for highly viscous medium mixing (helical, screw and anchor stirrers) are not recommended.

In a stirred tank reactor the primary gas dispersion occurs at the immediate vicinity of the stirrer blade. At its outer tips, trailing vortex systems are formed[91]. With increasing distance from the blade tips, the trailing vortex (pseudoturbulence with large eddies and periodical character) transfer into microturbulence which yields a higher energy dissipation rate and a higher gas dispersion degree than the macroturbulence. However, with increasing distance from the blade the relative turbulence intensity diminishes and thus, in a radial direction, the energy dissipation rate passes through a maximum.

The smallest bubble size is controlled by the dynamic turbulence pressure in this maximum energy dissipation range.

As soon as the bubbles leave this range, the dynamic turbulence pressure rapidly diminishes in its surroundings and the dynamic equilibrium bubble size quickly increases.

In coalescence-promoting media of low viscosity the dynamic equilibrium bubble size is quickly established. In coalescence suppressing media this coalescence is considerably inhibited. In highly viscous media it is necessary to distinguish between "small" bubbles, which have a very low tendency to coalesce and "intermediate-to-large" bubbles, which display a high tendency to coalesce. Furthermore, in mechanically agitated reactors the gas is first collected on the down stream face of the blades, if highly viscous media are employed. This gas is sucked into the trailing vortices and dispersed. The volume of this gas trail increases with rising medium viscosity and volumetric gas flow rate. At high gas flow rates and in highly viscous media the volumes of these gas trails become so large that the entire space between the blades is filled with gas, i.e. the stirrer is flooded by gas (Fig. 15). In highly viscous media the dynamic equilibrium bubble size is rapidly established with regard to the "intermediate-to-large" bubbles. These bubbles rapidly grow due to their coalescence. Large bubbles have high buoyancy forces; they quickly ascend in the liquid and cannot be recirculated to the stirrer blade vicinity by the liquid loop. The "intermediate-to-large" bubble gas mean residence time is short, its longitudinal dispersion low, and its value strongly influenced by the gas flow rate.

In the immediate stirrer blade vicinity, high dynamic turbulence pressures prevail which disperse the gas phase. The higher this dynamic turbulence pressure, τ_T, the smaller the bubbles. However, small bubbles are also formed by break-up during the interactions of two larger bubbles far from the stirrer. These small bubbles have low buoyancy forces due to their small size. They slowly ascend in the liquid and are recirculated by the liquid flow. The mean gas residence time due to these small bubbles is long (or very long). Since the small bubble redispersion frequency is low, the gas phase due to the small bubbles behaves like a macro- mixed liquid in a well stirred tank. Because of the long residence times of small bubbles, their fraction gradually increases and can attain very high values.

With increasing bubble age the enrichment and alignment of the polymers at the interface becomes considerable[39, 104]. This changes the interfacial behavior in such a way that the surface elasticity increases. This further reduces the coalescence probability of those bubbles. Thus, small bubbles with high age are no longer able to coalesce. Furthermore, if they are attached to the surface of large bubbles they remain separated.

4.1 Apparatus and Instruments

An aerated stirred tank reactor usually consists of a tank with vanes, an agitator, and gas and liquid provisions. To keep the medium temperature constant, a heating/cooling jacket or coils are employed. To measure the hydrodynamical properties (gas hold-up, E_G and Sauter bubble diameter, d_S, mixing time, power input, P, oxygen transfer .rate, OTR, volumetric mass transfer coefficient, k_La, and heat transfer coefficient) special equipment is necessary. To determine the relative gas hold-up, E_G, in stirred tank reactors, one can measure the height of the bubbling layer, H, and compare it with the height of the bubble free liquid, H_L. E_G can be calculated by

$$E_G = \frac{H - H_L}{H} . \tag{21}$$

This method, however, becomes erroneous as the tank scale increases. Therefore, H is usually kept constant and the difference $H - H_L$ is measured by other methods, e.g. the tank is filled by the liquid with a volume V to a definite height, H'. After starting the aeration, a suction pump withdraws the two-phase volume which is jutted out beyond H'. The volume of the removed liquid, V_1', is equal to the gas hold-up present in the two-phase system. The relative gas hold-up is given by[95]:

$$E_G = \frac{V_1'}{V} . \tag{49}$$

It is difficult to measure the bubble size distribution and d_S in highly viscous media. The photographic method can only be applied to the wall region[43] and the electrooptical method is not suitable for such measurements either[48, 50]. Only electrical conductivity probes can be employed[51]. However, the evaluation of these measurements is quite difficult.

To determine the mixing time, Θ, a tracer is usually added to the medium instantaneously and its concentration is measured as a function of time. As a tracer, a salt solution (change of electrical conductivity is measured), a dye solution (color change is followed), an acid or a base (change of pH is measured) or a cold or hot liquid (change of temperature is followed) etc., can be employed. It is important to place the detector in the reactor in a position in which the representative tracer concentration can be measured[96]. To determine the power input, the rotational speed, N, and the torque on the shaft, M_N, are measured and P is calculated by (27). To determine OTR and/or $k_L a$ the dissolved oxygen concentration is usually measured in the medium by polarographic oxygen electrodes. In highly viscous media the mass transfer resistance due to the liquid boundary layer at the electrode membrane surface can seriously increase the time constant of the instrument. If the non-steady state method is used for measuring $k_L a$, the position of the electrode must be carefully chosen. In some laboratories a small stirrer is employed to keep the boundary layer at the electrode membrane thin[94]. To determine the heat transfer coefficient in a well mixed batch reactor it is sufficient to measure the temperature of the medium and the flow rate, as well as the entrance and exit temperature of the cooling and heating medium.

4.2 Mathematical Models

If one assumes a perfect mixing in both phases, the simple stirred tank model can be employed for calculating the performance of the reactor. The liquid as well as the gas phase properties are described by lumped parameters. In batch operated systems, some process variables (composition and sometimes temperature of the liquid) are a function of time while others (flow rate, composition and temperature of the gas) are constant. In steady-state operated reactors, all process variables are constant within the reactor, but the distribution of liquid residence times in steady-state stirred tank reactors using highly viscous media indicates some non-idealities (cf.[135–137]). The jet formed by the stream entering the tank has a fairly strong effect on the velocity distribution within the tank. The interaction of the rotation impeller blades with the feed jet seems to influence the residence time distribution regardless of the system's geometry[137]. Such residence time distributions can be represented by mixing models consisting of a plug flow region in series with a complete mixing region and a parallel by-pass stream. A dead space region can also be included (see e.g.[134]). Another model assumes that the reactor volume can be divided into two well mixed regions at the level of the impeller and that the transfer of material between these regions is limited. It is assumed that nutrient solutions and cells both enter and leave from the upper portion of the culture. Thus, the lower region has dead space character[138]. Since the parameters of these models are determined by fitting the calculated to the measured data, all of them can describe the experimental results fairly well.

4.3 Hydrodynamic Properties

Very little data have been published on the relative gas hold-up, E_G[95,97,113]. In general, E_G increases with increasing N and Q_G. E_G and its dependence on N diminishes if the medium viscosity is increased (Figs. 16 and 17). The author is not

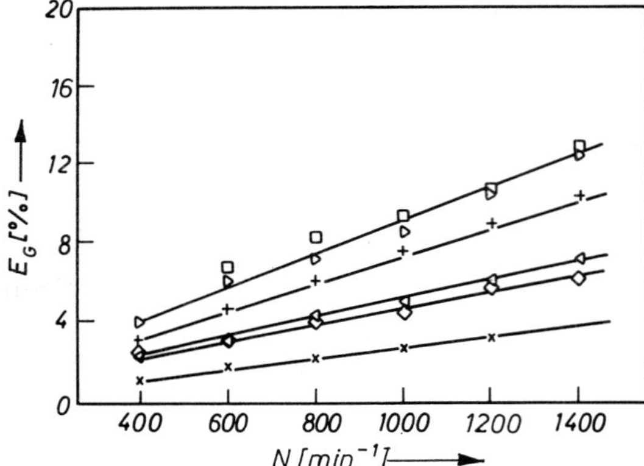

Fig. 16 Relative gas hold-up E_G in a stirred tank loop reactor ($H = 600$ mm, $D_t = 450$ mm) as a function of turbine stirrer speed N and gas flow rate Q_G; 0.25% CMC solution[95)]

w_{SG}	(cm s^{-1})	w_{SG}	(cm s^{-1})
□	11.04	◁	3.25
▷	7.86	◇	2.18
+	5.43	×	1.08

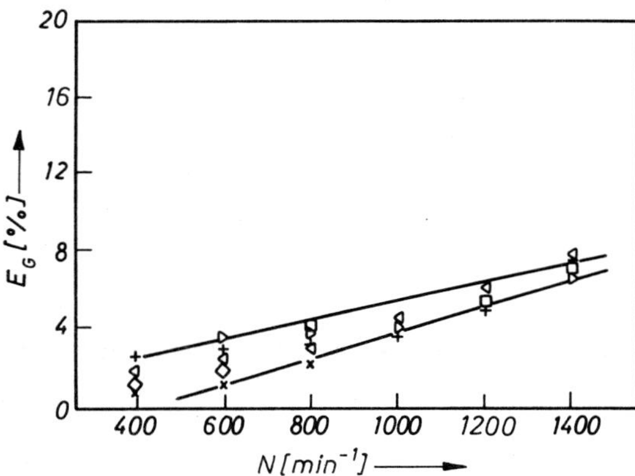

Fig. 17 Relative gas hold-up E_G in a stirred tank loop reactor ($H = 600$ mm, $D_t = 450$ mm) as a function of turbine stirrer speed N and gas flow rate Q_G; 1.5% CMC solution[95)] (for symbols see Fig. 16)

aware of bubble size distributions and/or bubble Sauter diameters, d_S, which were measured in aerated stirred tank reactors employing highly viscous media.

Results of mixing time measurements in aerated highly viscous media are scarce[110)]. Páca et al.[110)] employed pseudoplastic colloidal starch solutions $\eta_{\text{apparent}} \simeq 0.02$ to 0.2 Pa · s) in a 300 l vaned stirred tank ($D_1 = 508$ mm, $H_t/D_t = 3$) with a three-stage turbine stirrer ($d/D_t = 0.33$). At low viscosities ($\leqq 0.02$ Pa · 2s) no impeller speed effect on θ was found. At higher viscosities (0.2 Pa · s), θ diminishes with increasing N, e.g. from $\theta = 50$ s (at $N = 150$ min^{-1}) to $\theta = 25$ s (at $N = 500$ min^{-1}) if 0.88 *vvm* is used.

4.4 Power Input

In media with low viscosity and without aeration two ranges of the power input can be distinguished:

1) laminar flow $Re_N \leqq 10$ with $P = k_1 N^2 d_N^3 \eta$, (50a)

2) turbulent flow $Re_N \geqq 10^2$ with vanes
$Re_N \geqq 10^5$ without vanes

where $P = k_2 N^3 d_N^5 \varrho$. (51a)

If one introduces the Power or Newton number, *Ne*

$$Ne = \frac{P}{N^3 d_N^5 \varrho},$$

Eqs. (50) and (51) can be written as

$$Ne = k_1 \frac{1}{Re_N} \quad (50\,b)$$

for the laminar and

$$Ne = k_2 \quad (51\,nb)$$

for the turbulent region.

For the transient region $10^2 \leq Re_N \leq 10^4$, Eq. (52) is valid:

$$Ne = k_3\, Re_N^{-1/3}, \quad (52)$$

where k_1, k_2 and k_3 are constants and

$Re_N = \frac{N d_N^2}{\nu}$ the stirrer Reynolds number.

To apply these relationships to non-Newtonian media the representative viscosity η_r is introduced:

$$\eta_r = \frac{\tau}{\left(\overline{\frac{dv}{dx}}\right)}, \quad (53)$$

where τ is the shear stress and

$\left(\overline{\frac{dv}{dx}}\right)$ the mean shear velocity.

For the calculation of the mean shear velocity two approximations can be used.

According to the approximation of Metzner and Otto[98]:

$$\left(\overline{\frac{dv}{dx}}\right) = q^* N, \quad (44)$$

where q^* is an empirical constant, which slightly depends on the stirrer type and the system's geometry. The q^* values are tabulated in[86].

According to Calderbank and Moo-Young[99]

$$\left(\overline{\frac{dv}{dx}}\right) = q^* N \left(\frac{4n}{3n+1}\right)^{n/(1-n)}. \quad (54)$$

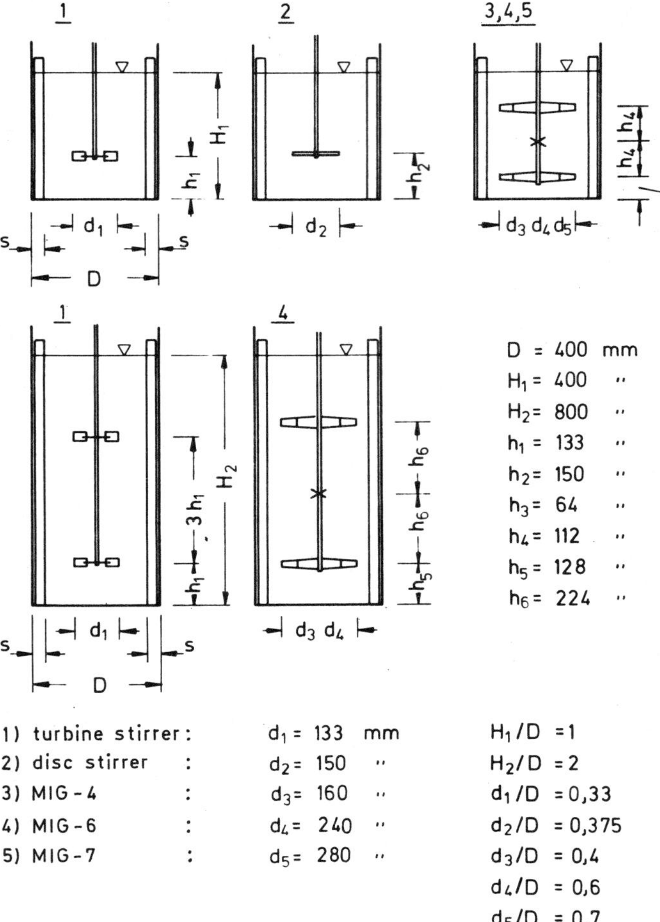

Fig. 18 Some data on the equipments employed by Höcker[93)]

This yield for η_r

$$\eta_r = \frac{K}{(q^*N)^{1-n}} \left(\frac{3n+1}{4n}\right)^n . \tag{55}$$

When working with non-Newtonian media, η is replaced by η_r in the Reynolds number.

For non-aerated systems, Ne is usually plotted as a function of Re_N. For aerated systems in Ne, P represents the power input during the aeration. In this case Ne is plotted as a function of the aeration number, $Q_G^* = Q_G/N\, d_N^3$, with the Froude number

$$Fr = \frac{N^2 \mathrm{d}_N}{g},$$

Table 2. Properties of the media employed by Höcker[92,93]

No.	Glycerol solutions		CMC solutions		PAA solutions	
	η	ϱ	K	n	K	n
	Pa × s	kg × m^{-3}	Pa × s^n	—	Pa × s^n	—
1	0.001	1000	0.001	1	0.001	1
2	0.003	1095	0.004	0.82	0.010	0.71
3	0.010	1150	0.091	0.80	0.019	0.67
4	0.060	1200	0.570	0.66	0.038	0.63
5	0.110	1225	1.180	0.60	0.106	0.51
6	0.160	1230	2.630	0.58	0.142	0.50
7	0.210	1235	4.260	0.54	0.540	0.44
8	0.400	1245			1.350	0.38
9	1.050	1255			3.450	0.34

Table 3. Constants k_1 and k_2 of Eqs. (50) and (51) for Newtonian liquids[93]

H/D	Stirrer	k_1	k_2
1	Turbine stirrer	75.85	5.0
	Disc stirrer	70.0	0.48
	MIG-4	63.1	0.68
	MIG-6	52.5	0.55
	MIG-7	47.0	0.525
2	Turbine stirrer	180.0	10.0
	MIG-6	72.4	0.63

as a parameter and *Ne* as a function of the Galilei number

$$Ga = \frac{Re_N}{Fr} = \frac{d_N^3 g}{\nu^2},$$

with the aeration number Q_G^* as a parameter.

The number of publications on $Ne(Q_G, Fr)$ relationships is scarce[92–95,100,101,178,179]. The most thorough investigations were carried out by Höcker[93,178,179]. He employed a turbine stirrer, a disc stirrer and three different types of MIG agitators (*M*ehrstufen-*I*mpuls-*G*egenstromrührer = multistage-impulse-countercurrent agitator[102]): MIG-4, MIG-6 and MIG-7, with different d_N/D_t ratios (Fig. 18).

Glycerol solutions as Newtonian, CMC solutions (CMC RK 5000, Wolf, Walsrode) as pseudoplastic and PAA solutions (Separan AP 30, Dow Chemical Co.) as viscoelastic model media were studied[93]. The properties of the non-Newtonian media are compiled in Table 2.

If Newtonian liquids are used the course of the $Ne\,(Re_N)$ functions can be described by Eqs. (50) and (51).

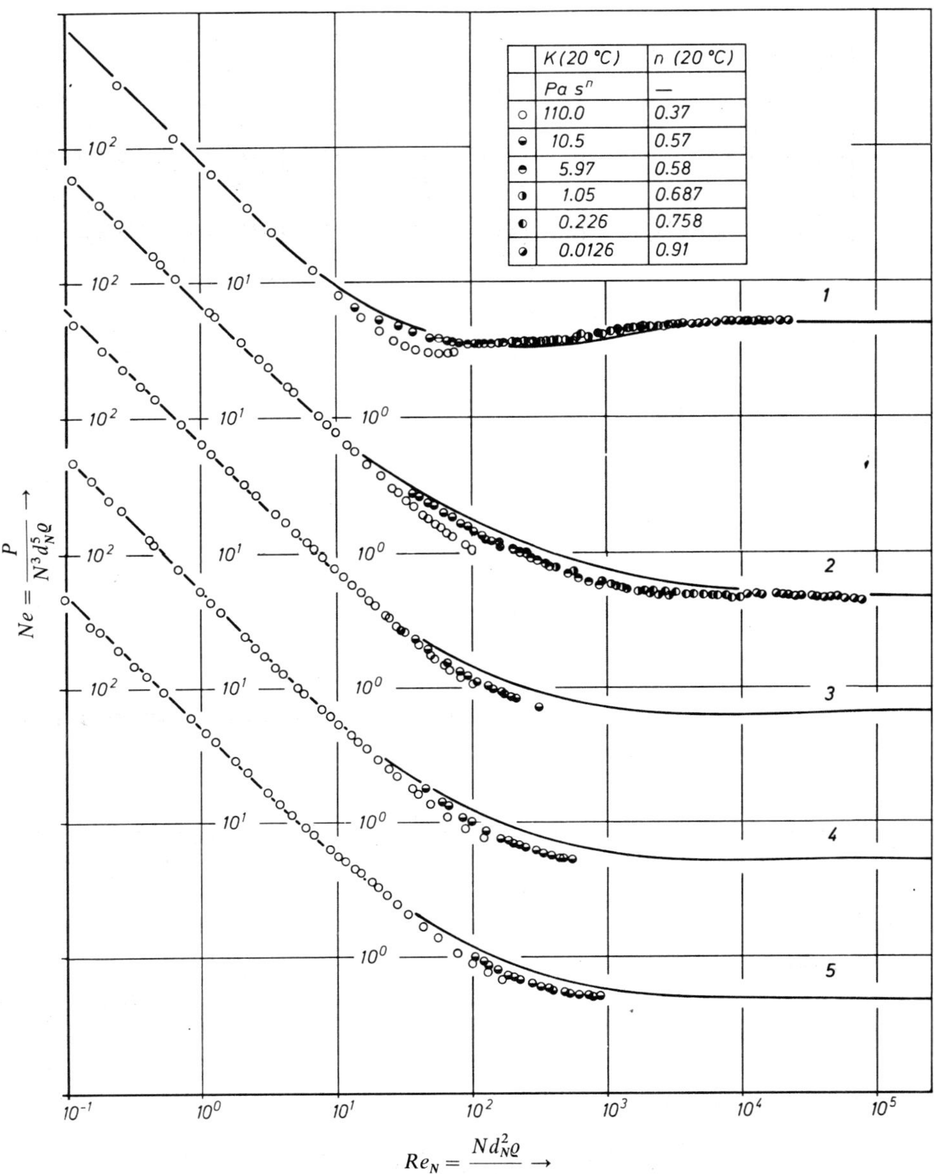

Fig. 19 Newton (power) number Ne as a function of Reynolds number Re_N for CMC solutions; $H/D_t = 1$ [93)]

①	turbine stirrer	④	MIG-6 agitator
②	disc stirrer	⑤	MIG-7 agitator
③	MIG-4 agitator	—	measured by Newtonian liquids

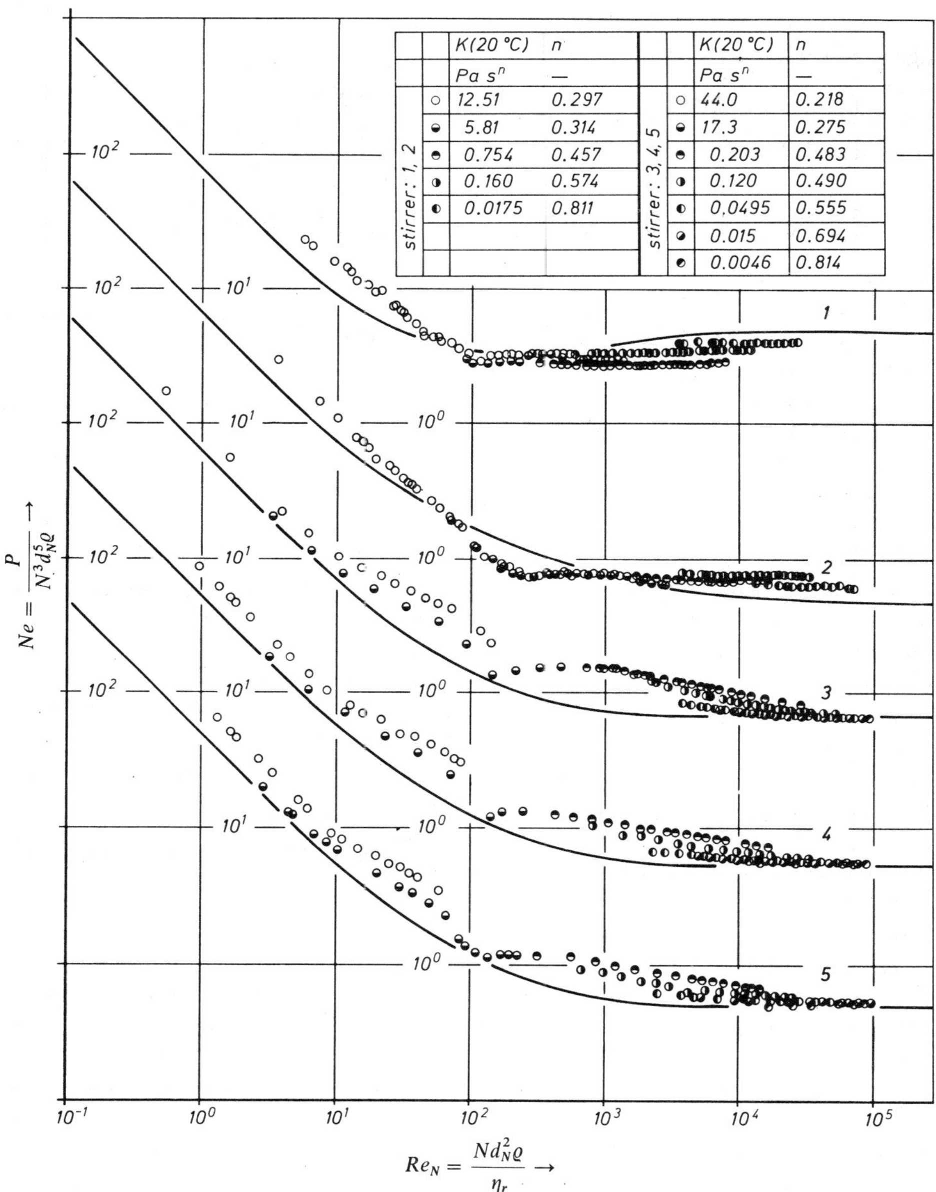

Fig. 20 Newton (power) number Ne as a function of Reynolds number Re_N for PAA solutions. $H/D_t = 1$ [93] (for symbols see Fig. 19)

In Table 3 the constants k_1 and k_2 of these relationships are compiled for the agitators employed by Höcker[93].

In Figs. 19 and 20, typical power input diagrams are shown for CMC and PAA solutions[93]. Fig. 19 indicates that the application of the representative viscosity according to Metzner and Otto

$$\eta_r = 11N$$

in the Re_N number yields excellent agreement between Newtonian and pseudoplastic liquids in the laminar region ($Re_N \leqq 10$), but there are significant deviations in the transient range (Fig. 19). Using pseudoplastic liquids the laminar-to-turbulent transition occurs at higher Re_N numbers than in Newtonian liquids. This phenomenon was explained by Metzner and Otto[98] by a considerably faster turbulence decay in non-Newtonian liquids than in Newtonian liquids with increasing distance from the stirrer. At high Re_N numbers the agreement between the calculated and measured Ne number is satisfactory. Employing PAA solutions no unequivocal laminar flow can be attained which prevails within a sufficiently large range of Re_N to determine q^* in Eq. (44). Therefore, it is assumed that for PAA solutions the same q^* as for pseudoplastic media can be employed. Thus, in Fig. 20 the curves were calculated by $q^* = 11$. One can see from this Fig. that in the range $Re_N = 1$ to 100 the measured Ne numbers are higher than the calculated ones. This is in agreement with the observation of other authors[105-107] and can be explained by the turbulence, which is stimulated by the viscoelasticity at lower Re_N numbers than in viscoinelastic media. This arises together with an additive energy dissipation[93]. In the transient and turbulent regions the Ne number is influenced by the stirrer type. For a turbine stirrer, the measured Ne values are lower than the Newtonian ones. Employing the disc and MIG agitators, the measured Ne values are higher than the corresponding Ne number in Newtonian media. This agitator effect is explained by the different flow patterns which prevail in a stirred tank using a turbine agitator vs. a disc and/or a MIG agitator[93].

The power input is considerably influenced by the aeration of the media.

In Figure 21 the Ne number is plotted as a function of the aeration number Q_G^*, with the Foude number Fr, as a parameter, employing glycerol solution and, for comparison, water. In liquids with low viscosity, Ne diminishes with increasing Q_G^* and Fr due to the rising relative gas hold-up, E_G. However, with rising viscosity this Q_G^* dependence gradually disappears. In highly viscous media Ne no longer depends on Q_G. In these system the Fr number effect is very significant. The course of the $Ne(Q_G^*)$ function in highly-viscous media can be explained by the formation of stable gas trails behind the stirrer blades (see also Fig. 15). This is coupled with a strong reduction of the power input due to aeration of the medium. Obviously, the energy dissipation is considerably reduced if the wake of the blades is large and stable bubble trails are formed. The volume of these gas trails depends on the Fr number (rotation speed). The course of the Ne (Q_G^*) function is quite similar if non-Newtonian media and a turbine stirrer are employed. For MIG agitators the angular distance between the blades is larger (180°) than that of the turbine stirrer with 6 blades (60°); therefore, the stirrer flooding occurs at much higher aeration rates than that for turbine stirrers. This can clearly be recognized

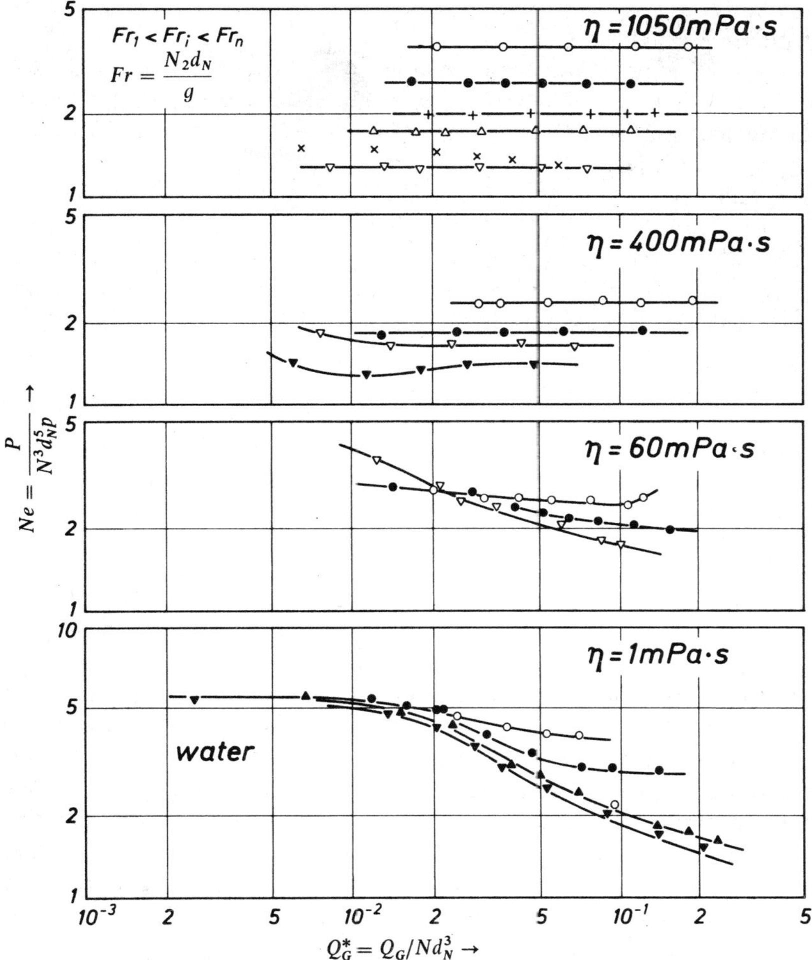

Fig. 21 Newton (power) number Ne as a function of aeration number Q_G^* for glycerol solutions and Froude number Fr as a parameter. Turbine stirrer; $H/D = 1$ [93].

	Fr		Fr
○	0.09	▲	0.64
●	0.17	▽	0.91
+	0.32	▼	1.64
△	0.46	×	2.2

from a comparison of these two stirrers under comparable conditions (Figs. 22 and 23).

As a consequence of this MIG agitator behavior, the Newton number, Ne, increases with increasing aeration number, Q_G, if highly viscous PAA solutions are applied.

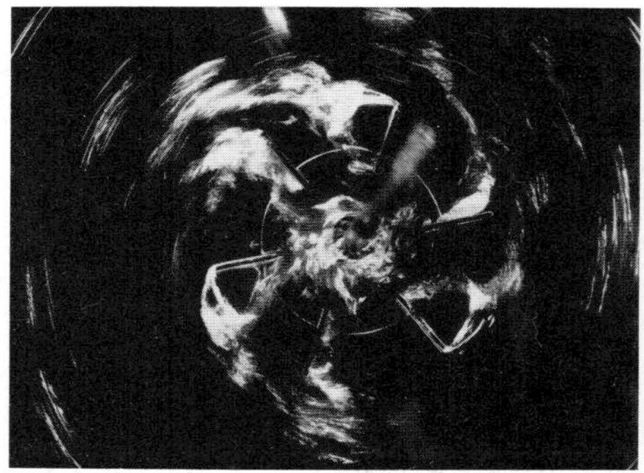

Fig. 22 Gas trails in the wake of the blades of turbine stirrers when highly viscous media are employed[108]

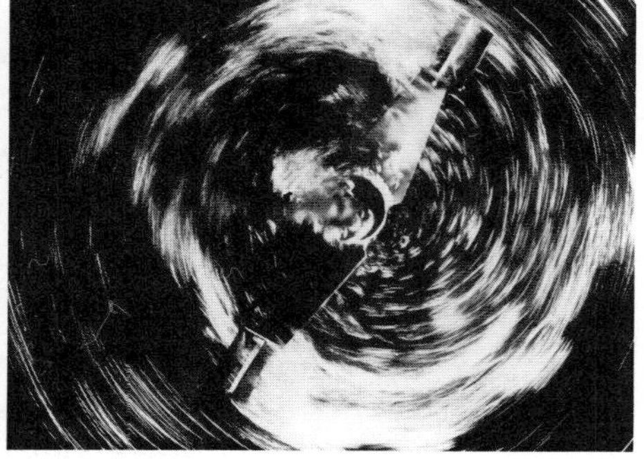

Fig. 23 MIG agitator in highly viscous aerated medium[108]

This remarkable phenomenon can be explained by the fact that at low stirrer speeds and aeration rates, fairly large gas trails prevail in the blade wakes. With increasing aeration rates the volume of these trails significantly diminishes.

The compare the behavior of the glycerol, CMC and PAA solutions it is useful to plot the *Ne* number as a function of the Galilei number, *Ga*, with the aeration number Q_G as a parameter at constant *Fr* number[93].

In Figure 24, this plot is shown for glycerol, CMC and PAA solutions at $Fr = 1$, employing a turbine stirrer.

In glycerol solutions and for $Q_G^* =$ const *Ne* does not depend on *Ga* as long as $(Ga)^{0.5} > 10^4$. With increasing viscosity (i.e. decreasing *Ga* number) *Ne* strongly diminishes as soon as $(Ga)^{0.5} < 10^4$, and at low *Ga* all curves join into one line. The behavior of the *Ne*(*Ga*) function for $(Ga)^{0.5} > 10^4$ is similar to that of water, i.e. the introduced gas can be dispersed by the stirrer. The reduction of *Ne* and the joining of the *Ne*(*Ga*) curves at low *Ga* is caused by the large stable gas trails in the blade wakes. This behavior of the aerated glycerol solutions also holds true of CMC solutions (Fig. 24).

In the aerated viscoelastic PAA solutions, however, an effective gas dispersion only occurs when $(Ga)^{0.5} > 10^5$. Only in this region is *Ne* constant and independent of *Ga*. In the range $(Ga)^{0.5} = 10^4$ to 10^5, *Ne* is considerably reduced by increasing viscosity (diminishing *Ga* number). In the range $(Ga)^{0.5} < 10^4$ all of the $Ne(Ga^{0.5})$ curves run together (Fig. 24). This behavior is caused by the viscoelastic property, which already yields stable gas trails at relatively low viscosities.

By using MIG agitators the dispersion mechanism is considerably changed. As has been mentioned already, the degree of gas dispersion is low at low gas flow rates and stirrer speeds. With rising gas flow rate and stirrer speed the dispersion effect increases. Accordingly, *Ne* does not depend on *Ga* and only slightly on Q_G^* if CMC solutions are employed[93]. In PAA solutions this behavior significantly changes as can be seen from Figure 25. With rising viscosity (decreasing *Ga* number) the *Ne* number increases and passes through a maximum at $Q_G^* = 0$. This course is also preserved for $Q_G^* > 0$. After starting the aeration, *Ne* considerably drops and then remains nearly constant, even with increasing Q_G^*.

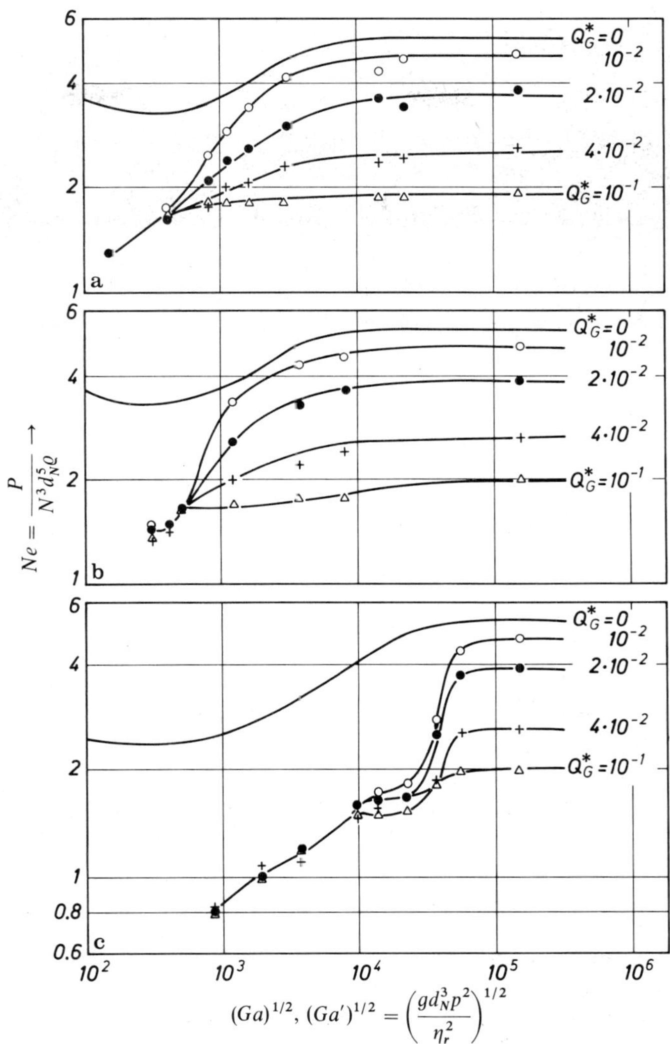

Fig. 24 Newton (power) number *Ne* as a function of Galilei number *Ga*, and aeration number Q_G^* as a parameter Turbine stirrer[93], $Fr = 1$. **(a)** Glycerol/water, **(b)** CMC/water, **(c)** PAA/water; ○ 0.01; ● 0.02; + 0.04; △ 0.20

It is important to stress the following properties of the MIG agitator[93]:

- in the range of the technical gas flow rates the power requirement does not depend on the gas flow rate either in water or in highly viscous media;
- due to the large angular distances of the stirrer blades, gas trail formation in the blade wake is suppressed. Thus, with increasing viscosity the power input increases in contrast to the turbine stirrer for which the power input is strongly reduced as the viscosity increases.

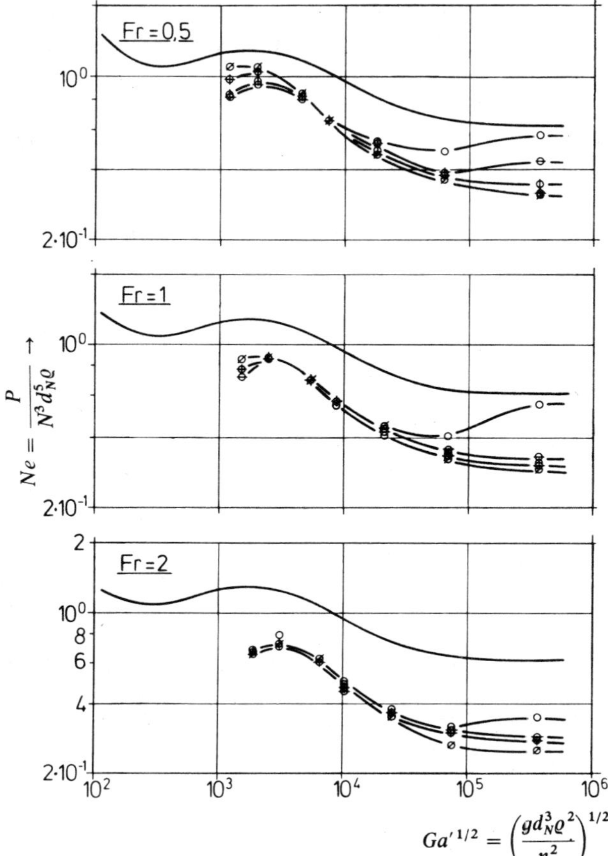

Fig. 25 Newton (power) number Ne as a function of Galilei number Ga and aeration number Q_G^* as a parameter[93]. $H/D_t = 2$, MIG-6 agitator, PAA solution

Symbol	Q_G^*	Symbol	Q_G^*
—	0	⦶	0.04
○	0.01	◇	0.08
⊖	0.02	⊘	0.2

4.5 Oxygen Transfer Rate, OTR, and Volumetric Mass Transfer Coefficient, $k_L a$

Only few papers have been published on OTR into highly viscous media employing stirred tanks. Three different methods are applied to the determination of $k_L a$.

- The dynamical method was employed for the determinion of $k_L a$ by Yagi and Yoshida (glycerol, CMC, PA Na-solutions, Millet-jelly water)[112], Páca et al. (pseudoplastic starch solutions)[110], Kiepke (CMC)[94], and Höcker (glycerol, CMC, PAA)[93, 178, 179];
- the steady-state method was used to measure $k_L a$ by Perez and Sandall [carbopol(carboxypolymethylene)934][109], Keitel (CMC)[95] and König et al. (glycerol, CMC, PAA)[64];
- the chemical method was used by Páca et al. (pseudoplastic starch solutions)[110], Ranade and Ulbrecht (PEO, PAA)[113], Loucaides et al. (three-phase system)[111] and Blakebrough et al. (three-phase system)[115].

The most comprehensive investigations were carried out by Yagi et al.[112] and Höcker[93, 178, 179]. Yagi et al. characterized the non-Newtonian liquids (CMC and PA Na) according to Prest et al.[116] by a constant λ. It is defined as the reciprocal of the shear rate at which the reduced complex viscosity, i.e. the ratio of the representative viscosity η_r to the zero-shear viscosity η_0, is 0.67. The representative viscosity is defined as

$$\eta_r = \frac{\tau}{\left(\dfrac{\overline{dv}}{dx}\right)}, \tag{53}$$

where according to Metzner and Otto[98]

$$\left(\frac{\overline{dv}}{dx}\right) = 11.5N. \tag{44}$$

However, it is not possible to determine η_0 with high accuracy and therefore this viscosity characterization of the medium is unsuitable.

Perez and Sandall[109] applied the method of Calderbank and Moo-Young[99] to the characterization of the medium viscosity:

$$\eta_r = \frac{K}{(11N)^{1-n}}\left(\frac{3n+1}{4n}\right)^n. \tag{53}$$

Höcker[93], Keitel[95], Kiepke[94] and König et al.[64] used the Metzner-Otto relationship (44) to define the medium viscosity.

Under steady-state conditions, two stress functions can be evaluated from the rotational flow of a liquid between a cone and a plate. The torque M_i can be converted to the shear stress by:

$$\tau = \frac{3M}{2R^3}, \tag{56}$$

where R is the radius of the cone, and the axial thrust T between the cone and the plate can be used to compute the primary normal stress difference

$$\sigma_1 = \frac{2T}{\pi R^2}, \tag{57}$$

which is one of the measures of viscoelasticity. The following relationships are assumed[113]:

$$\tau = K(D)^n, \tag{14a}$$

$$\sigma_1 = h(D)^{\lambda_1}. \tag{14b}$$

Furthermore, the characteristic time of the liquid can be described by t_{cr}, defined as

$$t_{cr} = \frac{\sigma_1}{D^2}, \tag{58}$$

where $\eta = K(D)^{n-1}$.

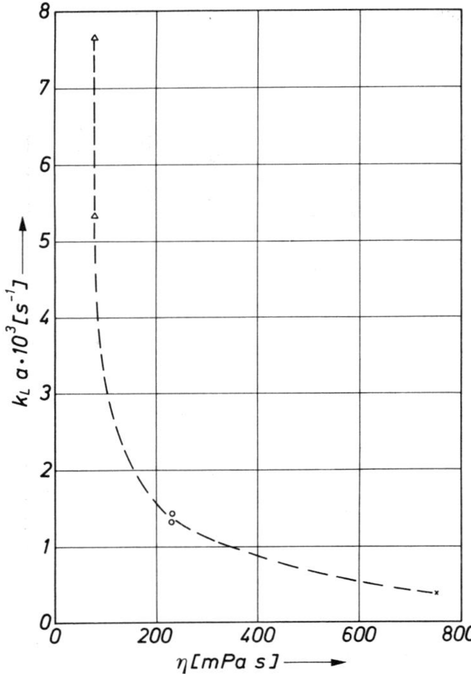

Fig. 26 Volumetric mass transfer coefficient $k_L a$ as a function of viscosity at different gas flow rates Q_G. Self-aerating tube stirrer $D_t = 19$ cm, $H_t = 34$ cm, $N = 2800$ rpm. Glycerol solutions[64]. △ 80 m Pa · s, ○ 230 m Pa · s, × 755 Pa · s

Substituting η and Eq. (14b) into Eq. (58) one obtains

$$t_{cr} = \frac{h(D)^{\lambda_1}}{D^2 K D^{n-1}} = \frac{h}{K} D^{\lambda_1 - n - 1} . \tag{59}$$

Ranade and Ulbrecht[113] applied the modified Deborah number, *De*

$$De = N t_{cr} \tag{60}$$

to the characterization of the viscoelasticity of the medium, using Eqs. (56) to (60).

Employing a self-aerating tube stirrer at high rotational speed, ($N = 2800$ rpm which is characteristic of these stirrers), the aeration rate does not influence $k_L a$ in glycerol solutions[64]. Therefore, $k_L a$ can be plotted as a function of the medium viscosity (Fig. 26). With increasing viscosity, $k_L a$ considerably diminishes.

A stirred tank loop reactor can be operated in a totally filled mode, with the liquid level above the upper edge of the draft tube, or in the oberflow mode, with the liquid level in the draft tube below its upper edge[95]. In Figure 27, $k_L a$ is plotted as a function of N for these two different modes employing a 1% CMC solution. One can recognize that no significant difference exists between them. In both of these modes $k_L a$ increases with rising N and w_{SG}; this gas flow rate effect, however, is fairly slight. The same also holds true of 1.5% CMC solutions (Fig. 28). With increasing CMC concentration, $k_L a$ at first diminishes but changes only slightly in the range 0.75 to 1.5%.

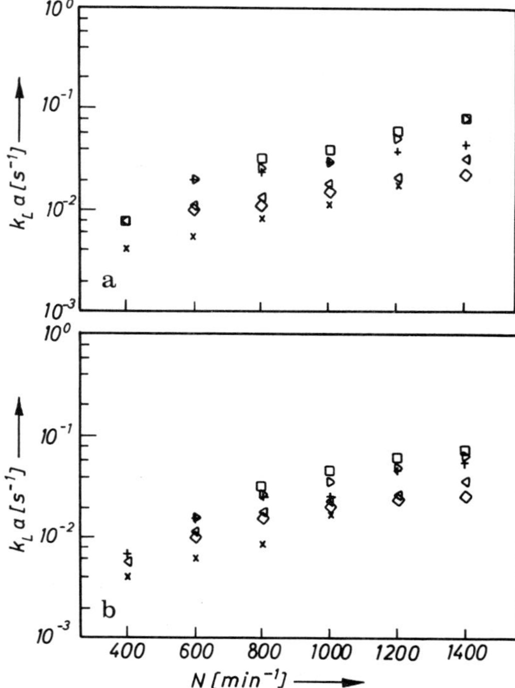

Fig. 27 Volumetric mass transfer coefficient $k_L a$ as a function of impeller speed N and superficial gas velocity w_{SG} as a parameter. Stirred tank loop reactor. $D_t = 45$ cm, draft tube hight $H_D = 60$ cm, 1% CMC solution[95]. a) totally filled reactor; b) with overflow (for symbols see Fig. 16)

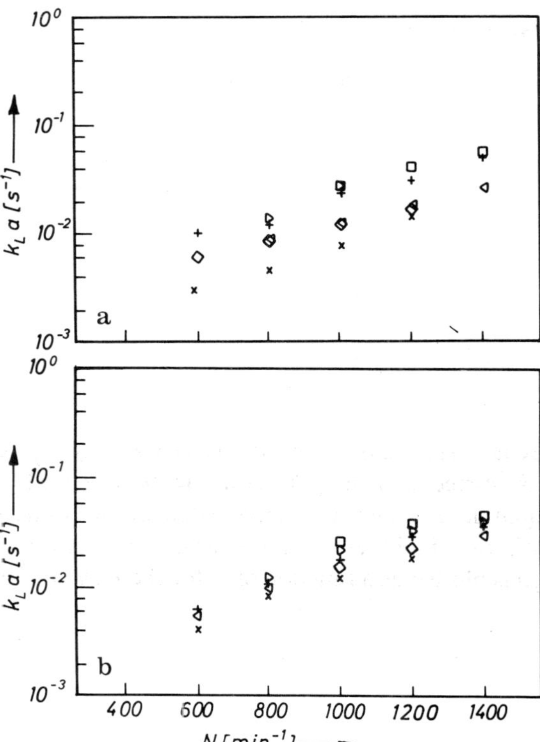

Fig. 28 Volumetric mass transfer coefficient $k_L a$ as a function of impeller speed N with superficial gas velocity w_{SG} as a parameter. Stirred tank loop reactor $D_t = 45$ cm, $H_D = 60$ cm, 1.5% CMC solution[95]. a) totally filled reactor; b) with overflow (for symbols see Fig. 16)

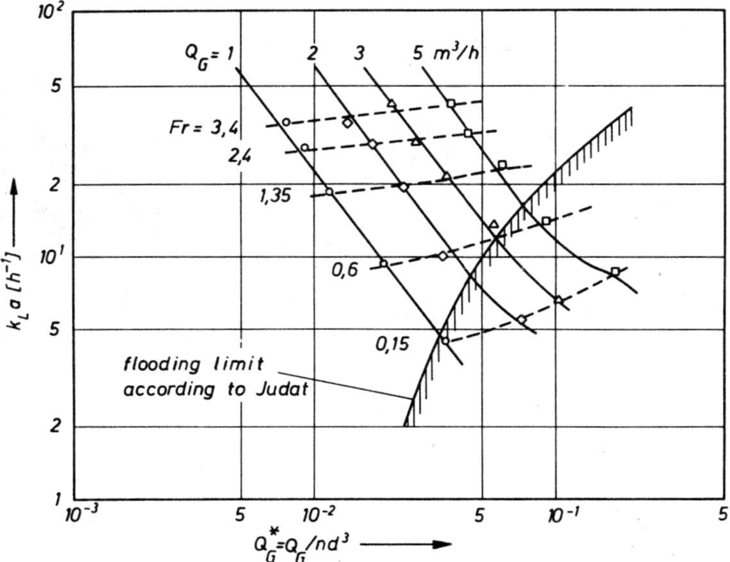

Fig. 29 Volumetric mass transfer coefficient k_La as a function of aeration number Q_G^* with Froude number Fr as a parameter. Stirred tank $D_t = 40$ cm, $H_t = 40$ cm, turbine stirrer $d_N = 13.2$ cm, CMC solution $\eta_r = 560$ m Pa · s [94]

In a stirred tank with a turbine stirrer, k_La increases only slightly with rising gas flow rate, Q_G if a CMC solution is applied[94]. The effect of the Froude number (rotational speed) is considerable[94] (Fig. 29). However, at very low rotational speed, N, k_La is independent of N[109] (Fig. 30).

The results of the k_La measurements can be correlated in different ways. Two types of presentations have been reported:

Type I: $Sh_N = f(Re, Sc, ...)$ (62)

and

Type II: $k_La\left(\frac{V}{Q_G}\right) = f\left(\frac{P}{Q_G}\right)^*, Sc\,.$ (63)

To type I belong the relationships recommended by Yagi et al.[112], Perez et al.[109] and Ranada et al.[101].

Type II was suggested by Zlokarnik[117], Höcker[93], Kiepke[94], Keitel[95] and Henzler[114].

A) Type I:
Yagi and Yoshida[112] recommended relationship (64):

$$Sh'_N = f_1\, Re_N^{n_1}\, Fr_N^{n_2}\, Sc^{n_3}\left(\frac{w_{SG}\eta}{\sigma}\right)^{n_4}\left(\frac{Nd_N}{w_{SG}}\right)^{n_5}(1 + 2.0\, De^{0.5})^{n_6}\,, \tag{64}$$

where $Sh'_N = \frac{k_Lad_N^2}{D_m}$ a modified stirrer Sherwood number.

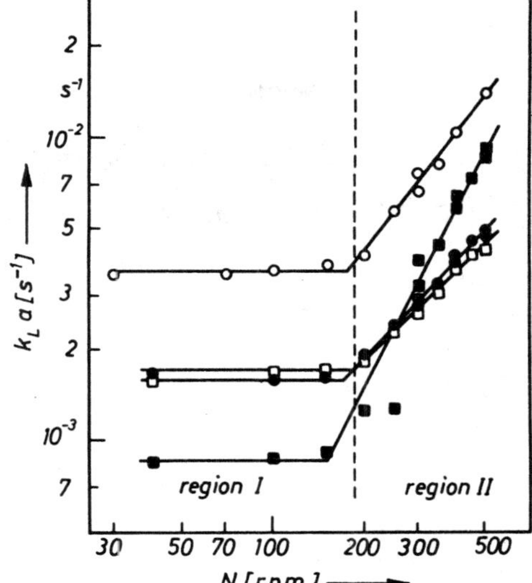

Fig. 30 Volumetric mass transfer coefficient k_La as a function of impeller speed N. Turbine stirrer. $D_t = 15.2$ cm, $H_t = 15.2$ cm, $w_{SG} = 0.162$ cm s^{-1} [109].
○ water
□ 0.25% Carbopol ($n = 0.916$, $K = 0.0428$)
● 0.75% Carbopol ($n = 0.773$, $K = 0.507$)
■ 1.00% Carbopol ($n = 0.594$, $K = 5.29$)

The empirically determined constants have the following values:

$f_1 = 0.06, n_1 = 1.5, n_2 = 0.19, n_3 = 0.50, n_4 = 0.6, n_5 = 0.32, n_6 = -0.67.$

The validity range of Eq. (64) is

$0.2 \leqq w_{SG} \leqq 10 \text{ cms}^{-1}$
$4 \leqq N \leqq 10 \text{ rps}$
$0.01 \leqq k_La \leqq 0.1 \text{ s}^{-1}$
$30 \leqq Sh'_N, Sc^{-0.5} \leqq 3 \times 10^3$
$0.017 \leqq n \leqq 1$
$0.0017 \leqq t_{cr} \leqq 8.3 \text{ s}$
$0.117 \leqq K \leqq 5.2 \text{ Pa} \times \text{s}^n$.

Perez and Sandall[109] recommended Eq. (65):

$$Sh'_N = f_2\, Re_N^{n_1}\, Sc^{n_3} \left(\frac{d_N w_{SG}}{\sigma}\right)^{n_7} \cdot \left(\frac{\eta_g}{\eta}\right)^{n_8} . \qquad (65)$$

The constants have the values:

$f_2 = 21.2, n_1 = 1.11, n_3 = 0.5, n_7 = 0.447, n_8 = 0.694.$

The vailidity range of Eq. includes (65):

$0.162 \leqq w_{SG} \leqq 0.466 \text{ cm s}^{-1}$
$2800 \leqq Re_N \leqq 26700$
$200 \leqq N \leqq 500 \text{ rpm}$
$0.009 \leqq K \leqq 0.04 \text{ Pa} \times \text{s}^n$
$0.916 \leqq n \leqq 1.0$
$455 \leqq Sc \leqq 1490$
$2340 \leqq Sh'_N \leqq 20900.$

Ranade and Ulbrecht[101)] have applied Eq. (66)

$$Sh'_N = f_3\, Re_N^{n_1} \left(\frac{\eta}{\eta_w}\right)^{n_9} (1 + De)^{n_{10}}, \tag{66}$$

where η_w is the dynamic viscosity of water.

The constants have the values:

$$f_3 = 2.5 \times 10^{-4},\ n_1 = 1.98,\ n_9 = 1.39,\ n_{10} = -0.67.$$

The validity range of Eq. (66):

$$2 \times 10^4 \leqq Re_N \leqq 2 \times 10^5$$
$$4 \times 10^{-3} \leqq k_L a \leqq 10^{-1}\ \text{s}^{-1}$$
$$1.27 \times 10^{-3} \leqq K \leqq 5 \times 7 \times 10^{-2}\ \text{Pa} \cdot \text{s}^n$$
$$0.55 \leqq n \leqq 1.0$$
$$70 \times 10^{-3} \leqq h\ 750 \times 10^{-3}$$
$$0.77 \leqq \lambda_1 \leqq 0.975.$$

B) Type II:

Höcker[93)] plotted his data according to Zlokarnik[117)], i.e. $k_L a \left(\frac{V}{Q_G}\right)$ as a function of $(P/Q_G)^*$. In Fig. 31, the CMC values are correlated utilizing this plot. For a given CMC concentration the experimental points can be described by parallel straight lines on this plot. With increasing CMC concentration these lines are shifted to lower $k_L a(V/Q_G)$ values. Only the line for water has a different slope.

To compare the influence of different stirrer types on Fig. 32, $k_L a(V/Q_G)$ was plotted as a function of (P/Q_G) for a 0.75% CMC solution employing a turbine stirrer, MIG-4 and MIG-6 agitators and a disc stirrer[93)]. All of these points can be represented by a straight line; when using PAA solutions and different stirrer types the $k_L a(V/Q_G)$ vs. $(P/Q_G)^*$ plot yields two straight lines. The upper represents the data for MIG agitators and the lower those for turbine and disc stirrers[93)] (Fig. 33).

Recently, Henzler[114)] have suggested some relationships of type B: For CMC solutions (Fig. 34):

$$k_L a(V/Q_G) = 0.082\ (P/Q_G)^{*0.6}\ Sc^{*\,-0.3}\ , \tag{67}$$

where $(P/Q_G)^* = (P/Q_G)\,[\varrho_L (\nu_g)^{2/3}]^{-1}$ is the dimensionless specific power input with regard to the gas flow rate.

Validity range of Eq. (67):

$$8 \times 10^3 \leqq Sc^* \leqq 1.5 \times 10^5\ ,$$
$$2 \times 10^2 \leqq (P/Q_G)^* \leqq 2 \times 10^5\ .$$

Relationship (67) is based on the data of Yagi et al.[112)], Höcker[93)] and Kiepke[94)].

Henzler also recommended a relationship for the Newtonian glucose and glycerol solutions[114)] (Fig. 35):

$$k_L a(V/Q_G) = 0.045 (P/Q_G)^{*0.5}\ Sc^{-0.3}\ . \tag{68}$$

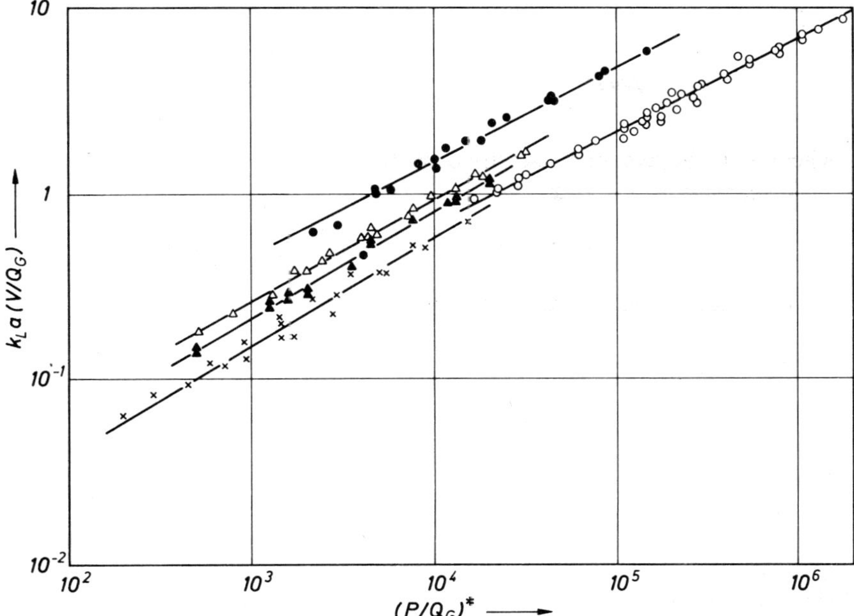

Fig. 31 Dimensionless volumetric mass transfer coefficient $k_L a(V/Q_G)$ as a function of dimensionless specific power input $(P/Q_G)^*$ for water and CMC solutions. $D_t = 40$ cm, $H_t = 40$ cm, turbine stirrer[93].

$$Ga'' = (\varrho/K)^{\frac{2}{2-n}} d^{\frac{2n}{2-n}} g\,, \qquad Sc = \frac{\eta_s/\varrho}{D}\,, \qquad \frac{k_L a}{Q/V} = f[(P/Q)^*,\ SC]$$

		K (Pa · s^n)	n (—)
○	water	0.001	0.1 ?
●	0.3% CMC	0.050	0.8
△	0.75% CMC	0.210	0.74
▲	1.0% CMC	0.582	0.64
×	1.5% CMC	1.470	0.60

Validity range:

$$4 \times 10^2 \leqq Sc \leqq 1.5 \times 10^6\,,$$
$$10^3 \leqq (P/Q_G)^* \leqq 2 \times 10^6\,.$$

Eq. (68) is based on the data of Höcker[93] and Yagi et al.[112].

A similar relationship was suggested for millet jelly water[114] based on the data of Yagi et al.[112] (Fig. 35):

$$k_L a(V/Q_G) = 0.0125\ (P/Q_G)^{*\,0.6}\ Sc^{*\ -0.17}\,. \tag{69}$$

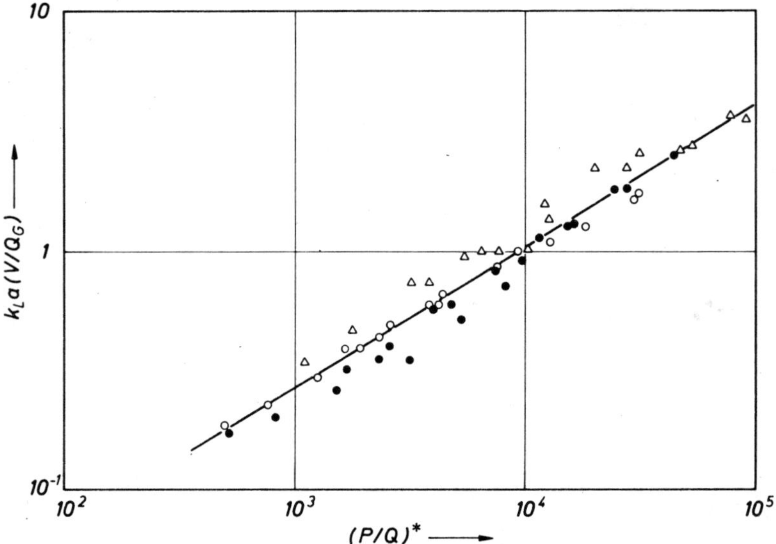

Fig. 32 Dimensionless volumetric mass transfer coefficient $k_L a(V/Q_G)$ as a function of dimensionless specific power input $(P/Q_G)^*$ for 0.75% CMC solution[93]. $D_t = 40$ cm, $H_t = 40$ cm, $K = 0.21$ Pa · s^n ($n = 0.75$). ○ turbine stirrer; ● MIG-4 agitator; △ disc stirrer

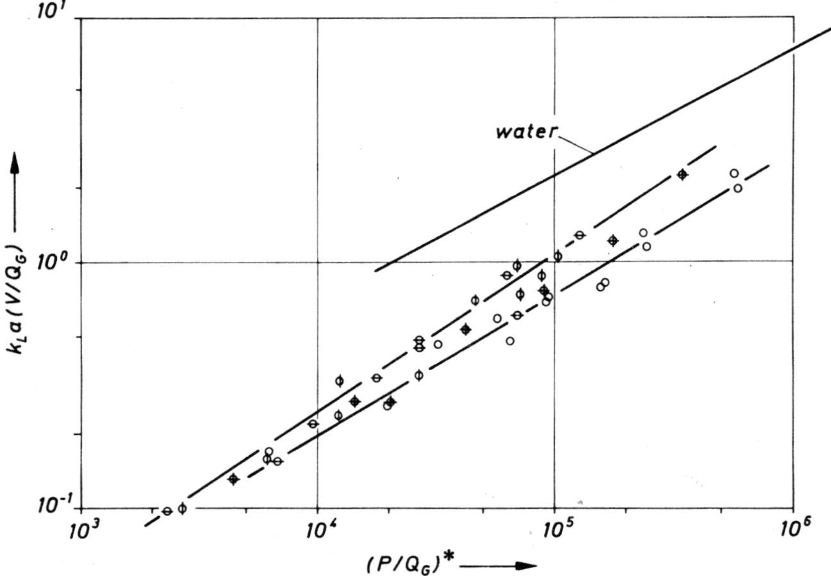

Fig. 33 Dimensionless volumetric mass transfer coefficient $k_L a(V/Q_G)$ as a function of dimensionless specific power input $(P/Q_G)^*$ for 0.1% PAA solution. $D_t = 40$ cm, $H_t = 40$ cm[93], $K = 0.24$ Pa · s^n ($n = 0.51$). ○ turbine stirrer; ⊖ MIG-4; ⦶ MIG-6; ⊕ disc agitator

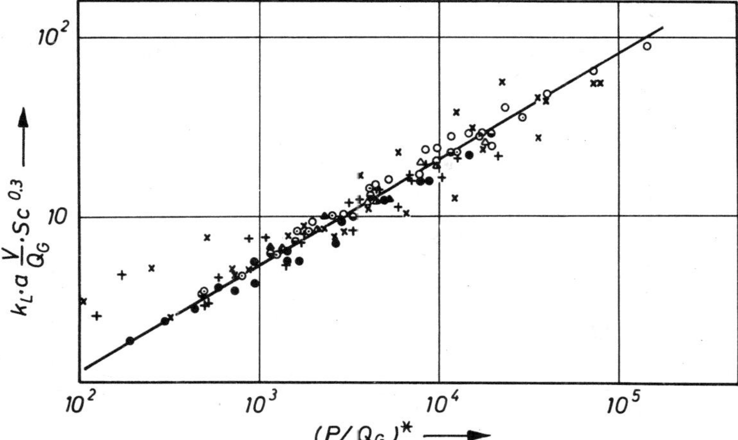

Fig. 34 $St_G\ Sc^{*0.3}$ as a function of the dimensionless specific power input $(P/Q_G)^*$ for CMC solutions. Turbine stirrer. $H_t/D_t = 1$, $d_N/D_t = 0.25$—0.33 [114, 166].

	η_r (m Pa · s)	Ref.
△	51—58	
◬	112—125	112)
▲	262—302	
○	16—24	
⊙	48—81	93)
◒	75—156	
●	152—342	
× +	220—1500	94)

Validity range:

$$5 \times 10^2 \leqq Sc^* \leqq 5 \times 10^5 \,,$$
$$10^3 \leqq (P/Q_G)^* \leqq 2.5 \times 10^5 \,.$$

Henzler[114] did not succeed to develop a similar relationship for viscoelastic media (PAA solutions).

4.6 Heat Transfer

Only few papers are concerned with heat transfer processes employing aerated stirred tanks and highly viscous liquids[118–123].

A comparison of the heat transfer coefficients, α, between medium and jacket in non-erated and aerated systems indicates that at low rotational speeds, aeration increases α whereas at high rotational speeds it is diminished (Fig. 36). The influence of the viscosity on α_j is shown in Fig. 37. With increasing viscosity α_j diminishes. This reduction of α_j is especially significant at high aeration rates and for $\eta > 500$ mPa · s.

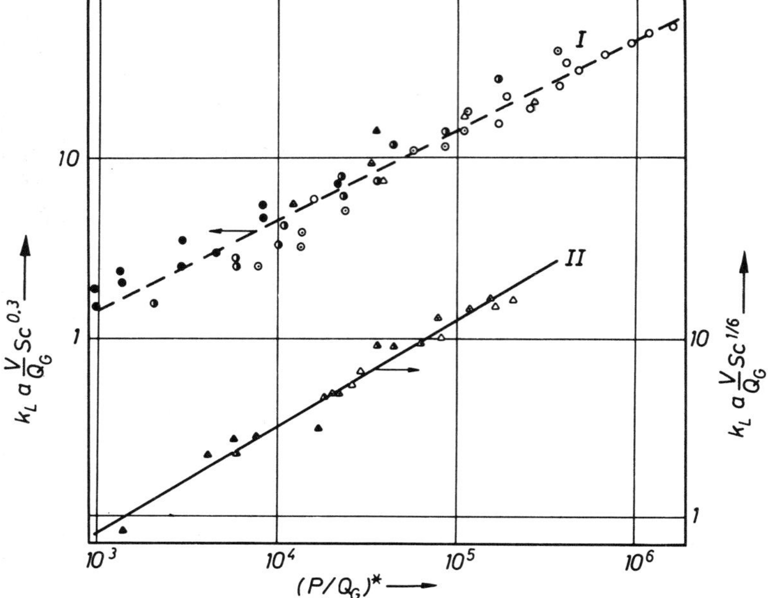

Fig. 35 $St_G\ Sc^{*0.3}$ as a function of dimensionless specific power input $(P/Q_G)^*$ [114, 166]. I glucose and glycerol solutions; II millet jelly water

	η_r (m Pa · s)		Ref.		η_r (m Pa · s)
	millet jelly water		[112]		
○	water	0.9	[177]	△	1.3
⊙		12		◬	2.1
◑	glucose	51	[93]	◭	13.3
●	solution	267		▲	70.2
△	glycerol	0.9	[112]		
◬	solution	5.1	[112]		

Figure 38 indicates that α_j increases with the specific power input. The data with 0.5 and 1.0% CMC solution relating to different aeration rates can be represented by a single relationship. Only α_j values measured in 1.5% CMC solutions cannot be described by this common relationship.

A comparison of the α_j values, evaluated employing different stirrer types, is shown in Fig. 39. One recognizes that the two-stage INTERMIG stirrer with the larger d_N/D ratio yields at the same specific power input much higher α_j values than the turbine stirrer. To attain the same α_j value this turbine stirrer needs ten times as high a specific power input than the two-stage INTERMIG stirrer. This improvement is mainly due to the higher d_N/D_t ratio of the INTERMIG stirrer. In pseudoplastic media the power input by the stirrer is mainly used to pump the medium in the immediate stirrer vicinity. At the jacket wall this stirrer effect is fairly low. By increasing the d_N/D_t ratio the medium velocity at the jacket wall can significantly

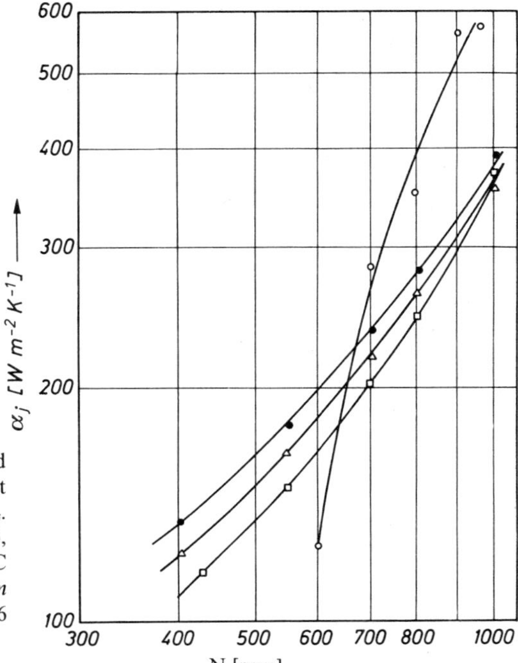

Fig. 36 Influence of stirrer speed N and aeration rate on the heat transfer coefficient between the medium and jacket wall, α_j. Stirred tank (D_t = 40 cm, H_t = 40 cm), turbine stirrer d_N/D_t = 0.33; 1.5% CMC (Tylose 10000)[121]; CMC = 1.5%. *vvm* = Q_G/V (min^{-1}); ○ 0; □ 0.5; △ 1; ● 1.6

be raised[121]. In the range $Re_N = 10^3$ to 10^6 all of the data relating to water and to pseudoplastic CMC solution can be represented by a simple relationship:

$$Nu_N = f_1 \, Re_N^{n_1} \, Pr^{1/3} \left(\frac{\eta}{\eta w}\right)^{0.14} . \tag{70}$$

Even by neglecting the viscosity term because of the small exponent (0.14) a satisfactory correlation can be attained (Fig. 40). However, in the range $Re_N < 10^3$ there is a deviation from the Newtonian relationship due to strong pseudoplastic behavior of CMC solutions at high concentrations[121]. Kahilainen et al.[122] and Steiff[123] have compared the relationships published in the literature. The equations of Steiff[118] and Nagata[120] can only be applied to fairly low viscous media (η_r < 100 mPa · s). Kahilainen et al.[122] employed cellulose suspensions exhibiting pseudoplastic behavior. The relationship recommended by them is valid for the range n_r = 35 to 205 mPa · s:

$$Nu^* = f_1 \, Re_N^{*n_1} \, Pr^{*1/2} \, Q_G^{*n_3} , \tag{71}$$

where $Nu^* = \dfrac{\alpha D_t}{\lambda^*}$ is the modified Nusselt number

$$\lambda^* = E_G \lambda_g + (1 - E_G) \, \lambda_L ,$$

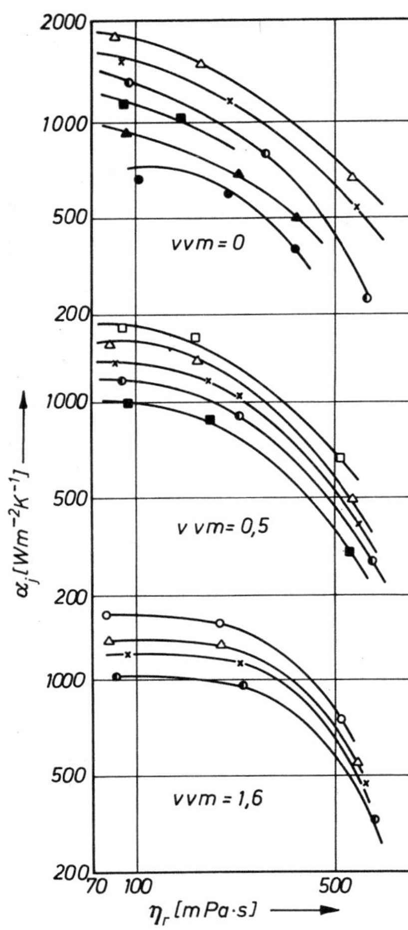

Fig. 37 Influence of viscosity η_r on heat transfer coefficient α_j at different stirrer speeds N and aeration rates $vvm = Q_G/V$ (min^{-1}). Stirrer tank (D_t = 40 cm, H_t = 40 cm), turbine stirrer (d_G/D_t = 0.33). CMC solutions[121)]

N	(rpm)	N	(rpm)
●	300	×	700
▲	400	△	800
■	500	□	900
◐	600	○	1000

λ_g, λ_L = heat conductivity of the gas and/or liquid

$$Re_N^* = \frac{Nd_N^2}{\nu_r} = \text{modified Reynolds number}$$

$$\nu_r = \eta_r/\varrho_L$$

$$\eta_r = K(q^*N)^{n-1}\left(\frac{3n+1}{4n}\right)^n$$

$$Pr^* = \frac{\eta_r C_L}{\lambda^*} = \text{modified Prandtl number}$$

C_L = specific heat of the liquid.

$n_3 \simeq 0$.

The constants of Eq. (71) are given in Table 4[122)]:

Table 4. Constants of Eq. (71)[122]
Stirred tank (D_t = 590 mm, H_t/D_t = 2.88, H/D_t = 1.78)
turbine stirrer (d_N/D_t = 0.33)

Heat exchanger	Number of stages	f_1	n_1	n_2
Coil (A_h = 0.56 m^2)	1	3.00	0.319	0.128
	2	0.50	0.384	0.365
Tube (A_h = 0.56 m^2)	1	8.32	0.430	−0.320
	2	0.034	0.660	0.340
Plate (A_h = 0.028 m^2)	1	1.978	0.409	−0.030
	2	0.057	0.541	0.412

For the geometrical data of the heat exchanger see [122]

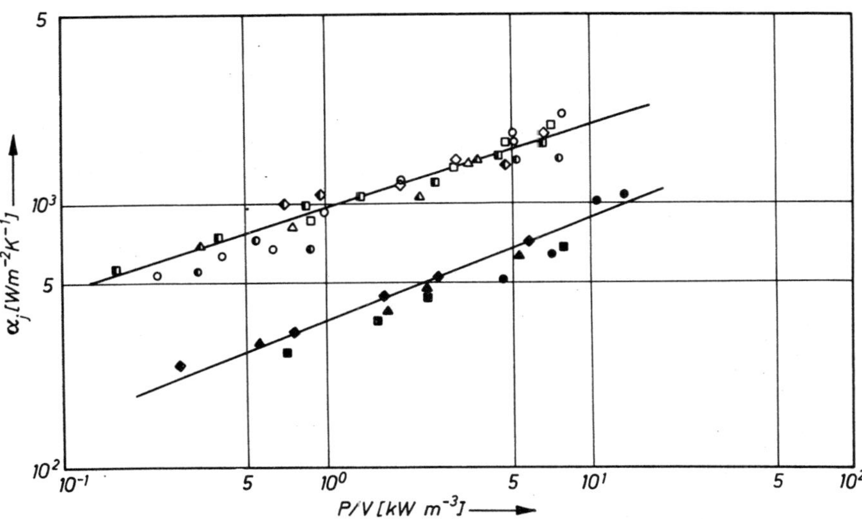

Fig. 38 Heat transfer coefficient α_j as a function of specific power input P/V. Stirred tank (D_t = 40 cm, H_t = 40 cm), turbine stirrer (d_N/D_t = 0.33), CMC solutions[121].

CMC (%)		0.5	1.0	1.5
vvm	0	○	◐	●
	0.5	□	◧	■
	1.0	△	◭	▲
	1.6	◇	⬖	◆

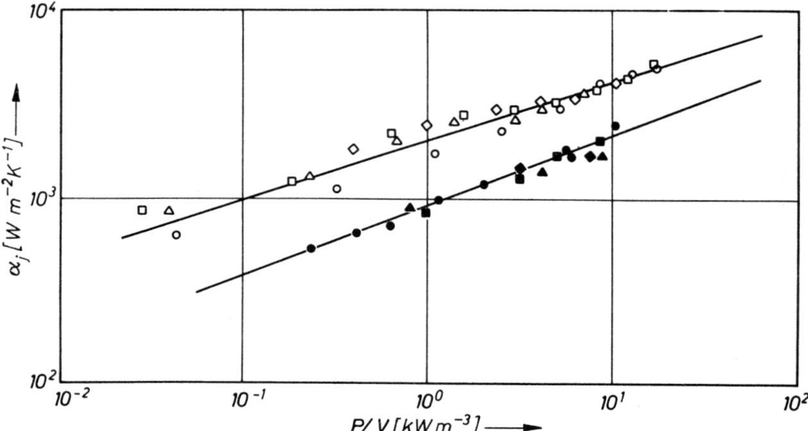

Fig. 39 Heat transfer coefficients α_j as a function of specific power input P/V. Influence of the stirrer type. Stirred tank (D_t = 40 cm, H_t = 40 cm), *TS* (turbine stirrer), $d_N/D_t = 0.33$. 2 IMIG (two-stage INTERMIG stirrer) ($d_N/D_t = 0.60$) 0.50% CMC solution[121].

vvm	2 IMIG	*TS*
0	○	●
0.5	□	■
1.0	△	▲
1.6	◇	◆

The validity range of Eq. (71) is:

$$11 \leqq Nu^* \leqq 148$$
$$10^3 \leqq Re_N^* \leqq 1.6 \times 10^4$$
$$230 \leqq Pr^* \leqq 1360$$
$$5 \leqq N \leqq 12.6 \text{ rps}$$
$$0.20 \leqq n \leqq 0.58$$
$$26 \leqq K \leqq 440 \text{ mPa} \cdot \text{s}^n$$
$$35 \leqq \eta_r \leqq 205 \text{ mPa} \cdot \text{s}.$$

Other recent relationship were recommended by Suryanarayanan et al.[124] who evaluated Nu_j using a tank with heated jacket, a cooled coil and a turbine stirrer:

$$Nu_j = 0.22\,(Re_N^*)^{0.63}\,Pr^{*0.33}\left(\frac{d_N}{D_t}\right)^{0.14}\left(\frac{d_a}{D_t}\right)^{0.09}\left(\frac{D_{Co}}{D_t}\right)^{-0.21}\left(\frac{d_o}{D_t}\right)^{-0.35}, \quad (72)$$

where

$$Nu_j = \frac{\alpha_j D_t}{\lambda} = \text{jacket Nusselt number}$$

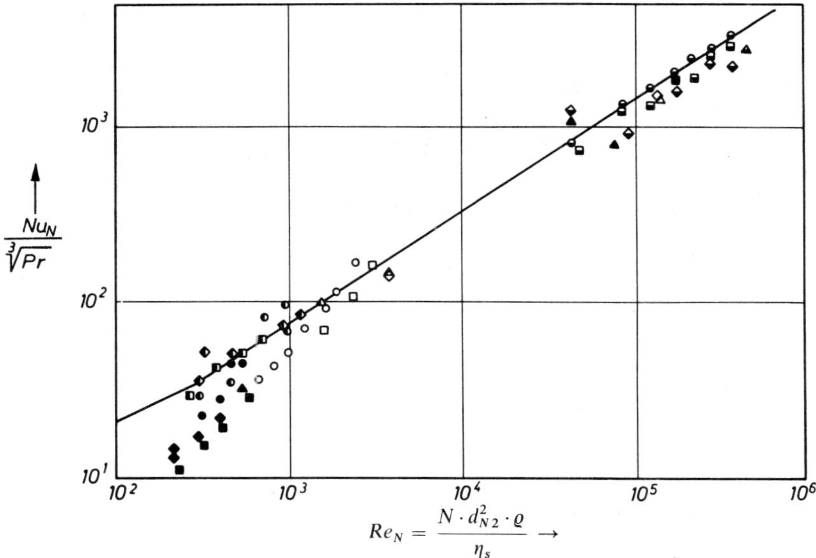

Fig. 40 Nu_N — Re_N relationship for Newtonian and non-Newtonian media. Stirred tank ($D_t = 40$ cm, $H_t = 40$ cm), turbine stirrer ($d_N/D_t = 0.33$), two-stage INTERMIG stirrer ($d_N/D_t = 0.6$), water and CMC solutions[121)]

vvm	H_2O	0.5% CMC	1.0% CMC	1.5% CMC
0	◒	○	◐	●
0.5	⬓	□	◧	■
1.0	▲	△	◭	▲
1.6	⬙	◇	⬖	◆

α_j = heat transfer coefficient at the jacket wall
d_a = depth of agitator from the tank bottom
D_{Co} = mean diameter of the coil helix
d_o = coiled tube outside the diameter

$$Re_N^* = \frac{d_N^2 N \varrho_L}{\eta_r}$$

$$Pr^* = \frac{\eta_r C_L}{\lambda_L}$$

λ_L = liquid heat conductivity
η = (see Eq. (53))

$$Nu_{Co} = 0.21\, Re_N^{*0.66}\, Pr^{*0.33} \left(\frac{d_N}{D_t}\right)^{0.17} \left(\frac{d_a}{D_t}\right)^{0.13} \left(\frac{D_{Co}}{D_t}\right)^{-0.29} \left(\frac{d_o}{D_t}\right)^{-0.45}, \quad (73)$$

where $Nu_{Co} = \frac{\alpha_{Co} D_t}{\lambda_L}$ is the coil Nusselt number and α_{jo} = the heat transfer coefficient at the outer coil surface.

Relationships were evaluated in a flat-bottomed, cylindrical, jacketed tank (D_L = 356 mm) equipped with a helical coil (D_{Co} = 16.9 to 27.8 mm) and a four flat-bladed turbine agitator (d_N = 7.8 to 15.2 mm). To prevent vortex formation four equally spaced baffles of 30 mm width were installed vertically at the vessel wall.

As test liquids 1% and 2% CMC Na (sodium CMC) and 1% and 2.5% SA (sodium alginate) solutions were employed.

The validity ranges of Eqs. (72) and (73) are:

$$180 \leqq N \leqq 670 \text{ rpm}$$
$$0.47 \leqq n \leqq 1.0$$
$$1.24 \leqq K \left(\frac{3n+1}{4n}\right)^n \leqq 100 \text{ mPas}^n$$
$$200 \leqq Re_N^* \leqq 21\,700$$
$$49 \leqq PR^* \leqq 1220.$$

A comparison of Eq. (71) (employing the constants in Table 4) with Eqs. (72) and (73) and similar relationships (Fig. 41) indicate a considerable deviation of Eq. (71) from the others. This departure may be explained by the fact that in the Nu-Re_N^* relationships (*a*) to (*d*) in Fig. 41 the value $^2/_3$ was assumed for the exponent of the R_N number. Furthermore, Eq. (71) was evaluated for cellulose suspensions, the other equations for CMC and CMC Na solutions.

5 Single Stage Tower Reactors

Bubble-column reactors are popular in the chemical industry because of their versatile use and economical advantages, i.e. low investment costs due to simple construction and low variable costs of production due to the low energy requirement of their operation[127]. They belong to the novel reactor types which will profitably be employed for special biotechnological production processes (e.g. for SCP

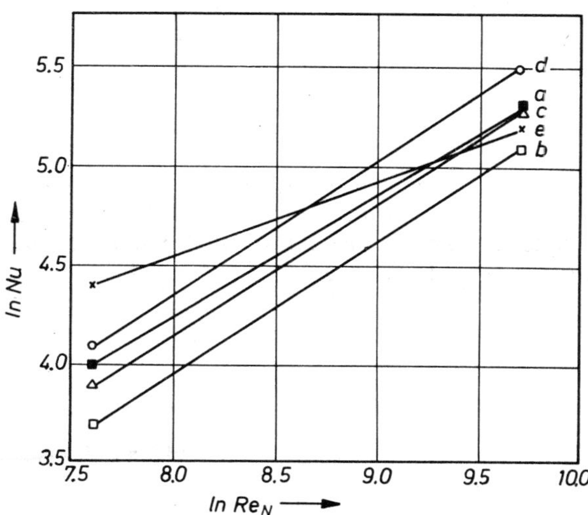

Fig. 41 Comparison of Nu—Re_N relationships of different authors[122]. ln Nu = ln A^N + $b \cdot$ ln Re_N. (a) Raja Rao[124]; (b) Skelland et al.[125]; (c) Noorud-din M. Raya[126]; (d) Edney et al.[119]; (e) Kahilainen et al.[122]

production[128]) in the near future, if enough know-how their design and operation becomes available.

5.1 Apparatus and Instruments

The apparatus consists of a column and usually a static aerator, a gas and a liquid supply. The gas enters the column at the bottom and leaves it at its head. In concurrent columns, the liquid enters the column at the bottom and leaves it at the top as does the gas. In countercurrent columns, the liquid enters the column at the top and leaves it at the bottom. Bubble column reactors are practically employed in several modifications[129,130]. In this chapter, however, only simple tower reactors will be considered.

To determine the relative gas hold-up, E_G, the bubbling layer height H and the bubble-free liquid layer height H_L are measured and E_G is calculated by

$$E_G = \frac{H - H_L}{H}. \tag{21}$$

H_L can be measured by the static pressure at the bottom of the column, e.g. by use of a stand pipe.

Since in slender columns the system properties reveal a significant position dependence, it is more difficult to make representative measurements in these columns than in stirred tanks. It is usually necessary to employ more than one detector, e.g. to measure the mixing time, the OTR and/or k_La etc.

To measure the power input it is generally sufficient to measure the gas flow rate and the compression energy needed for the aeration by means of pressure gauges. For the determination of the longitudinal liquid dispersion coefficient, E, either a series of probes along the columns are required (if the steady-state tracer technique is used, cf.[72]) or two probes (one near the entrance and the other near the exit) if the non-steady state method (distribution of residence time) is applied (e.g.[71])). In the latter case, a data logger (for storage) and computer (data evaluation) or process computer (for on-line evaluation[131,132]) are needed. The determination of the bubble size distribution requires a plane parallel window (for flash photography) or special probes for bubble detection[7,43,49]. In highly viscous systems, the electrooptical probe cannot be employed[48,50] and the electric conductivity probe must be modified[51].

For the k_La determination several oxygen electrodes are needed if the non-steady state method is applied and if representative values are desired. If the steady-state is used, two columns are necessary and in the investigated column several oxygen electrodes must be distributed along the column[84,133].

5.2 Mathematical Models

A slender bubble volumn can only be described by a model with distributed parameters. Two commonly employed models are the dispersion model and the cell model.

The dispersion model has already been considered in Ref.[127] and is therefore not treated here in detail. The cell model assumes that the column consists of several perfectly mixed cells. In the simplest type of cell model the cells can only be transferred by the liquid in one direction. The intensity of longitudinal dispersion is characterized by the number of cells (cf.[134]), whereas in the advanced cell models a back flow exists (cf.[71]).

5.3 Hydrodynamic Properties

It is necessary to make some general statements on the behavior of two-phase flow in single-stage bubble columns:

- When a narrow bubble size distribution occurs, the two-phase flow is referred to as a bubble flow, a laminar flow or a homogeneous flow.
- If bubbles of very different sizes are present, they have different ascending velocities and the two-phase flow becomes unstable. This state is called turbulent flow or heterogeneous flow.

"Homogeneous" flow prevails if small primary bubbles are formed in coalescence-suppressing media at relatively low gas flow rates. Since the small primary bubbles are mostly preserved, the bubble size distribution is narrow and the bubble ascending velocity is uniform. The two-phase flow in bubble columns with $d_p \simeq d_e$ in coalescence-promoting and/or-hindering media is "homogeneous" as long as the gas velocity is low. In these systems a narrow bubble size distribution occurs due to $d_p \simeq d_e$, i.e. the coalescence cannot increase the bubble size.

In this "homogeneous" flow the properties of the two-phase flow are uniform in every cross section of the column[7] and the longitudinal liquid dispersion intensity is low[72].

By increasing the gas flow rate above a critical value, the two-phase flow becomes "heterogeneous". Two types of heterogeneous flow can be distinguished.

- At low gas flow rates "heterogeneous" flow results, if $d_p \ll d_e$ and the coalescence of the bubbles is unhindered, i.e. large bubbles with diameter d_e are formed by coalescence. These bubbles have the smallest drag in the center of the column and consequently move into the region where they increase the flow velocity. Since a fraction of the liquid is transported to the top of the bubbling layer very quickly the rest must move backwards along the column wall. The resulting non-uniform velocity profile in the liquid accelerates the coalescence, since non-spherical large bubbles suffer a higher local drag the further they are from the center. This non-uniform drag tilts and drives the bubbles into the center more quickly than would occur with a uniform velocity profile. The enrichment of large bubbles in the column center further increases the non-uniformity of the liquid flow velocity. Large bubbles are very effective in collecting small bubbles due to the diminished flow resistance in the wake. The large concentration of bubbles in the column center increases the coalescence rate because more bubbles come into a position favorable for coalescence.

This type of "heterogeneous" flow mainly occurs in highly viscous media. In low viscosity media another type of "heterogeneous" flow exists.

- At high gas flow rates the bubble velocity is lower than the effective gas velocity, which is maintained by large high-turbulent eddies consisting of different sized bubbles (bubble collectives). These bubble collectives grow due to their high rise velocity and spinning movement which cause small bubbles to be attracted and entrapped[139) 140)]. The large difference between the velocities of the bubble collectives and the rest of the two phase flow causes large-scale turbulence in the center of the column. In these system, the two-phase system largely exhibits radial non-uniformity and the intensity of the liquid longitudinal dispersion is high.

Such "heterogeneous" flow can also occur at low gas flow rates, if very small bubbles are formed in strong coalescence-suppressing media. These bubbles have a very low ascending velocity. This is much lower than the effective gas velocity necessary to maintain the gas flow rate. To increase the bubble ascending velocity, many bubbles form agglomerates which have a higher ascending velocity than that of single bubbles. Also, these bubble collectives grow during their rise through the column due to their spinning movement. Again, the large differences between the ascending velocities of these turbulent eddies and the surrounding environment causes the "heterogeneous" flow.

For the determination of the longitudinal liquid dispersion coefficient, *E*, in viscous systems, only few investigations have been carried out. Most of them employed slightly viscous systems[141)], e.g. cane sugar solutions up to 50 *wt*-% ($\leq$19.2 mPa · s)[145)], sugar solutions up to 50 *wt*-% ($\leq$12.7 mPa · s)[144)], glycerol solutions up to 61 *wt*-% ($\leq$11 mPa · s)[143)] and up to 67 *wt*-% ($\leq$14 mPa · s)[142)]. The longitudinal liquid dispersion coefficients, *E*, were determined by fitting the calculated longitudinal concentration profiles of the dissolved oxygen to the measured ones using, according to Niebeschütz[73)] in highly viscous Newtonian systems (glycerol solutions).

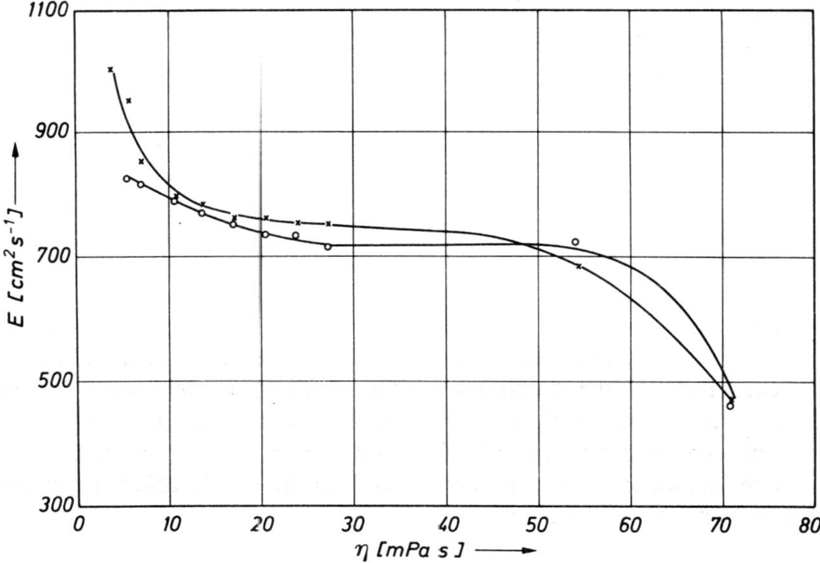

Fig. 42 Longitudinal liquid dispersion coefficient E as a function of viscosity η employing glycerol solutions and two aerators; $w_{SL} = 1.21$ cm/s, $D_c = 14$ cm, $H_c = 280$ cm, $w_{SG} = 4.28$ cm s^{-1} [73)]. ○ porous plate; × injector nozzle

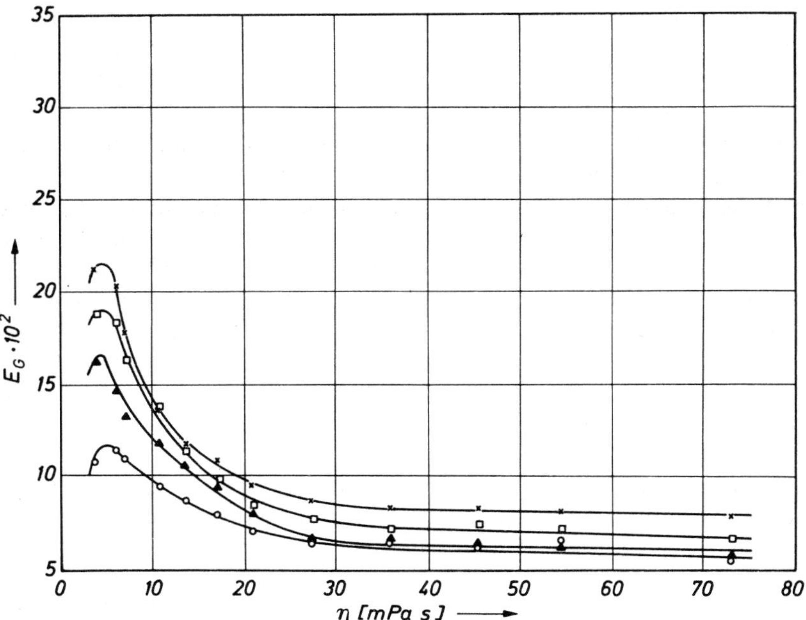

Fig. 43 Mean relative gas hold-up E_G as a function of dynamic viscosity using a porous plate gas distributor and glycerol solutions; D_c = 14 cm, H = 280 cm, w_{SL} = 1.21 cm s^{-1} [133]. ○ w_{SG} = 1.07 cm $^{-1}$; ▲ w_{SG} = 2.14 cm s^{-1}; □ w_{SG} = 3.21 cm s^{-1}; × w_{SG} = 4.28 cm s^{-1}

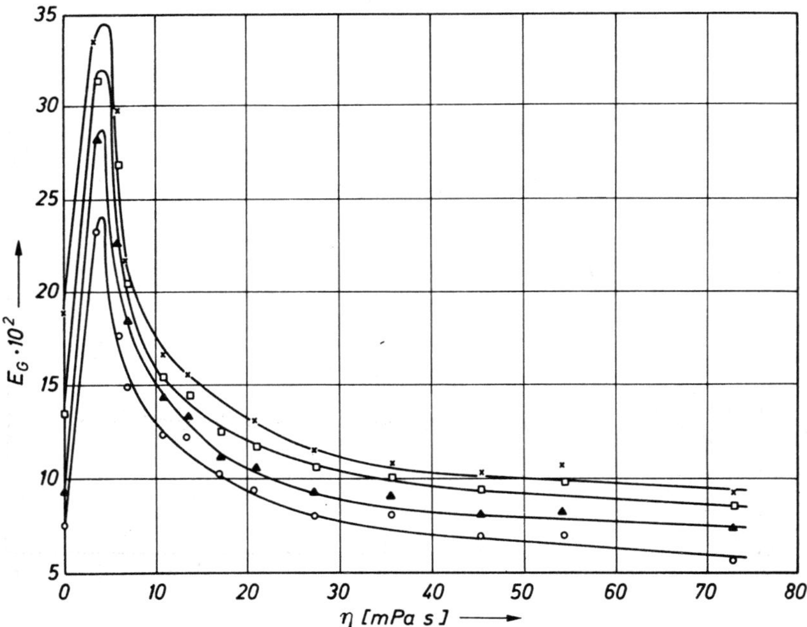

Fig. 44 Mean relative gas hold-up E_G as a function of dynamic viscosity η employing an injector nozzle and glycerol solutions; D_c = 14 cm, H = 280 cm, w_{SL} = 1.21 cm s^{-1} [133] (for symbols see Fig. 43)

In Fig. 42, E is plotted as a function of viscosity, employing two different aerators. One can recognize that in the viscosity range $20 \leq \eta \leq 55$ mPa · s, E is nearly constant. Neither the viscosity nor the superficial gas velocity[73)] influences E in this region significantly. At low η values, E is higher and at high η values it is lower than the value: $E \simeq 725\ \mathrm{cm^2 s^{-1}}$. This behavior of E holds true of all investigated superficial gas velocities.

However, these E values have fairly low accuracy, because the quality of fitting depends only slightly on the E values. E values were evaluated in a similar way employing CMC solutions by Buchholz, but these have not been published[150)].

Relative gas hold-up data measured in single-stage bubble columns using highly viscous media are also scarce.

Glycerol solutions were employed by Eissa, Schügerl[146)], Bach, Pilhofer[147, 148)], Niebeschütz[73)] and Buchholz et al.[133)]. CMC solutions were used by Buchholz et al.[84, 133, 149)].

In glycerol solutions E_G passes through a maximum with increasing viscosity[73, 146–148)]. This behavior of E_G is observed with porous plate gas distributors (Fig. 43) and with injector nozzle gas distributors (Fig. 44). From these figures one can see the slight influence of the superficial gas velocity on E_G in the range $\eta \geqq 20$ mPa · s (or cP). In the range $\eta > 40$ mPa · s, the viscosity influence on E_G is also slight.

The corresponding relative bubble swarm velocity, w_S, with regard to the liquid, at first increases with rising viscosity. In the viscosity range $30 < \eta < 60$ mPa · s does not depend on η if a porous plate gas distributor is employed (Fig. 45) and only slightly depends on η using an injector nozzle gas distributor (Fig. 46). With increasing superficial gas velocity, w_S strongly increases. No model or empirial relationships have been published yet which can describe this behavior of E_G and/or w_S.

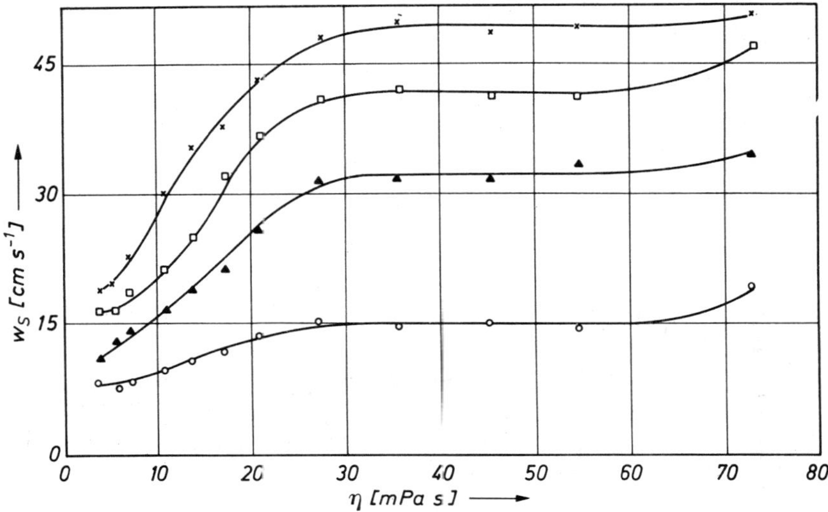

Fig. 45 Relative bubble-swarm velocity w_S with regard to the liquid as a function of dynamic viscosity η using a porous plate gas distributor and glycerol solutions; $D_c = 14$ cm, $H_c = 280$ cm, $w_{SL} = 1.24\ \mathrm{cm\ s^{-1}}$ [133)] (for symbols see Fig. 43)

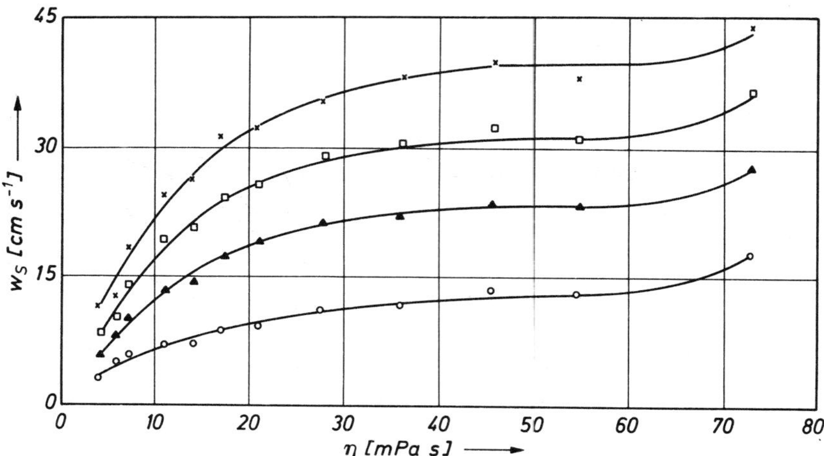

Fig. 46 Relative bubble-swarm velocity with regard to the liquid as a function of dynamic viscosity η applying an injector nozzle gas distributor and glycerol solutions $D_c = 14$ cm, $H_c = 280$ cm, $w_{SL} = 1.21$ cm s^{-1} [133] (for symbols see Fig. 43)

Relative gas hold-up values were determined applying CMC solutions by Buchholz[84, 133, 149, 150].

In Figure 47, E_G is plotted as a function of w_{SG} emplyoing CMC solutions of different concentrations. The rheological behavior of these pseudoplastic media can be described according to the Oswald de Waele relationship (Eq. 30)) by the fluid consistency index K and flow behavior index n. In Fig. 48, these constants are plotted as a function of the CMC concentration for the solutions used in Fig. 47.

One recognizes from Fig. 47 that at low CMC concentration E_G passes through a maximum, then through a minimum and finally increases again with increasing w_{SG}. At higher CMC concentrations, E_G increases, then passes a plateau and finally increases again as w_{SG} rises. However, this course of $E_G(w_{SG})$ is somewhat misleading. The increase of E_G at high w_{SG} values is caused by very large bubble formation. If the volume of these bubbles is substracted from E_G, a corrected value $E_{G\,\mathrm{corr}}$,

$$E_{G,\,\mathrm{corr}} = E_G - E_{G,\,\mathrm{slug}}\,, \tag{74}$$

can be evaluated. With rising w_{SG}, $E_{G,\,\mathrm{corr}}$ runs through a maximum and diminishes. No increase occurs at high w_{SG}[151]. E_G increases with increasing flow behavior index, n, and dropping fluid consistency index, K (Fig. 49).

Several photographs were taken of single bubbles[8–9, 74–78, 152–155]; however, only few of bubble swarms. Otake et al.[10] investigated the bubble size distributions in glycerol solutions and Buchholz et al.[84, 133] in CMC solutions, both of them employing a column 14 cm in diameter; however, Otake et al. used a 130 cm high and Buchholz et al. 391 cm high column. Furthermore, Buchholz et al. took photographs at three different distances from the aerator, namely at 49.5 cm, 124.5 cm and 180 cm, while Otake et al. took motion pictures along the column.

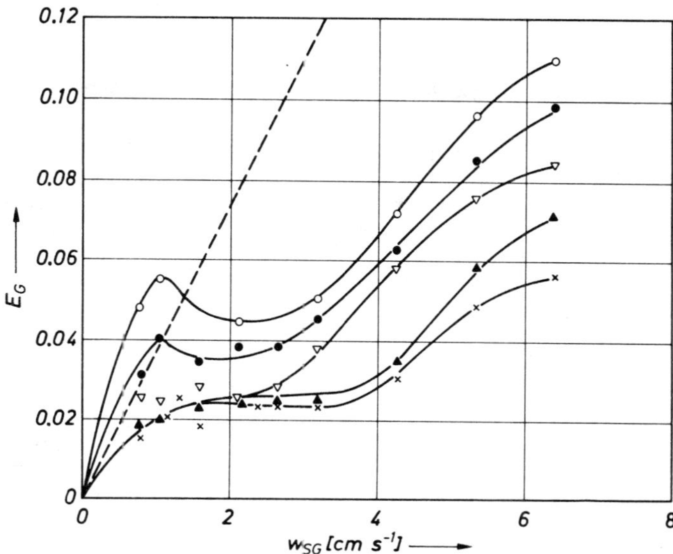

Fig. 47 Mean relative gas hold-up E_G as a function of superficial gas velocity w_{SG} in solutions of different CMC concentrations; D_c = 14 cm, H = 391 cm, w_{SL} = 1.3 to 1.5 cm s^{-1} [84]; ○ 1.0 wt-%; ● 1.2 wt-%; ▽ 1.4 wt-%; ▲ 1.56 wt-%; × 1.7 wt-% CMC solutions, ——— H_2O dist

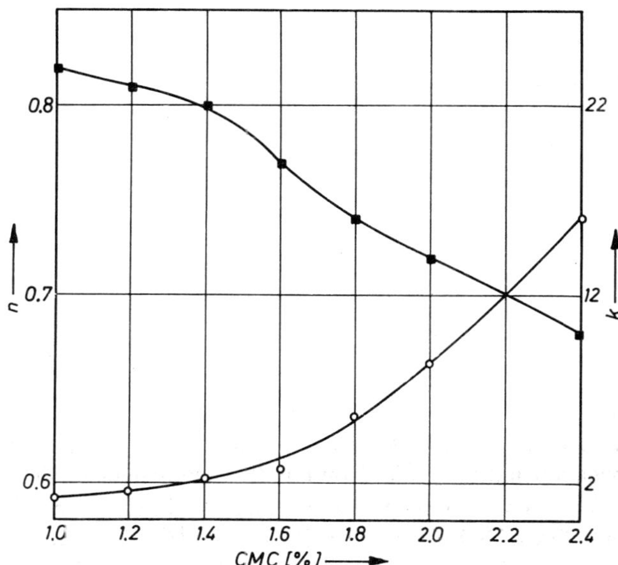

Fig. 48 Flow behavior index n and fluid consistency index K as a function of the CMC concentration for media applied by H. Buchholz[84].

○ K, ■ n

$$\tau = k\left(\frac{dc}{dx}\right)^n$$

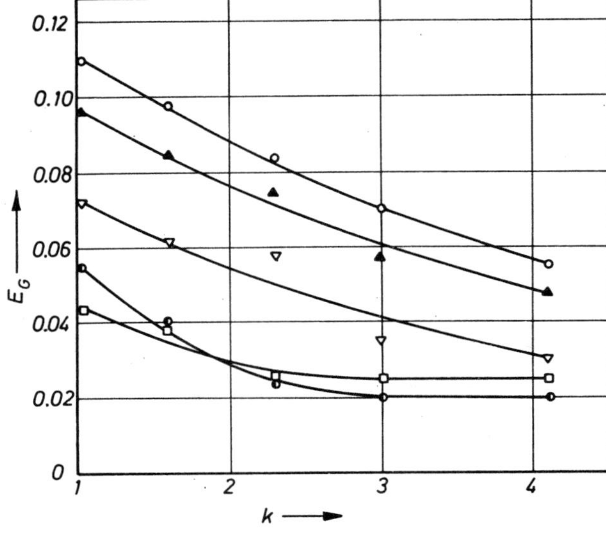

Fig. 49 Mean relative gas hold-up E_G as a function of fluid consistency index K at different superficial gas velocities. CMC solutions. $D_c = 14$ cm, $H = 291$ cm, $w_{SL} = 1{,}3$ to 1.5 cm s^{-1} [150].
○ $w_{SG} = 6.4$ cm s^{-1};
▲ $w_{SG} = 5.3$ cm s^{-1};
▽ $w_{SG} = 4.2$ cm s^{-1};
□ $w_{SG} = 2.1$ cm s^{-1};
⊙ $w_{SG} = 1.1$ cm s^{-1}

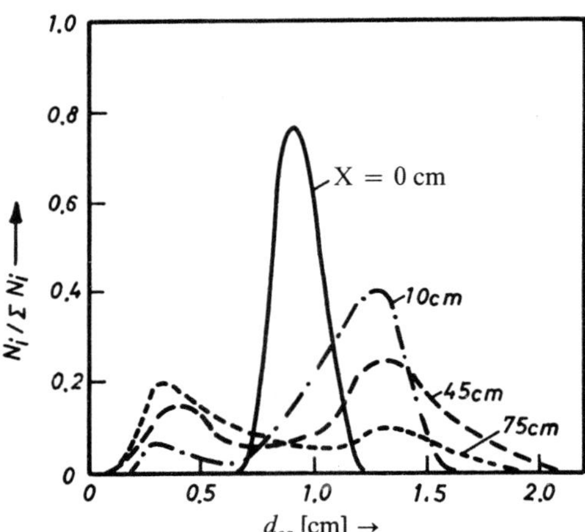

Fig. 50 Bubble size distribution at different heights X from the aerator. 62 wt-% glycerol solution. $D_c = 14$ cm; $H = 130$ cm; $w_{SG} = 0.08$ cm s^{-1} [10]

Figure 50 clearly indicates that the primary bubbles (at $X = 0$ cm, 0.8 cm in diameter) are destroyed with increasing height whereby small ($d_{eq} \simeq 0.6$ cm) and large ($d_{eq} \simeq 1.3$ cm) bubbles are formed. One can also observe from Figure 51 how the small bubble fraction increases with rising distance, X.

Keeping the location of the measurements constant but increasing the gas flow rate, a similar shift of bubble size distribution can be observed (Fig. 11). With increasing w_{SG} the "small bubble" fraction ($d_{eq} \simeq 1.0$ cm) and "large bubble" fraction ($d_{eq} \simeq 2$ to 3 mm) increases, while the "medium bubble" fraction ($d_{eq} \simeq 1$ to 2 mm) diminishes. With rising w_{SG} the distance, X, at which very large bubbles are formed, is shifted from 40 to 20 cm.

Therefore, it is not possible to characterize these system by a uniform bubble size distribution. Two or three bubble collectives are present and their fractions vary

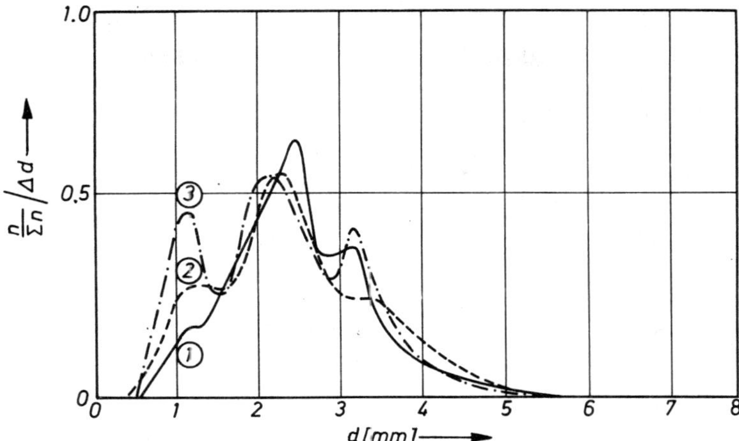

Fig. 51 Bubble size distribution at different heights X from the aerator; 1% CMC solution, $D_c = 14$ cm, $H_c = 391$ cm, $w_{SL} = 1$ cm s^{-1}, $w_{SG} = 2.67$ cm s^{-1} [84]. ① $X = 49.5$ cm; ② $X = 124.5$ cm; ③ $X = 180$ cm

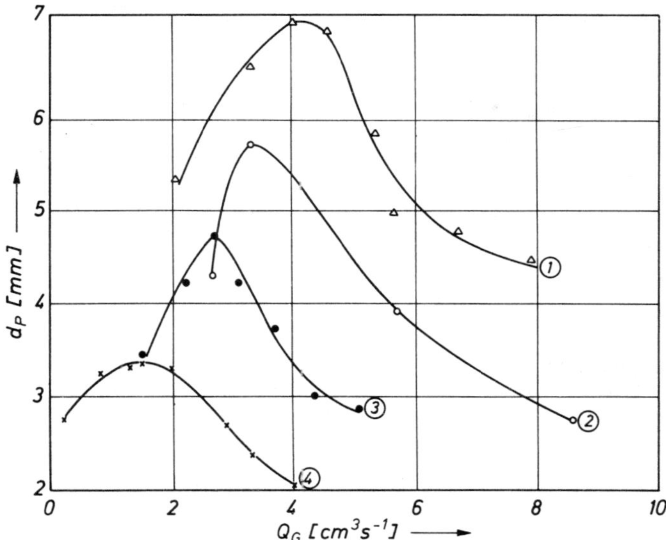

Fig. 52 Equivalent primary bubble diameter d_p as a function of gas flow rate Q_G through the orifice. Orifice diameter $D_ö$ as a parameter. 68% glycerol solution[75,154]. ① $D_ö = 472$ µm; ② $D_ö = 382$ µm; ③ $D_ö = 300$ µm; ④ $D_ö = 186$ µm; ⑤ $D_ö = 100$ µm; ⑥ $D_ö = 54$ µm

with the position in the column. The coalescence/redispersion process was investigated by Otake et al.[10] and has already been discussed in Sect. 3.1.

Hallensleben et al.[75], Bhavaraju et al.[156] investigated the bubble behavior in the immediate vicinity of the gas distributor. Hallensleben et al. varied the orifice diameter between 0.054 mm and 0.472 mm and the gas flow rate, Q_G, through the orifice up to 10 cm^3 s^{-1}. In all of the investigated systems, the primary equivalent bubble diameter, d_p, at first increases with Q_G, runs through, a maximum value of d_p increases from 2.4 mm to 7 mm with rising orifice diameter from 186 μm employing 68% glycerol solution (Fig. 52)[75,157]. The maximum value of d_p corresponds to that which has been calculated by the model of Meister and Scheele[158]. The reduction of d_p from this maximum with growing gas flow rate was explained by a turbulence mechanism[75].

According to Bhavaraju et al.[156] the bubble diameter in aerator region I, which is practically identical to d_p, increases when using nozzle diameters of 800 to 2000 μm with increasing gas flow rate in the range up to $Q_G = 80\ \text{cm}^3\ \text{s}^{-1}$. A decrease of d_e occurs only in region II with rising Q_G due to turbulence. They could not find a bubble break-up region II in highly viscous media even for

$$Re_0 = \frac{4Q_G}{D_{\ddot{o}}\nu_G} \gg 2000\,.$$

Hallensleben et al.[75] took high speed motion pictures (6000 s^{-1}) which clearly indicate that the bubble break-up can be a primary process if Q_G is high enough. These results therefore refect a clear contradiction.

It is possible that the gas distributor type considerably influences this bubble break-up process. Hallensleben et al. employed orifices for bubble formation while Bhavaraju et al. used nozzles.

5.4 OTR and k_La

Only a few investigations have been reported on k_La measured in single-stage bubble columns employing highly viscous media: Akita and Yoshida[159,160] Niebeschütz[73], Buchholz et al.[133] employed glycerol solutions, Buchholz[150], Buchholz et al.[84,133,149] CMC solutions and Baykara and Ulbrecht[161] PAA and PEO solutions.

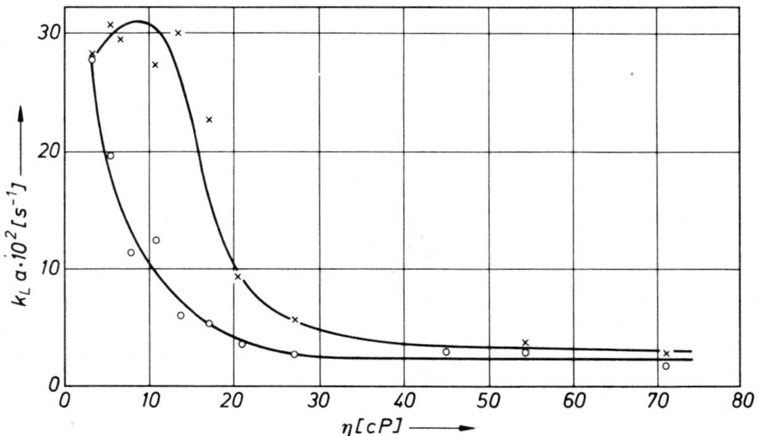

Fig. 53 Volumetric mass transfer coefficient k_La as a function of dynamic viscosity η using glycerol solution. $D_c = 14$ cm, $H = 380$ cm, $w_{SL} = 1.21\ \text{cm s}^{-1}$, $w_{SG} = 2.14\ \text{cm s}^{-1}$ [133]. ○ porous plate; × injector nozzle

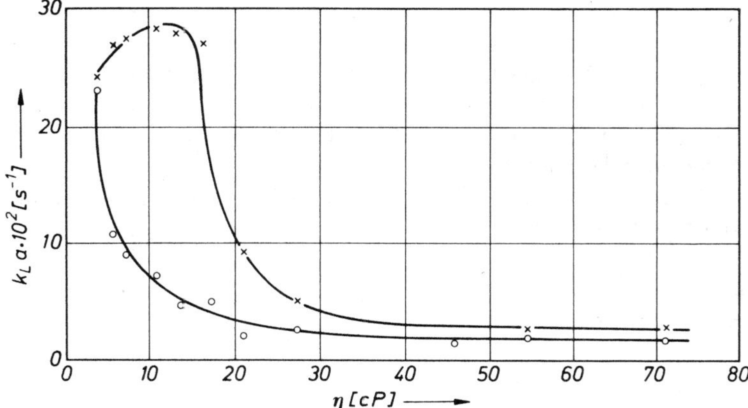

Fig. 54 Volumetric mass transfer coefficient $k_L a$ as a function of dynamic viscosity employing glycerol solution. $D_c = 14$ cm, $H = 380$ cm, $w_{SL} = 1.21$ cm s^{-1}, $w_{SG} = 4.28$ cm s^{-1} [133]. ○ porous plate; × injector nozzle

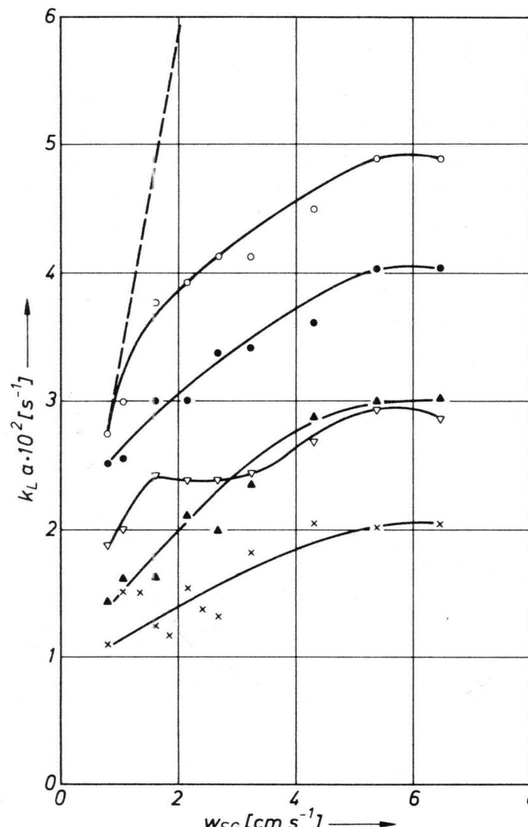

Fig. 55 Volumetric mass transfer coefficient $k_L a$ as a function of superficial gas velocity w_{SG} at different CMC concentrations. $D_c = 14$ cm, $H = 391$ cm, $w_{SL} = 1.3$ to 1.5 cm s^{-1} [84]. ○ 1.0% CMC; ● 1.2% CMC; ▽ 1.4% CMC; ▲ 1.56% CMC; × 1.7% CMC solution; ——— H_2O dist.

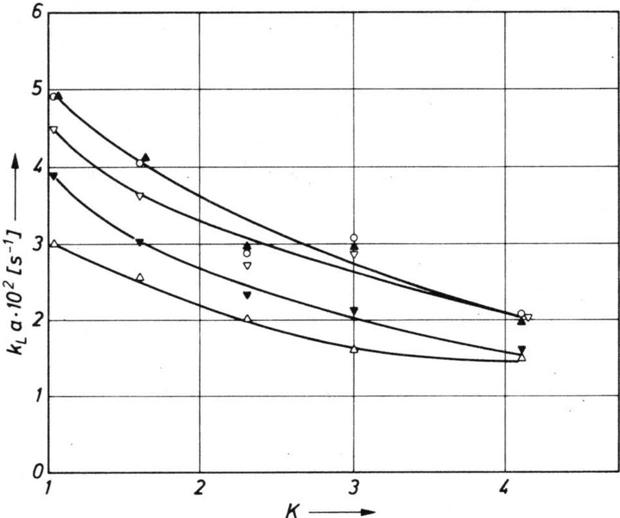

Fig. 56 Volumetric mass transfer coefficient k_La as a function of fluid consistency index K at different superficial gas velocities w_{SG}. D_c = 14 cm, H = 391 cm, w_{SL} = 1.3—1.5 cm^{-1} [84]. ○ w_{SG} = 6.4 cm^{-1}; ▲ w_{SG} = 5.3 cm s^{-1}; ▽ w_{SG} = 4.3 cm $^{-1}$; ▼ w_{SG} = 2.1 cm s^{-1}; △ w_{SG} = 1.1 cm s^{-1}

When using glycerol solutions, k_La depends only slightly on the superficial gas velocity, w_{SG}. With increasing viscosity, k_La rapidly diminishes and for $\eta > 30$ mPa · s it approaches a constant value if a porous plate is applied (Fig. 53).

Using an injector nozzle with increasing superficial gas velocity, k_La increases, passes through a maximum at about 10 mPa · s, then rapidly diminishes and approaches a constant value for $\eta > 40$ mPa · s (Fig. 53). By w_{SG} the k_La (η) curves change only slightly (Fig. 54). No model or experimental relationship which can describe this behavior has been reported.

Baykara and Ulbrecht applied only slightly viscous PAA and PEO solutions to the evaluation of the volumetric mass transfer voefficients[161]. Again, with increasing viscosity, k_La diminishes.

Table 5. Energy requirement for the formation of the measured volumetric mass transfer coefficient D_c = 14 cm, H = 391 cm and porous plate gas distributor[84]

CMC (%)	$k_La \times 10^2$ (l s^{-1})	P/V_L (kW m^{-3})
1.0	4.9	0.525
1.2	4.1	0.545
1.4	3.0	0.525
1.56	3.0	0.525
1.7	2.0	0.550

The effect of w_{SG} is fairly pronounced. With increasing w_{SG} the coefficient $k_L a$ increases. This both applies to PEO and PAA solutions in the range $\eta_{app} < 15$ mPa · s[161].

In contrast to glycerol solutions, $k_L a$ considerably increases with growing w_{SG} and, similar to glycerol, diminishes with rising concentration of CMC (Fig. 55). Similar to E_G, the volumetric mass transfer coefficient increases with growing flow behavior index, n, and with diminishing fluid consistency index K (Figs. 56).

In Table 5 the specific power inputs P/V_L, due to the air compression, are compiled. These are needed to produce the $k_L a$ values in a bubble column employing CMC solutions.

The $k_L a$ values determined by Buchholz[84] employing CMC solutions were correlated by Henzler (Fig. 57)[114, 166]

$$\frac{k_L a}{w_{SG}}\left(\frac{v_r^2}{g}\right)^{1/3} = 0.06\left[\frac{w_{SG}}{(g v_r)^{1/3}}\right]^{-0.9} Sc^{*-0.5}, \tag{75}$$

for the validity range ($D_c = 14$ cm, $H = 391$ cm, porous plate):

$$8 \times 10^{-2} \leqq \frac{k_L a}{w_{SG}}\left(\frac{v_r^2}{g}\right)^{1/3} Sc^{0.5} \leqq 0.8,$$

$$2 \times 10^{-3} < \left[\frac{w_{SG}}{(v_r g)^{1/3}}\right]^2 < 8 \times 10^{-1}.$$

In Eq. (75), v_r, the representative kinematic viscosity, was calculated analogously to Metzner and Otto[98] using relationship (76) to define a representative shear rate:

$$\left(\frac{dv}{dx}\right) = 15 w_{SG}, \tag{76}$$

hence

$$\eta_r = \frac{\tau}{\left(\frac{dv}{dx}\right)} = \frac{\tau}{15 w_{SG}}. \tag{77}$$

6 Multistage Tower Reactors

If a single-stage column is separated into several compartments by the application of trays, the bubbles are newly formed at every tray. This may be necessary if the bubble coalescence rate is high[163].

Few papers have been published on multistage tower reactors employing highly viscous liquids. These include: the aeration of glycerol[11, 50], CMC-[50, 103] and PAA-[51] solutions investigated in a single-stage short column with stagnant liquids, CMC solutions in a five-stage column in concurrent operation[84, 150] as well as the

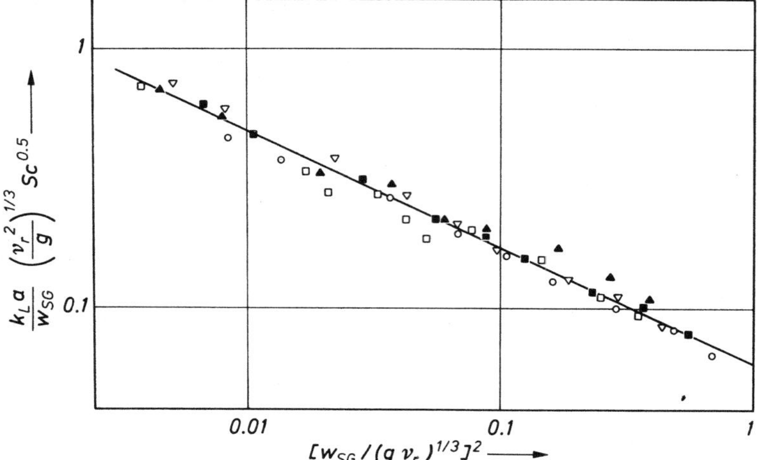

Fig. 57 Correlation according to Henzler for the volumetric mass transfer coefficients in single-stage bubble column (D_c = 14 cm, H = 391 cm) employing CMC solutions and porous plates[114, 166)]

symbol	η_r [m Pa · s]
○	46— 65
■	64— 95
▽	92—140
▲	109—172
□	130—217

aeration of glycerol-[36, 63, 162)] and PAA[37, 162, 164)] solutions in the upper (first stage of a countercurrent multistage column.

6.1 Apparatus and Instrumentation

In countercurrent columns the liquid phase flows from top to bottom and stage to stage by means of downcomers. The height of the layers can be varied by the length of the overflow and air is introduced at the bottow of the column. Below the perforated plates, air layers prevail. If the compartment separating trays are suitable, no liquid leak through the perforated plates occurs. In the investigations considered three perforated plates were employed as trays (Table 6).

For the determination of the mean relative gas hold-up, the bubble-free liquid hight, H_L, has to be measured in each of the stages. (The height of the bubbling layer, H_s, is fixed by the overflow height.) E_G is calculated again by the application of the relationship

$$E_G = \frac{H_s - H_L}{H_s} . \tag{21}$$

OTR and $k_L a$ can be determined in a twin bubble column. In a countercurrent multistage column, O_2 can be absorbed and in a second concurrent multistage

Table 6. Characterization of perforated trays used in multistage columns[36, 37]

A_H (fraction of free cross sectional area)	d_H (hole diameter) mm	N_H (number of holes)
0.35	0.5	558
0.62	1.0	248
4.82	3.0	214

column, it can be desorbed by purging the liquid with N_2. The liquid, after it has passed the first multistage column, saturated by O_2, is pumped from the bottom of the first column to the bottom of the second concurrent multistage column, in which it is saturated with N_2. At the top of this column the liquid is oxygen-free and is fed back into the top of the absorption column. The heat generated by pumping must be removed by a cooler to maintain a constant temperature. Through this liquid recirculation a steady-state sorption is maintained. In all of stages the dissolved oxygen concentration has to be measured if the OTR and k_La are to be determined in them. However, in the first (upper) stage of the first column the accuracy of the k_La determination is the highest. Therefore, it is suitable to determine k_La only in this stage. In this case, the dissolved oxygen concentration has to be measured in this stage as well as in the liquid which is fed into this stage. The dissolved oxygen concentration is usually measured by oxygen electrodes.

If the bubble size distribution is to be measured special stages are needed. To avoid a distortion of the bubbles on the photographs the stages must be provided with plane parallel windows.

6.2 Mathematical Models

Since the measurements of the mixing times in the stages indicate that these times are usually much shorter than the corresponding mean residence times of the liquids, one can assume that perfect mixing prevails in each of the compartments. Thus, each compartment can be treated as a perfectly mixed stirred tank reactor.

The Stanton number, St, can be calculated by Eq. (78) by means of the oxygen balance in the liquid, e.g. for the first stage:

$$St = \frac{C\bar{P}_B - C_E P_0}{(1 - C)\,\bar{P}_B}, \qquad (78)$$

where $St = k_La\,\tau_M$ = Stanton number
τ_M mean residence of the liquid in the first stage
C^* relative concentration of the dissolved oxygen with regard to the saturation
C_E^* C^* in the feed
P_0 pressure of the air at the entrance of the first stage
$\bar{P}_B$ mean pressure of the air in the first stage.

6.3 Hydrodynamic Properties

Only few investigations on the mean relative gas hold-up, E_G, in short bubble columns employing highly viscous media have been published. Franz[11, 50] determined E_G as well as E_{GK} and E_{GG} due to "small" and/or "intermediate-to-large" bubbles in glycerol solutions. A fairly long time is required to attain steady-state E_G and E_{GK} values (Fig. 4). In 50% glycerol solution the E_{GK} fraction is slight. (Fig. 5). With rising glycerol concentration, E_{GK} considerably increases (Fig. 6 to 8). E_{GK} depends only slightly on w_{SG} in contrast to E_{GG}, which increases with growing superficial gas velocity. In the viscosity range $\eta > 100$ mPa · s, a linear relationship exists between E_{GG} and w_{SG}:

$$E_{GG} = b w_{SG} \, . \tag{79}$$

In a 90% glycérol solution, employing a perforated plate with

$$\begin{aligned} d_H &= 0.5\,; & b &= 0.0166 \text{ s cm}^{-1} \\ d_H &= 1.0\,; & b &= 0.0153 \text{ s cm}^{-1} \\ d_H &= 3.0\,; & b &= 0.0142 \text{ s cm}^{-1} \end{aligned}$$

or in a 95% glycerol solution with

$$d_H = 0{,}5\,, \quad 1.0 \quad \text{and} \quad 3.0\,; \qquad b = 0.0166 \text{ s cm}^{-1}\,,$$

these relationships are valid for $w_{SG} \leqq 8 \text{ m s}^{-1}$.

Voigt[63, 165] as well as Franz[50, 103] determined E_G and E_{GG} values in CMC solutions. As can be seen from Fig. 58, the applied compartment separating tray

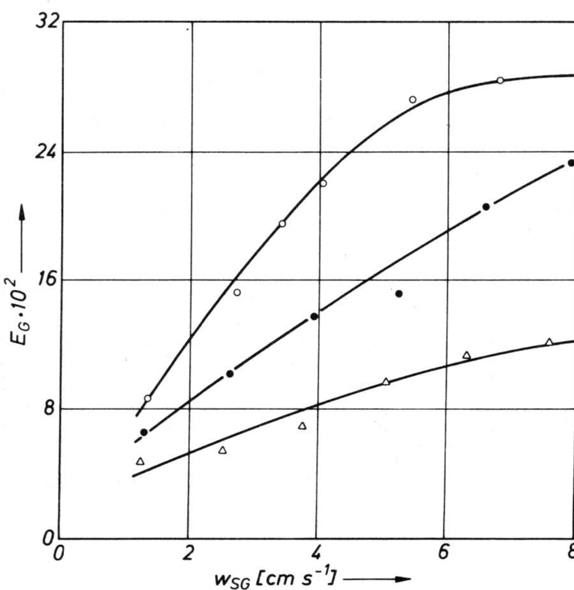

Fig. 58 Mean relative gas hold-up E_G as a function of w_{SG} in the first stage of a multistage counter-current column (D_c = 20 cm, H_s = 30 cm). 1% CMC solution. Perforated plate-compartment separating trays[63]. △ d_H = 3 mm; ● d_H = 1 mm; ○ d_H = 0.5 mm

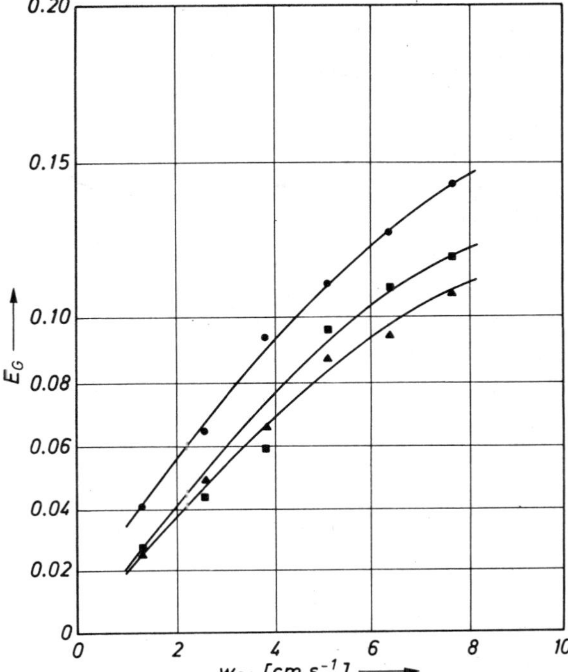

Fig. 59 E_G as a function of w_{SG}. D_c = 14 cm, $H_s \simeq$ 35 cm, 1.4% CMC solution. Perforated plates[103]. ● d_H = 0.5 mm; ■ d_H = 1.0 mm; ▲ d_H = 3.0 mm

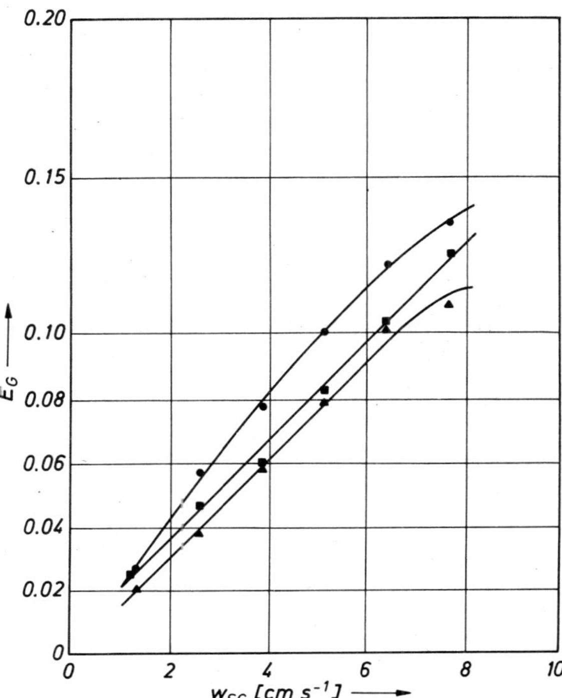

Fig. 60 E_G as a function of w_{SG}. D_c = 14 cm, $H_s \simeq$ 35 cm, 1.6% CMC solution. Perforated plates (for symbols see Fig. 59)[103]

considerably influences E_G. With increasing CMC concentration this influence diminishes (Figs. 59 to 60).

Peschke[51] determined, E_G, E_{GG} and E_{GK} values using PAA solutions.

As in the case of glycerol and CMC solutions, E_{GK} and E_G vary with time after the aeration was started. Again E_{GG} does not depend on time (Figs. 61 and 62). With increasing aeration rate E_{GK}, E_{GG} and E_G increase and a longer time is required to attain the steady-state E_{GK} and E_G values (Fig. 61).

With decreasing perforated plate hole diameter from 3.0 mm to 0.5 mm, E_{GK} and E_G significantly increase and E_{GG} varies only slightly (Fig. 62). The higher E_{GK}, the longer a time is necessary to attain its steady-state value again. If the aeration is stopped, the "intermediate-to-large" bubbles rapidly leave the two-phase system. However, a fairly long time is needed to remove the "small" bubbles from the liquid. In a 0.4% PAA solution employing a perforated plate with $d_H = 1.0$ mm, E_{GK} is for example reduced below 0.01% after 11 min (with $w_{SG} = 1.8$ cm s^{-1}), 17 min ($w_{SG} = 2.4$ cm s^{-1}), 21 min ($w_{SG} = 3.6$ cm s^{-1}) and 24 min ($w_{SG} = 4.5$ and 5.4 cm s^{-1}). With decreasing hole diameter this "small" bubble residence time significantly increases.

Again, only few investigations on bubble size distributions in highly viscous media have been published employing glycerol[11,50], CMC[50,103] and PAA[51] solutions.

The "small" bubble size distribution was evaluated by turning off the gas flow rate and taking photographs 15 s thereafter. These photographs were evaluated by a semiautomatic particle analyzer (TGZ 3 Leitz). The small bubbles are spherical and have usually a diameter of less than 1 mm. In Figure 63 such small bubbles with a mean Sauter diameter, $d_{SK} = 0.4$ mm, are shown. Figure 64 describes "inter-

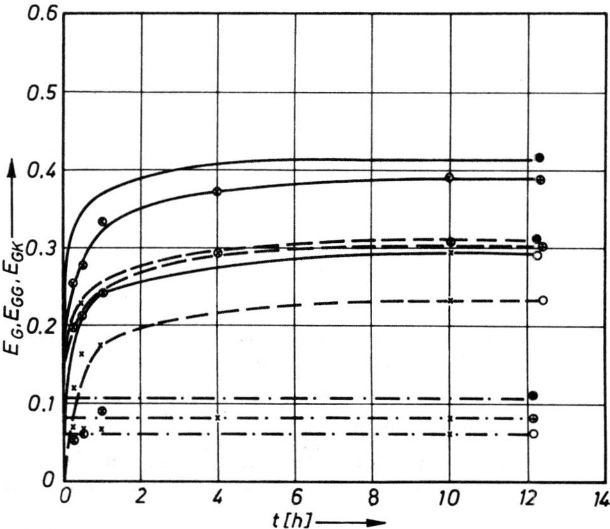

Fig. 61 E_G, E_{GG} and E_{GK} as a function of time t after starting aeration. 0.6% PAA solution, $d_H = 1.0$ mm, $D_c = 14$ cm, $H_s = 35$ cm[51]. ——— E_G; – – – – E_{GK}; —·—·—·— E_{GG}; ○ $w_{SG} = 1.8$ cm s^{-1}; ⊕ $w_{SG} = 3.6$ cm s^{-1}; ● $w_{SG} = 5.4$ cm s^{-1}

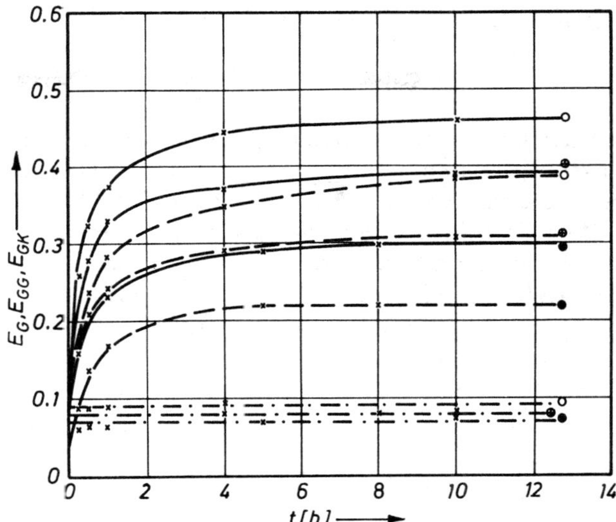

Fig. 62 E_G, E_{GG} and E_{GK} as a function of time t after starting aeration, 0.6% PAA solution, $w_{SG} = 3.6$ cm s^{-1}, $D_c = 14$ cm, $H_s \simeq 35$ cm[51]. ———— E_G; – – – – – E_{GK}; —·—·—·— E_{GG}; ○ $d_H = 0.5$ mm; ⊕ $d_H = 1.0$ mm; ● $d_H = 3.0$ mm

mediate" bubbles of ellipsoidal shape which were photographed in a 70% glycerol solution at low w_{SG}. Under these conditions, the bubble size is relatively uniform. The mean Sauter diameter is small, $d_{SG} = 3.5$ mm. "Large bubbles" are depicted in Fig. 65. They exhibit an irregular shape; therefore, it is very difficult to determine their mean Sauter diameter.

The distribution of small bubbles considerably changes after starting aeration. It becomes narrower with increasing time until their steady state is attained (Fig. 66). In Fig. 67, the mean Sauter diameter is plotted as a function of w_{SG} employing different glycerol concentrations and perforated plates as gas distributors. With rising glycerol concentration, d_{SG} becomes larger. The aerator effect is fairly slight, especially at high concentrations.

In short bubble columns using CMC solutions the Sauter mean bubble diameter, d_{SG}, increases with increasing superficial gas velocity and CMC concentration, as well as hole diameter, d_H, of the perforated plates (Fig. 68).

Peschke[51] determined d_{SK} and d_{SG} in PAA solutions. The parameter d_{SK} depends only slightly on the superficial gas velocity w_{SG}. d_{SG} increases with rising w_{SG}; however, with diminishing d_H, this dependence decreases (Fig. 69).

The increase in PAA concentration has the same effect on this $d_{SG}(w_{SG})$ curve. At higher PAA concentrations, d_{SG} is independent on w_{SG}.

By means of E_{GK} and d_{SK}, the specific interfacial area, a_K, due to the "small" bubbles, can be calculated by use of Eq. (1). In Fig. 70, a_K is plotted as a function of the glycerol concentration applying the perforated plates with $d_H = 1.0$ mm. Using perforated plates with $d_H = 0.5$ and 3.0 mm, similar curves are obtained. According to Fig. 70, a_K considerably increases with rising glycerol concentration and attains extremely high values. This is in contrast to the k_La measurements (see below): with

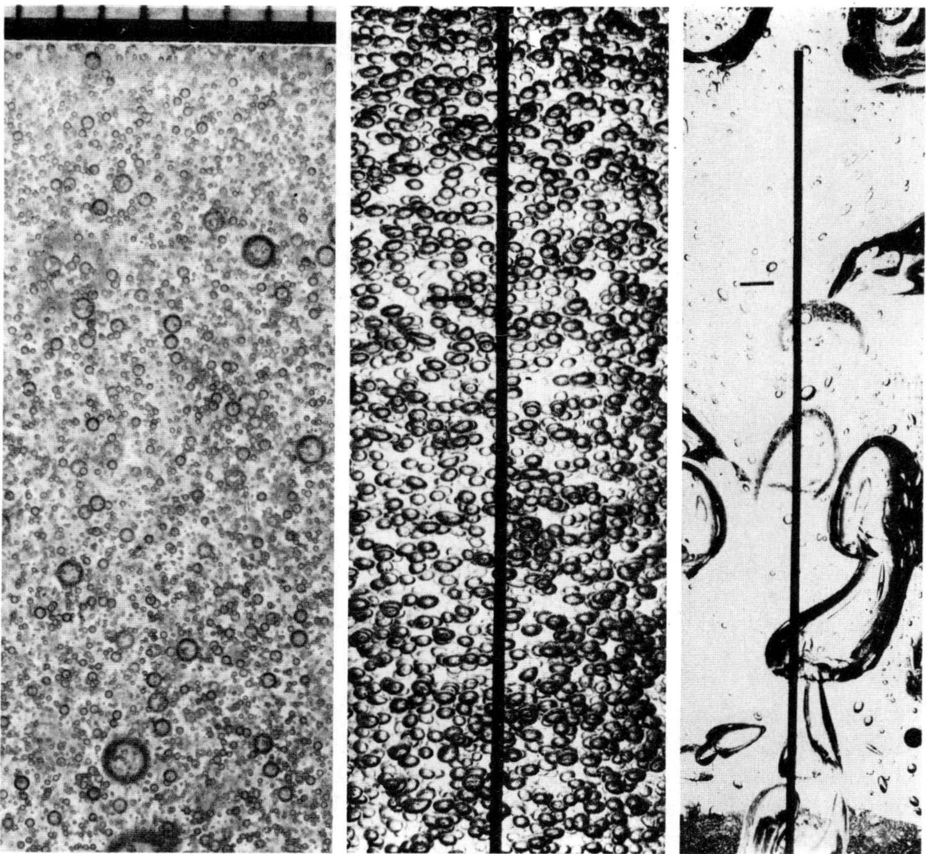

Fig. 63 "Small" bubbles in a 70% glycerol solution. D_c = 14 cm, H_s = 35 cm, d_H = 3.0 mm, w_{SG} = 6.8 cm s^{-1}. The scale on the top is given in mm. $d_{SK} \simeq$ 0.4 mm [50)]

Fig. 64 "Intermediate" bubbles in a 70% glycerol solution. D_c = 14 cm, H_s = 35 cm. d_H = 0.5 mm, w_{SG} = 1.3 cm s^{-1}, d_{SG} = 3.5 mm [50)]

Fig. 65 "Large" bubbles in a 95% glycerol solution. D_c = 14 cm, H_s = 35 cm, d_H = 1.0 mm, w_{SG} = 2.5 cm s^{-1}, d_{SG} = 31 mm [50)]

increasing glycerol concentration, k_La diminishes. Furthermore, the k_L values calculated by means of the relationship

$$k_L = \frac{k_La}{a_K} \tag{80}$$

are unrealistically small.

This indicates that a_K is a useless interfacial area for oxygen transfer. The residence time of the bubbles is obviously too high; their oxygen content is exhausted. Therefore, in the following only the specific surface area, $a_G = a$, due to the intermediate-to-large bubbles, is considered as an active oxygen transfer area. In Fig. 71, this specific interfacial area is plotted as a function of w_{SG}, employing glycerol solutions.

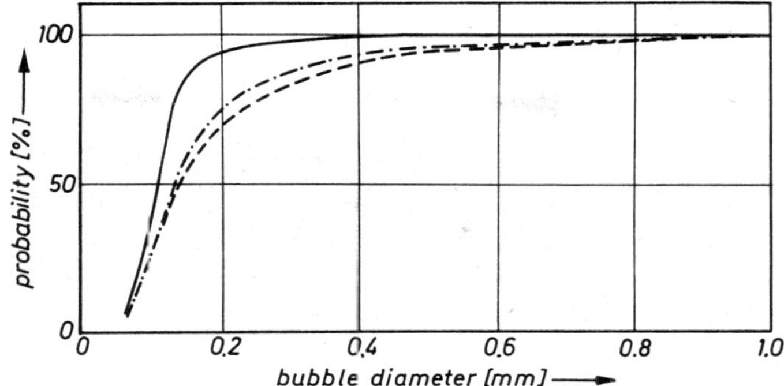

Fig. 66 Size distribution of "small" bubbles as a function of time after starting aeration. $D_c = 14$ cm, $H_s = 34$ cm, $d_H = 1.0$ mm, 70% glycerol solution. $w_{SG} = 1.3$ cm s^{-1} [11]. – – – – – – 20 min; —·—·—·— 60 min; ———— 4 h after starting aeration

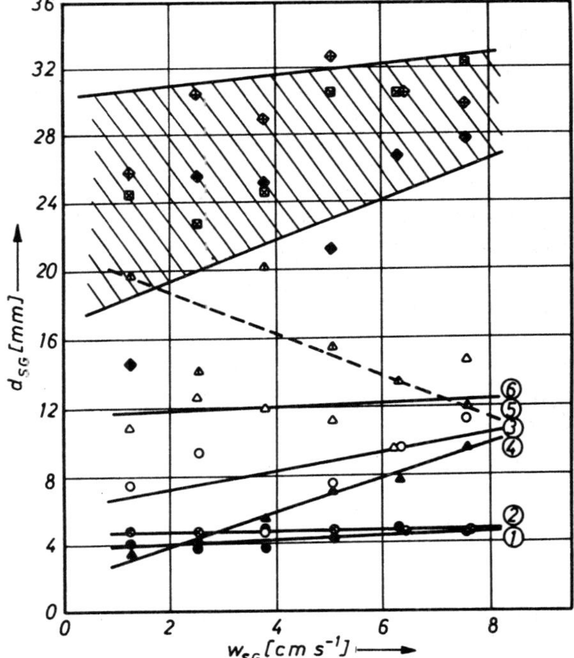

Fig. 67 Sauter diameter of "intermediate-to-large" bubbles d_{SG}, as a function of w_{SG}. $D_c = 14$ cm, $H_s = 34$ cm, perforated plate, Glycerol solutions[11])

glycerol [%]	d_H [mm]	symbol	
50	0.5	●	①
50	1.0	⊗	②
50	3.0	○	③
70	0.5	▲	④
70	1.0	⚠	⑤
70	3.0	△	⑥
90	1.0	□	
95	0.5	◆	
95	1.0	⊕	

One can recognize that with increasing glycerol concentration a considerably diminishes. At low concentrations, the gas flow rate and aerator type influence a. Parameter a increases with increasing w_{SG} and diminishes, if d_H is enlarged.

With increasing viscosity, the effects of gas flow rate and aerator type on a are considerably reduced.

The same applies to the specific interfacial areas, determined in CMC solutions (Fig. 72).

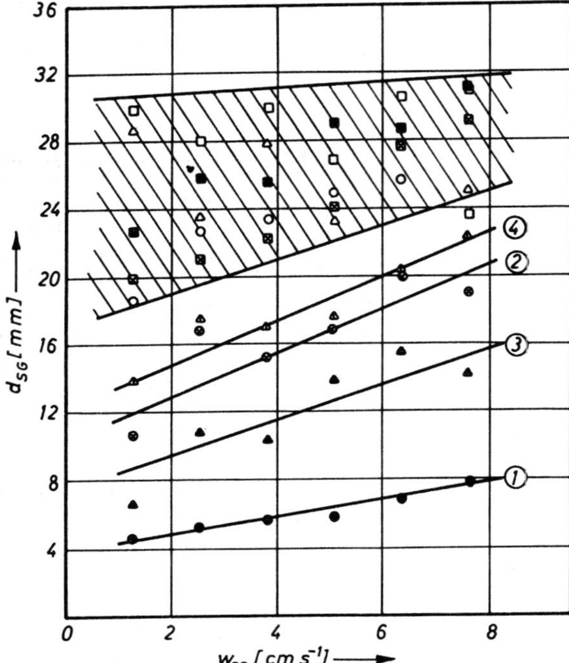

Fig. 68 Sauter diameter of "intermediate-to-large" bubbles, d_{SG} as a function of w_{SG}. D_c = 14 cm, H_s = 35 cm, perforated plate, CMC solutions[103)].

CMC (%)	d_H [mm]	symbol	
1	0.5	●	①
	1.0	⊗	②
	3.0	○	
1.4	0.5	▲	③
	1.0	⚠	④
	3.0	△	
2.0	0.5	■	
	1.0	⊠	
	3.0	□	

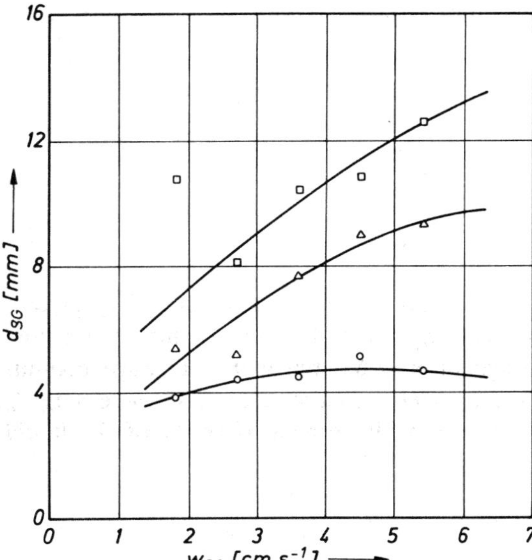

Fig. 69 Sauter diameter of "intermediate-to-large" bubbles d_{SG} as a function of w_{SG}. D_c = 14 cm, H_s = 34 cm, perforated plate, 0.2% PAA solution. ○ d_H = 0.5 mm; □ d_H = 1.0 mm; △ d_H = 3.0 mm

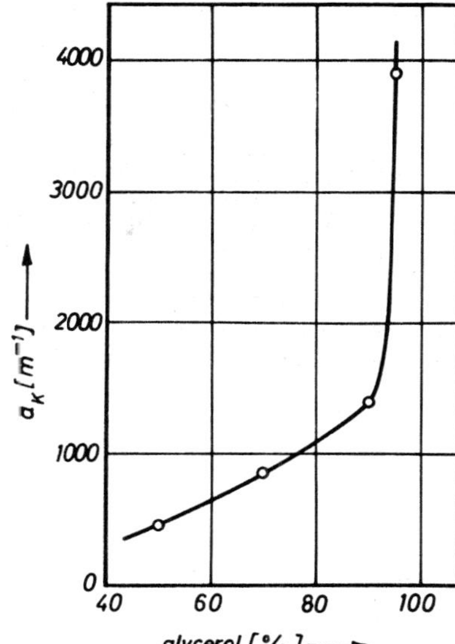

Fig. 70 Specific interfacial area a_K due to "small" bubbles as a function of glycerol concentration. D_c = 14 cm, H_s = 35 cm, perforated plate d_H = 1.0. w_{SG} = 1—8 cm s^{-1} [11]

Peschke[51] ascertained the specific interfacial areas a_K and a_G in PAA solutions. Again, the a_K values increase to extremely high values with increasing concentration. This is in opposition to k_La measurements since k_La diminishes with increasing PAA concentration. This again implies that a_K values are not decisive for k_La. Obviously, only the a_G value controls k_La. In Fig. 73, the a_G values are plotted as a function of the superficial gas velocity employing PAA solutions of different concentrations and perforated plates of various d_H values. With increasing superficial gas velocity, w_{SG}, a_G generally increases. The highest values are attained at lowest concentrations and with smallest d_H values. With increasing PAA concentration, the dependence of a_G on w_{SG} as well as on d_H gradually diminishes.

No general relationships are known for the calculation of E_G, E_{GG}, E_{GK}, d_s, d_K, and a_G in highly viscous media.

6.4 OTR and k_La

Recently studies on the determination of OTR and k_La using glycerol[36], CMC[36,63] and PAA[37] solutions hove been reported. In Table 1 the properties of these solutions are compiled.

In Figure 74, k_La is plotted as a function of w_{SG} 50,70 and 90% glycerol solutions and perforated plate trays with d_H = 0.5, 1.0 and 3.0 mm being used. k_La was measured in the upper stage of a countercurrent multistage column; this quantity diminishes with increasing glycerol concentration and increasing d_H. Both of these effects are gradually reduced with growing viscosity of the liquid. This tray effect on k_La is also pronounced if CMC solutions are applied (Fig. 75). With increasing bubbling layer height, H_S, k_La diminishes. At H_S = 40 cm, it approaches a constant value (Fig. 76). The CMC concentration markedly affects k_La (Fig. 77). With rising CMC concentration, k_La considerably diminishes.

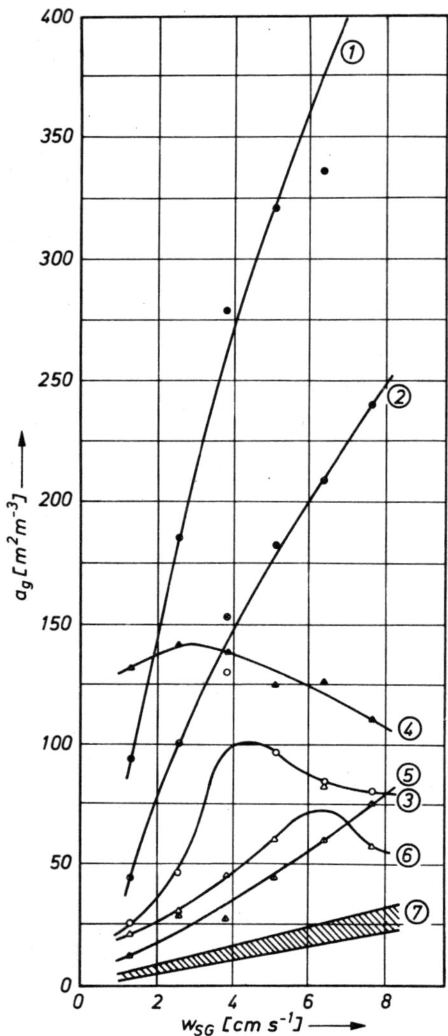

Fig. 71 Specific surface area a_G due to "intermediate-to-large" bubbles as a function of w_{SG} in glycerol solutions. D_c = 14 cm, H_s = 35 cm, perforated plate aerator[11].

glycerol (%)	d_H [mm]	symbol	
50	0.5	●	①
50	1.0	⊗	②
50	3.0	○	③
70	0.5	▲	④
70	1.0	⚠	⑤
70	3.0	△	⑥
90	1.0	}	⑦
95	0.5		
95	1.0		

Aerated PAA solutions behave similarly (Fig. 78). At low PAA concentration, a large tray effect exists. With increasing d_H and PAA concentration, k_La diminishes. At high PAA concentrations, only a slight tray effect is observed.

To illustrate the influence of the liquid rheological behavior on k_La in Fig. 79, k_La is plotted as a function of the dynamic viscosity η of glycerol solutions. With increasing η, k_La rapidly diminishes and for $\eta > 100$ Pa · s it approaches a constant value.

For pseudoplastic liquids, the fluid consistency index, K, approaximately corresponds to the viscosity. Therefore, in Figs. 80 and 81, k_La is plotted as a function of K employing CMC solutions. With increasing K, the volumetric mass transfer coefficient, k_La, diminishes. This effect is pronounced at high w_{SG} values and is reduced with decreasing superficial gas velocity. A comparison of Figs. 80 and 81

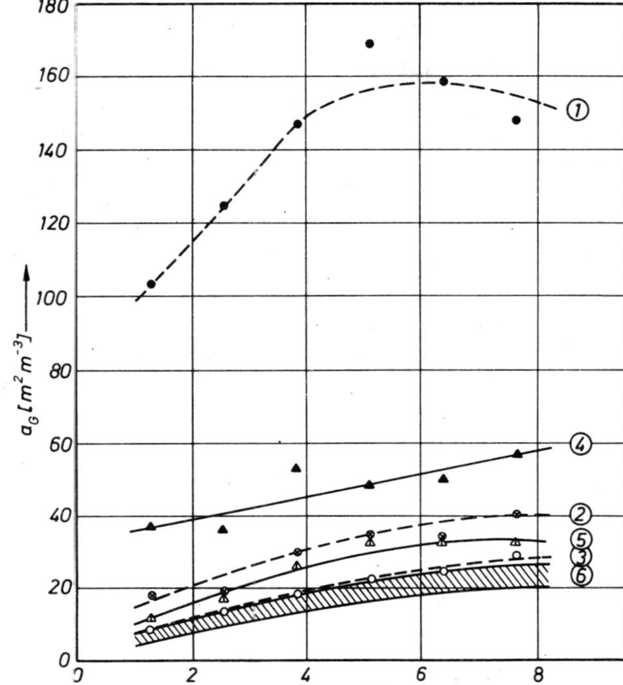

Fig. 72 Specific surface area a_G due to "intermediate-to-large" bubbles as a function of w_{SG} in CMC solutions. $D_c = 14$ cm, $H_s = 35$ cm, perforated plate aerator[103])

CMC (%)	d_H [mm]	symbol	
1	0.5	●	①
1	1.0	⊗	②
1	3.0	○	③
1.4	0.5	▲	④
1.4	1.0	△	⑤
1.4	3.0		⑥
2.0	0.5		⑥
2.0	1.0		⑥
2.0	3.0		⑥

reveals that with increasing d_H the *K*-effect also diminishes. The same also applies to PAA solutions (Fig. 82).

It is difficult to compare media of different rheological behavior. Figs. 83 to 85 qualitatively describe such comparisons where k_La is plotted for the same fluid consistency index, *K* (Fig. 83) and/or for the same shear velocities, *D* (Figs. 84 and 85) as a function of the superficial gas velocity, w_{SG}. All of these qualitative comparisons indicate that similar conditions (fluid consistency index *K*, shear rate *D*) in viscoelastic media the highest — and in Newtonian media the smallest — k_La values can be attained.

The low value of k_La determined in glycerol solutions can be explained by the low k_L values in glycerol due to the very low dissolved oxygen diffusivity in these solutions (see Table 1). Furthermore, the specific interfacial area, a_G, in a 90% glycerol solution is considerably lower than a_G in the corresponding 1% CMC solution (compare the set of curves (7) in Fig. 71 with curve (1) in Fig. 72).

The higher k_La values determined in PAA solutions rather than those obtained for CMC solutions can be explained by the higher mass transfer coefficients in the former due to viscoelasticity. According to Zana and Leal, mass transfer rates are significantly enhanced by viscoelasticity, which yields higher k_L values than those for purely viscous, pseudoplastic fluids[90]).

A comparison of Figs. 72 and 73 indicates that in PAA solutions higher a_G values can be attained than in comparable rheological CMC solutions under the same conditions (d_H, W_{SG}). This is due to the fact that in PAA solutions d_{SG} is

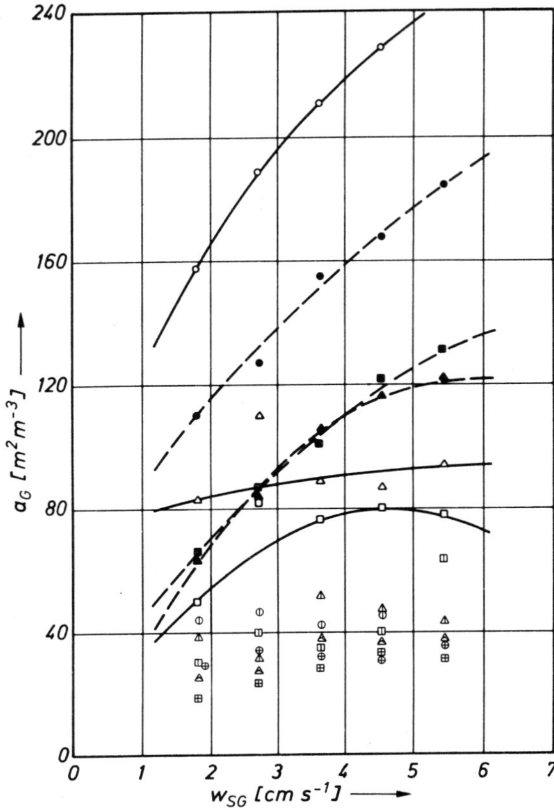

Fig. 73 Specific surface area a_G due ro "intermediate-to-large" bubbles as a function of w_{SG} in PAA solutions. D_c = 14 cm, H_s = 35 cm, perforated plate[51].

PAA concentrations (%):	0.2	0.4	0.6	1.0
d_H = 0.5 mm	○	●	⦶	⊕
d_H = 1.0 mm	□	■	◫	⊞
d_H = 3.0 mm	△	▲	△⃒	⨹

smaller and E_{GG} is larger than in comparable rheological CMC solutions under the conditions (compare the d_S values in Figs. 68 and 69).

The higher E_G values in viscoelastic solutions in comparison with pure pseudoplastic solutions are caused by the lower ascending bubble velocities in the former.

In slightly viscoelastic solutions, the gas/liquid interfacial area already exhibits an extremely significant viscoelastic property[175]. With increasing viscoelasticity, the conservation forces due to surface tension are supported and the drag coefficient is increased (Eq. (38)). The smaller Sauter diameter in viscoelastic solutions as compared with pseudoplastic solutions is probably attributed to the lower bubble coalescence rate.

All these effects together with comparable K and n values give rise to higher k_La values in viscoelastic liquids than in pure pseudoplastic liquids.

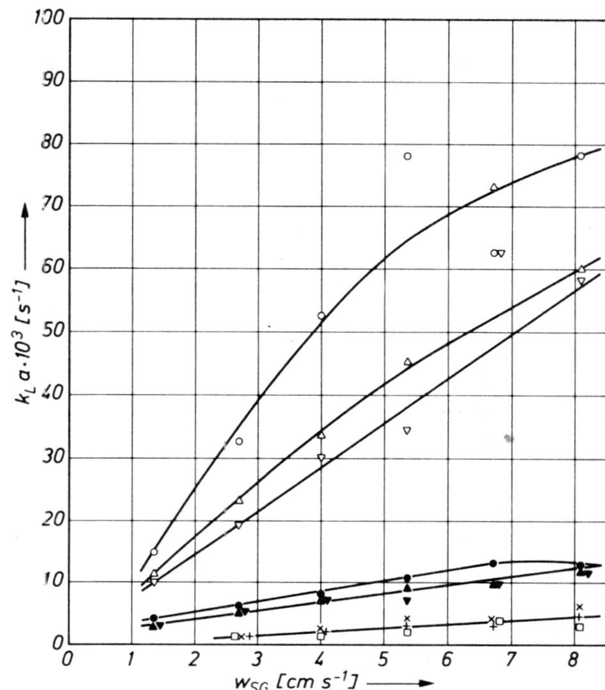

Fig. 74 Volumetric mass transfer coefficient k_La as a function of superficial gas velocity w_{SG}. First stage of a multistage countercurrent column, D_c = 20 cm, H_s = 30 cm. The perforated plate tray and glycerol concentrations are varied[36)]

glycerol (%)	d_H [mm]	symbol
50	0.5	○
50	1	△
50	3	▽
70	0.5	●
70	1	▲
70	3	▼
90	0.5	×
90	1	+
90	3	□

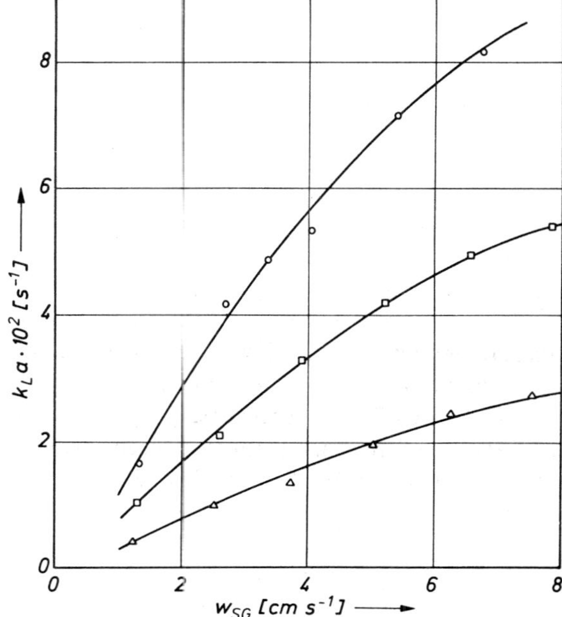

Fig. 75 Volumetric mass transfer coefficient k_La as a function of superficial gas velocity w_{SG}. First stage of a multistage countercurrent column. D_c = 20 cm, H_s = 30 cm, 1% CMC solution. Influence of the tray on k_La [36)]. d_H = 3.0 mm △; d_H = 1.0 □; d_H = 0.5 ○

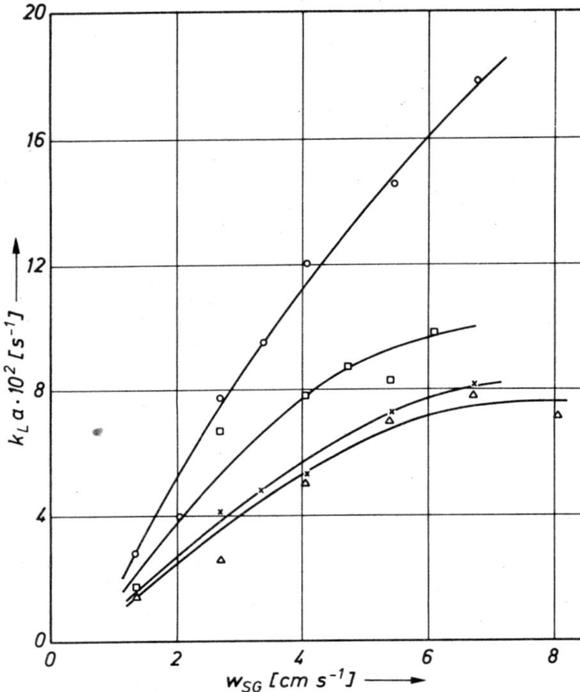

Fig. 76 Volumetric mass transfer coefficient k_La as a function of superficial gas velocity w_{SG}. First stage of a countercurrent multistage column. D_c = 20 cm, perforated plate, d_H = 0.5 mm. 1% CMC solution. Influence of bubbling layer height H_s on k_La[36]. H_s = 10 mm ○; H_s = 20 □; H_s = 30 ×; H_s = 40 △

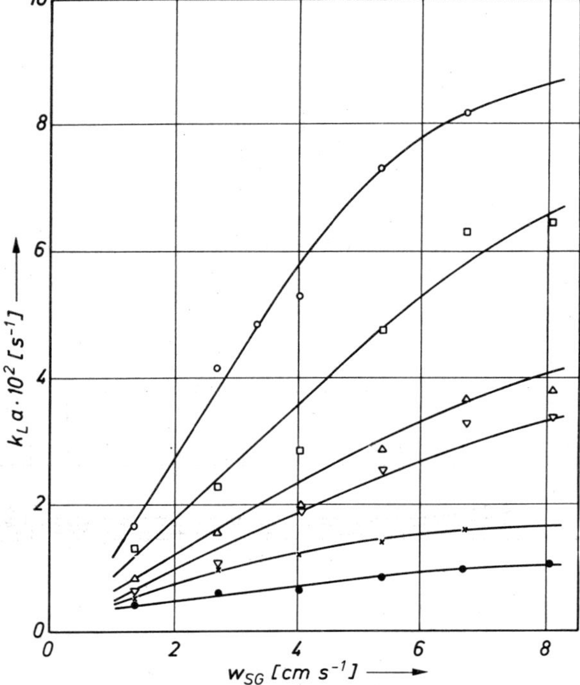

Fig. 77 Volumetric mass transfer coefficient k_La as a function of superficial gas velocity w_{SG}. First stage of a countercurrent multistage column. D_c = 20 cm, H_s = 30 cm, perforated plate d_H = 0.5 mm. Influence of the CMC concentration on k_La[36].

○ 1.0% CMC ▽ 1.6% CMC
□ 1.2% × 1.8%
△ 1.4% ● 2.0%

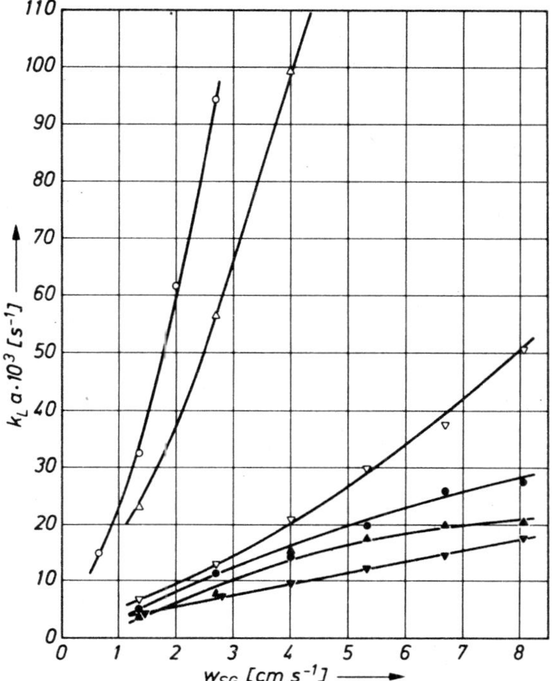

Fig. 78 Volumetric mass transfer coefficient $k_L a$ as a function of superficial gas velocity w_{SG}. First stage of a countercurrent multistage column. D_c = 20 cm, H_s = 30 cm; perforated plate trays. Influence of the tray type and the PAA concentration on $k_L a$ [37].

	0.2% PAA solution	0.5% PAA solution
d_H = 0.5 mm	○	●
1.0 mm	△	▲
3.0 mm	▽	▼

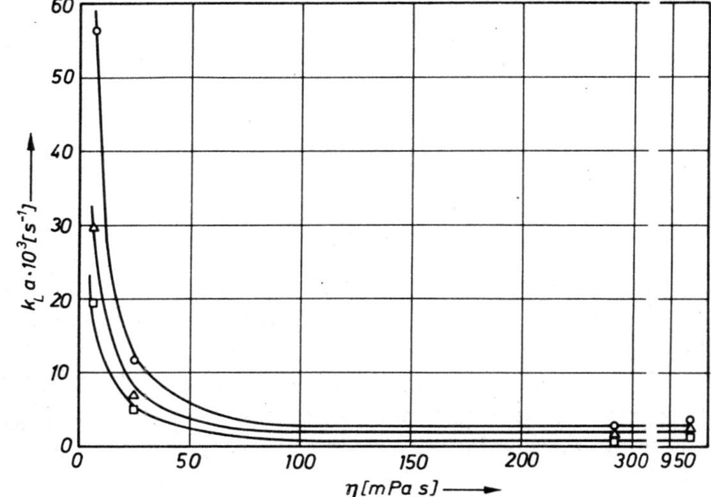

○ w_{SG} = 8 cm s^{-1}
△ w_{SG} = 4 cm s^{-1}
□ w_{SG} = 2.7 cm $*^{-1}$

Fig. 79 Influence of dynamic viscosity on $k_L a$. First stage of a countercurrent multistage column. D_c = 20 cm, H_s = 30 cm. Perforated plate trays; d_H = 3.0 mm. Glycerol solutions[36]

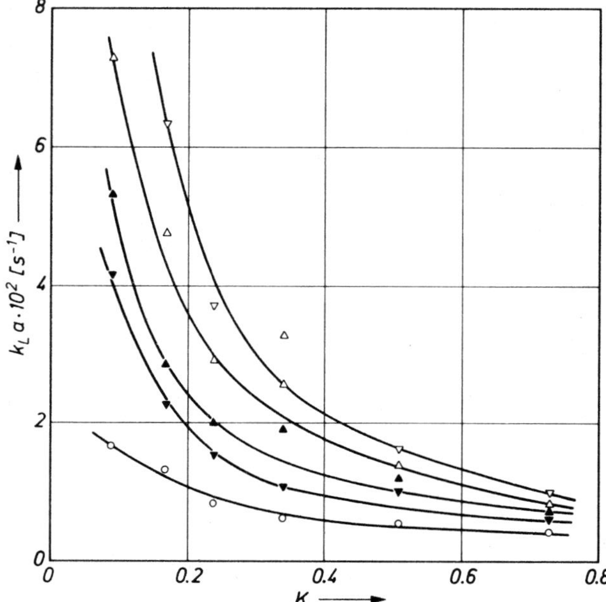

Fig. 80 Influence of fluid consistency index K on $k_L a$. First stage of a countercurrent multistage column. $D_c = 20$ cm, $H_s = 30$ cm, perforated plate trays, $d_H = 0.5$ mm, CMC solutions[36].

symbol	w_{SG} [cm s^{-1}]
⊙	1.35
▼	2.70
▲	4.05
△	5.40
▽	6.75

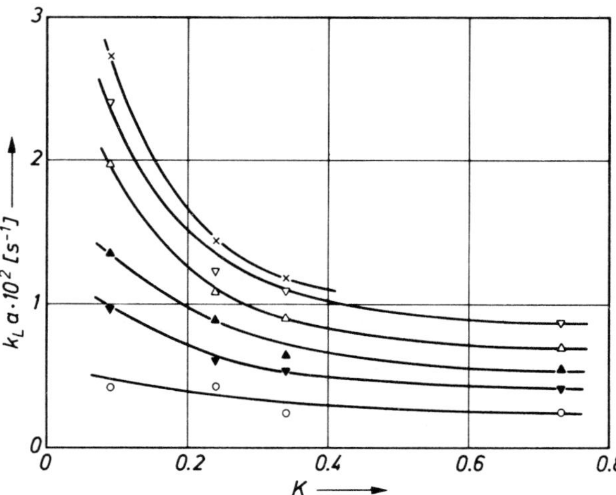

Fig. 81 Influence of fluid consistency index K on $k_L a$. First stage of a countercurrent multistage column. $D_c = 20$ cm, $H_s = 30$ cm, perforated plate trays, $d_H = 3.0$ mm, CMC solutions[36] (for symbols see Fig. 80). × $w_{SG} = 8.10$ cm s^{-1}

If the gas is dispersed by a turbulent mechanism between $k_L a$ and the specific power input, P/V, some relation ships must prevail.

In Figure 86 $k_L a$ is plotted as a function of P/V for glycerol solutions. One recognizes that with increasing specific power input, $k_L a$ increases and that for each concentration only one relationships exists regardless of the tray and height of the bubbling layer. Figure 87 shows the same plot for CMC solutions. In this medium $k_L a$ also markedly depends on P/V. However, at constant P/V, the volumetric mass transfer coefficient is higher if trays with small d_H are employed. With increasing d_H, $k_L a$ significantly diminishes.

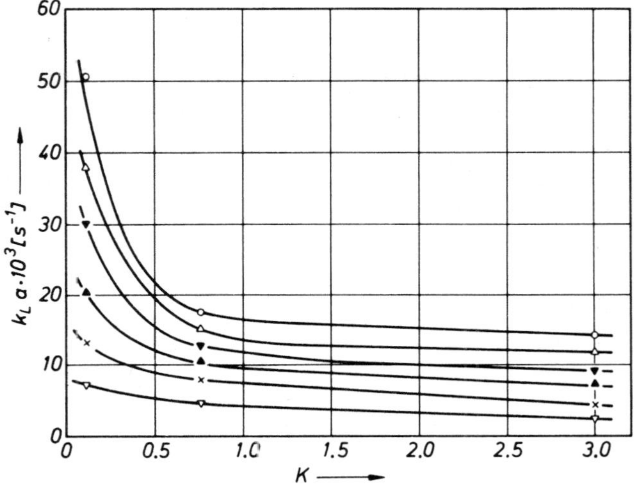

Fig. 82 Influence of fluid consistency index K on $k_L a$. First stage of a countercurrent multistage column. $D_c = 20$ cm, $H_s = 30$ cm; perforated plate tray, $d_H = 3.0$ mm, PAA solutions[37)]

symbol	w_{SG} [cm s^{-1}]	symbol	w_{SG} [cm s^{-1}]
○	8.0	▲	4.0
△	6.7	×	2.7
▼	5.3	▽	1.3

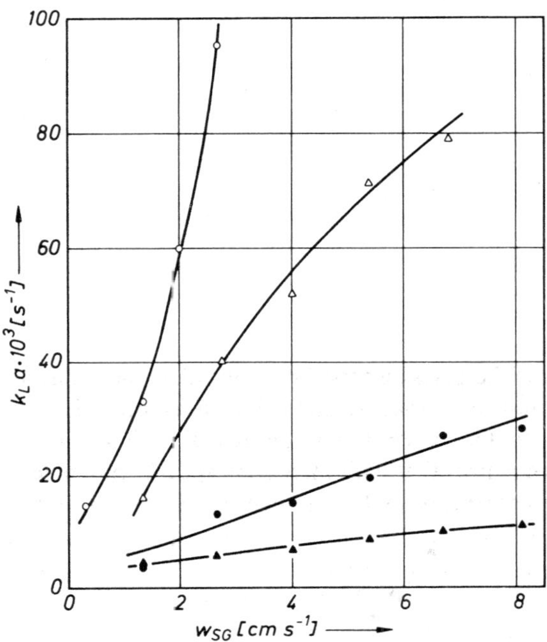

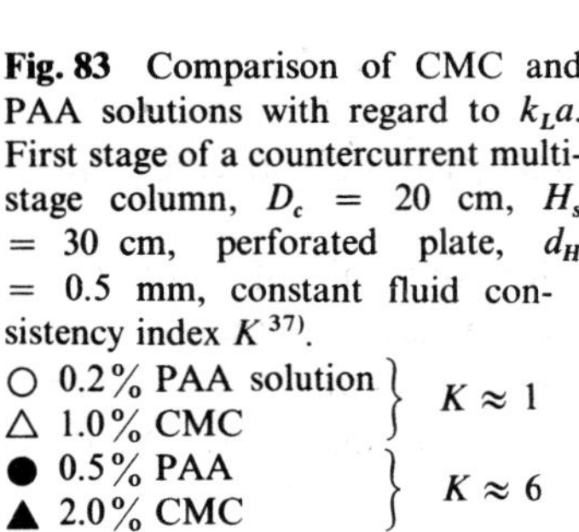

Fig. 83 Comparison of CMC and PAA solutions with regard to $k_L a$. First stage of a countercurrent multistage column, $D_c = 20$ cm, $H_s = 30$ cm, perforated plate, $d_H = 0.5$ mm, constant fluid consistency index K[37)].

○ 0.2% PAA solution
△ 1.0% CMC
} $K \approx 1$

● 0.5% PAA
▲ 2.0% CMC
} $K \approx 6$

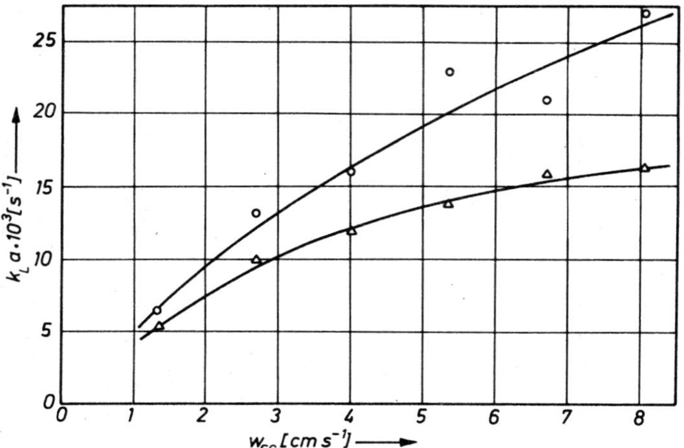

Fig. 84 Comparison of CMC and PAA solutions with regard to $k_L a$. First stage of a countercurrent multistage column. $D_c = 20$ cm, $H_s = 30$ cm, perforated plate, $d_H = 0.5$ mm, constant shear velocity $D = 100\ s^{-1}$ [37]. ○ 1% PAA solution; $\eta_D = 170$ m Pa · s; △ 1.8% CMC solution; $\eta_D = 160$ m Pa · s

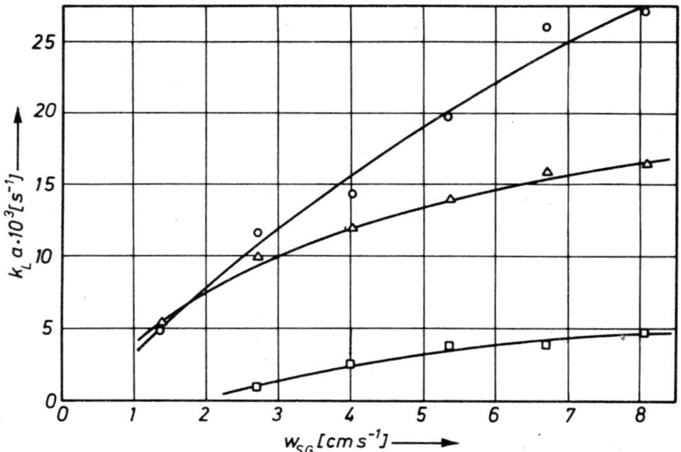

Fig. 85 Comparison of glycerol, CMC and PAA solutions with regard to $k_L a$. First stage of a countercurrent multistage column. $D_c = 20$ cm, $H_s = 30$ cm, perforated plate, $d_H = 0.5$ mm, constant shear velocity $D = 10\ s^{-1}$ [37]. ○ 0.5% PAA solution; $\eta_D = 230$ m Pa · s; △ 1.8% CMC; $\eta_D = 247$ m Pa · s; □ 90% glycerol; $\eta_D = 240$ m Pa · s

In Fig. 88 $k_L a$ is plotted as a function of P/V for PAA solutions. Again, $k_L a$ is enlarged with increasing P/V. At constant power input, the highest $k_L a$ value is attained by the tray with the smallest d_H.

This indicates that in CMC, as well as in PAA solutions, the turbulence microscale fraction influences the dispersion of the gas, in contrast to glycerol solutions where such

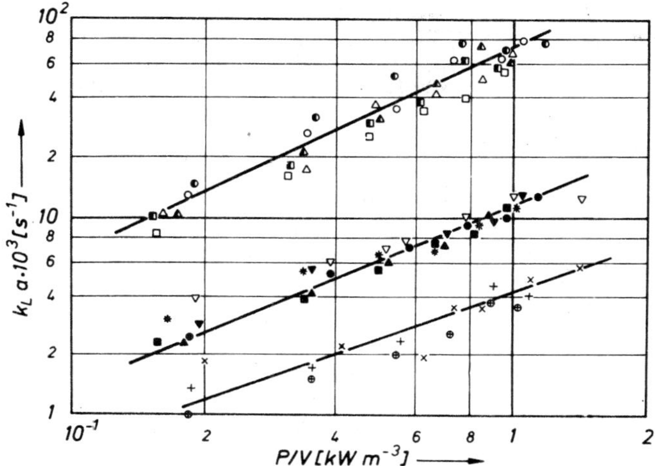

Fig. 86 Influence of specific power input P/V on k_La Glycerol solution[36].

50%	70%	90%	tray d_H [cm]	height H_s [cm]
○	●	×	0.5	40 cm
◐	▽	—	0.5	30 cm
△	▲	+	1.0	40 cm
◭	▼	—	1.0	30 cm
□	■	⊕	3.0	40 cm
◧	*	—	3.0	30 cm

an effect does not seen to exist. From Figs. 86 to 88 the constants β and γ of the function

$$k_La = \beta(P/V)^{\gamma} \tag{81}$$

can be evaluated: k_La [s^{-1}], P/V [kWm^{-3}].

For 50% glycerol: $\beta = 70 \times 10^{-3}$ $\gamma = 0.983$
70% glycerol: $\beta = 12.2 \times 10^{-3}$, $\gamma = 0.934$
90% glycerol: $\beta = 4.2 \times 10^{-3}$, $\gamma = 0.755$.

For 1% CMC and $d_H = 3.0$ mm $\beta = 51.5 \times 10^{-3}$, $\gamma = 1.04$
$d_H = 1.0$ mm $\beta = 74.0 \times 10^{-3}$, $\gamma = 0.947$
$d_H = 0.5$ mm $\beta = 134 \times 10^{-3}$, $\gamma = 0,.829$.

For 1% PAA, $H_s = 40$ cm and $d_H = 3.0$ mm $\beta = 9.2 \times 10^{-3}$, $\gamma = 0.756$
$d_H = 1.0$ mm $\beta = 17 \times 10^{-3}$ $\gamma = 0.894$
$d_H = 0.5$ mm $\beta = 26 \times 10^{-3}$ $\gamma = 0.813$
$H_s = 30$ cm and $d_H = 3.0$ mm $\beta = 50 \times 10^{-3}$, $\gamma = 1.22$
$d_H = 1.0$ mm $\beta = 200 \times 10^{-3}$, $\gamma = 1.30$
$d_H = 0.5$ mm $\beta = 500 \times 10^{-3}$, $\gamma = 1.49$.

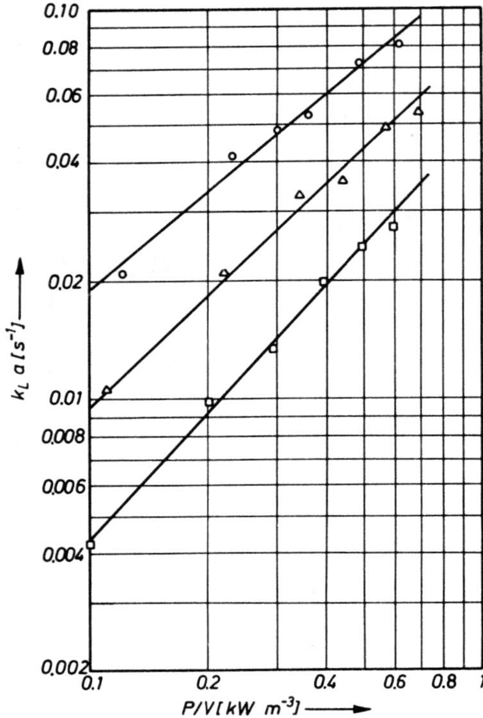

Fig. 87 Influence of specific power input P/V on k_La; CMC solution. $H_s = 30$ cm [165]. ○ $d_H = 0.5$ mm; △ $d_H = 1.0$ mm; □ $d_H = 3.0$ mm

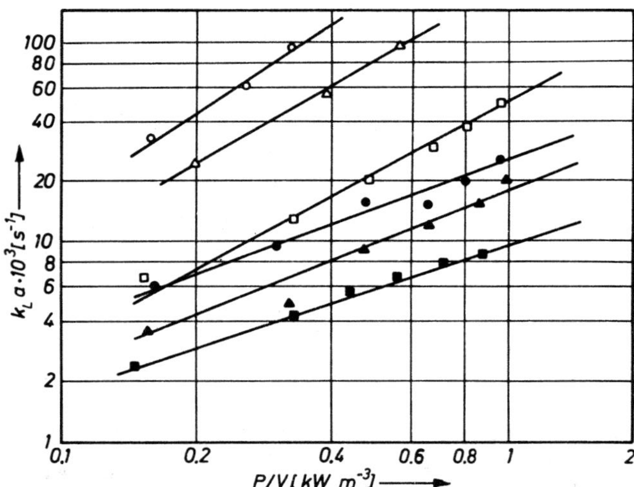

Fig. 88 Influence of specific power input P/V on k_La; PAA solution[37)]

0.2% PAA $H_s = 30$ cm	1.0% PAA $H_s = 40$ cm	tray, d_H [mm]
○	●	0.5
△	▲	1.0
□	■	3.0

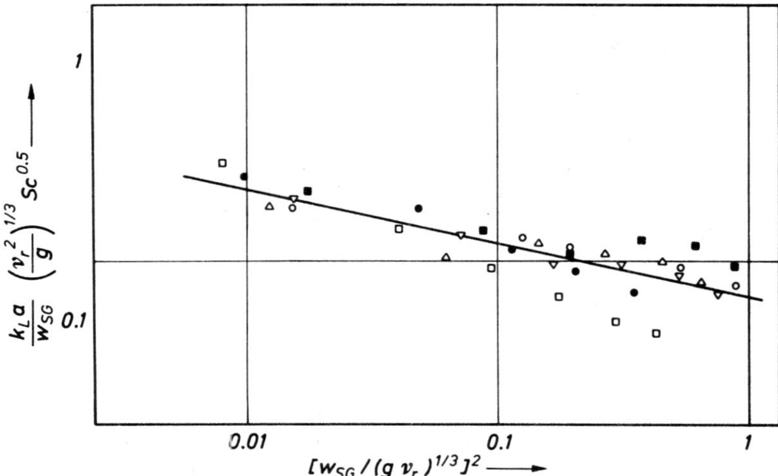

Fig. 89 Relationships for the calculation of $k_L a$; countercurrent multistage column; $D_c = 20$ cm, $H_s = 30$ cm, perforated plate trays, $d_H = 0.5$ mm, CMC solutions[166)]

symbol	η_r [m Pa · s]	symbol	η_r [m Pa · s]
○	37— 54	△	98—157
■	64— 94	●	146—228
▽	79—121	□	182—308

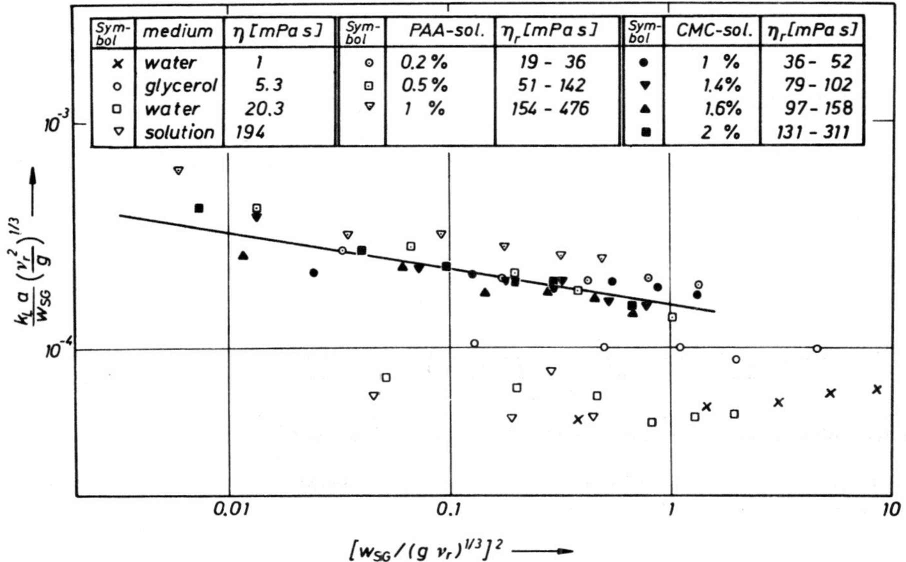

Fig. 90 Relationships for the calculation of $k_L a$; countercurrent multistage column; $D_c = 20$ cm, $H_s = 30$ cm, perforated plate trays, $d_H = 3.0$ mm [166)]

Based on these k_La data Henzler[166] established the following relationships for CMC solutions (Fig. 89):

$$\frac{k_La}{w_{SG}}\left(\frac{v_r^2}{g}\right)^{0.33} = 0.075\left[\frac{w_{SG}}{(gv_r)^{0.33}}\right]^{-0.40} Sc^{*-0.3}, \tag{82}$$

for $d_H = 0.5$ mm.

Validity range:

$$8\times 10^{-2} \leqq \frac{k_La}{w_{SG}}\left(\frac{v_r^2}{g}\right)^{0.33} Sc^{0,5} \leqq 2\times 10^{-1},$$

$$8\times 10^{-3} \leqq \left[\frac{w_{SG}}{(gv_r)^{0.33}}\right]^2 \leqq 1,$$

as well as for $d_H = 3.0$ mm (Fig. 90)

$$\frac{k_La}{w_{SG}}\left(\frac{v_r^2}{g}\right)^{0.33} = 1.60\times 10^{-4}\left[\frac{w_{SG}}{(gv_r)^{0.33}}\right]^{-0.32}$$

Validity range:

$$1.5\times 10^{-4} \leqq \frac{k_La}{w_{SG}}\left(\frac{v_r^2}{g}\right)^{0.33} \leqq 6\times 10^{-4}, \tag{83}$$

$$8\times 10^{-3} \leqq \left[\frac{w_{SG}}{(gv_r)^{0.33}}\right]^2 \leqq 1.0\,.$$

7 Other Reactor Types

Beyond the discussed stirred tank and bubble column reactors, several other reactor types are used in biotechnology if low viscosity media are employed. To these reactors belong the different types of loop reactors, plunging jet reactors. etc. Fluid dynamical properties in propeller loop reactors[167, 168] and in jet loop reactors[167, 169] were investigated employing non-aerated highly viscous media. A small wetted-wall column was used by Wasan et al.[170] for the determination of the mass transfer rate of oxygen into the medium, of the mass transfer coefficient and of the diffusivity of oxygen employing PEO, CMC, Carbopol and methocal solutions. However, the aim of these investigations was to characterize these solutions and not the wetted-wall column.

Static mixers can also be used to mix highly viscous media (e.g.[171]). However, the author is not aware of investigations in aerated static mixers employing highly viscous media. Also screw conveyers[171], ribbon screw impellers (e.g.[172]), anchor impellers (e.g.[173]) and helical impellers (e.g.[174]) are applied to mix non-aerated highly viscous media.

These constructions of the impellers are chosen to minimize the power input for a given mixing time rather than to maximize the gas dispersion and *OTR* for a given power input.

8 Comparison of Different Reactors. Recommendations

It is difficult to compare different reactors as only few comparable data have been published. In Table 7 some k_La values, which were measured at the same specific power input $P/V = 0.5$ kW m^{-3} (in different reactors), are compared.

One can recognize that in bubble columns much higher k_La values can be achieved at $P/V = 0.5$ kW m^{-3} than in stirred tank reactors. This applies to 1.5% CMC and 0.2% PAA solutions.

A comparison of single-stage and multistage bubble columns indicates that at low superficial gas velocities (low coalescence rate) in single-stage columns containing a more efficient gas distributor (porous plate), higher k_La values can be achieved than in multistage columns with a less effecitve gas distributor (perforated plates). At higher superficial gas velocities in multistage columns where the coalescence rate is higher, larger k_La values have been measured than in single-stage columns (this is not shown in Table 7). The k_La values in stirred tank reactors also increase with increasing P/V but they do not attain the high values found in single-stage bubble columns at low specific power input.

As long as the viscosity of the media is not very high ($\eta < 1000$ mPa · s) it is more economical to aerate the medium in the bubble column. At low superficial gas velocities, a single-stage solumn with a very effective aerator is recommended at high superficial gas velocities or, in the presence of antifoam agents, a multistage bubble column should be used as long as the reactor scale is similar to that employed in the investigations.

At viscosities >1000 mPa · s all of these reactors are very ineffective. However, while the single-stage bubble column cannot at all be recommended, the multistage

Table 7. Comparison of the volumetric mass transfer coefficients, k_La, obtained in different reactors using non-Newtonian media. Specific power input: $P/V = 0.5$ kW m^{-3}

Reactors	1.5% CMC solution		0.2% PAA solution	
	$k_La \times 10^3$ s^{-1}	vvm	$k_La \times 10^3$ s^{-1}	vvm
Stirred tank six-blade turbine[93]	2 to 5	0.37 to 2.75	2 to 5	0.37 to 2.75
Single-stage bubble column porous plate[84]	30	0.61 to 0.92	—	—
Six-stage bubble column perforated plate, $d_H = 1.0$ mm[36, 63]	23	1.33	80	1.0

Remarks: P/V for stirred tank reactor was calculated by Eq. (27); P/V for bubble column was calculated by Eq. (29). The compression energy for the stirred tank, the pumping energy for the six-stage column, and mechanical losses are neglected

bubble column and the stirred tank can still be employed. At still higher viscosities (>2000 mPa · s), other reactor types must be chosen.

Unfortunately, the author cannot recommend any new reactor type for this purpose and no data have been reported to scale up these reactors. It is necessary to further investigate highly viscous media in order to gain more and reliable data the application of which would result in an optimum reactor construction and a reliable scale-up with regard to $k_L a$.

9 Acknowledgement

The author gratefully acknowledges the financial support of the Ministry of Research and Technology of the Federal Republic of Germany, Bonn, and the cooperation of Dr. H. Buchholz, Dr. R. Buchholz, Dr. J. Voigt, K. Franz, V. Hecht, H. Niebeschütz, and G. Peschke.

10 List of Symbols

(M = mass, L = length, T = time, θ = temperature)

A	surface area	L^2
A_h	heat exchange surface area	L^2
a	specific gas/liquid interfacial area with regard to the liquid volume	L^{-1}
a_K	a due to small bubbles	L^{-1}
a_G	a due to "intermediate-to-large" bubbles	L^{-1}
C	concentration	ML^{-1}
C_D	drag coefficient	—
c	specific heat	L^2T^{-2}
$D = \mathrm{d}v/\mathrm{d}x$	shear velocity	T^{-1}
D_c	column diameter	L
$De = Nt_{cr}$	Deborah number	—
D_m	diffusivity	L^2T^{-1}
D_o	orifice diameter	L
D_t	tank diameter	L
d	bubble diameter	L
d_e	dynamic equilibrium bubble diameter	L
d_{eq}	diameter of the spherical bubble having the same volume as the bubble in question	L
d_H	hole diameter of perforated plate trays	L
d_N	impeller diameter	L
d_p	primary bubble diameter	L
d_S	Sauter mean bubble diameter	L
E	coefficient of longitudinal liquid dispersion	L^2T^{-1}
E_G	relative gas hold-up	—
E_{GG}	relative gas hold-up due to intermediate-to-large bubbles	—
E_{GK}	relative gas hold-up due to "small" bubbles	—

E_{GS}	relative gas hold-up due to slugs	—
$Fr = \frac{N^2 d_N}{g}$	impeller Froude number	—
$f(x)$	function of x	
$Ga = \frac{d_N{}^3 g}{\nu^2}$	impeller Galilei number	—
g	acceleration of gravity	LT^{-2}
H	height of the bubbling layer	L
H_L	height of the bubble free layer	L
H_i	constant in Eq. (17)	
H_s	height of the bubbling layer in multistage columns	L
H_t	height of the bubbling layer in stirred tank	L
h_1, h_2	coefficients in Eq. (14)	
h_i	constant in Eq. (16)	
I	ionic strength	ML^{-3}
i	i = 1,2,3 ...	
J	Joule	ML^2T^{-2}
K	fluid consistency index	$ML^{-1}T^{-n}$
K_S	Schenow constant	L^3M^{-1}
k	constant	
k_L	gas/liquid mass transfer coefficient	LT^{-1}
k_La	volumetric mass transfer coefficient	T^{-1}
$(k_La)^* = k_La(\nu/g^2)^{1/3}$	dimensionless volumetric mass transfer coefficient	—
L	cylinder length of the rotation viscosimeter	L
M_G	gas mass flow	MT^{-1}
M_i	torque exerted on the inner cylinder of the rotation viscosimeter	ML^2T^{-2}
M_N	torque exerted on the stirrer shaft	ML^2T^{-2}
N	impeller rotation speed	T^{-1}
$Ne = P/N^3 d_N{}^5 \varrho$	Power or Newton number	—
$Nu_j = \frac{\alpha_j D_t}{\lambda}$	jacket Nusselt number	—
$Nu_c = \frac{\alpha_c D_t}{\lambda}$	coil Nusselt number	—
n	flow behavior index	—
O_G	oxygen concentration in the gas phase	ML^{-3}
O_L	dissolved oxygen concentration in the bulk of liquid	ML^{-3}
O_L^*	dissolved oxygen concentration at the gas/liquid interface (at saturation)	ML^{-3}
OTR	oxygen transfer rate	MT^{-1}
P	power input	ML^2T^{-3}
$Pe_B = \frac{d_{eq} U}{D_m}$	bubble Peclet number	—

Symbol	Definition	Dimension
P/Q_G	specific power input with regard to Q_G	
$(P/Q_G)^* = (P/Q_G)[\varrho(gv)^{2/3}]^{-1}$	dimensionless specific power input with regard to Q_G	—
P/V	specific power input with regard to V	$ML^{-1}T^{-3}$
$(P/V)^* = (P/V)[\varrho(v^4g)^{1/3}]^{-1}$	dimensionless specific power input with regard to V	—
p	O_2 partial pressure in liquid	$ML^{-1}T^{-2}$
p	pressure	$ML^{-1}T^{-2}$
Q	flow rate	L^3T^{-1}
Q_G	gas flow rate	L^3T^{-1}
$Q_G^* = Q_G/Nd_N^3$	aeration number	—
Q_G/V	specific flow rate with regard to V	T^{-1}
$(Q_G/V)^* = (Q_G/V)(v^2g)^{1/23}$	dimensionless specific flow rate with regard to V	—
q^*	constant (Eq. (44))	
R	gas constant	$ML^2T^{-2}\Theta^{-1}$
R	equivalent bubble radius	L
$Re_B = \frac{d_{eq}U}{v}$	bubble Reynolds number	—
$Re_N = \frac{Nd_N^2}{v}$	impeller Reynolds number	—
$Re_N = \frac{Nd_N^2}{v_r}$	impeller Reynolds number for non-Newtonian liquids	
r	radius	L
$Sc = \frac{v}{D_m}$	Schmidt number	—
$Sc = \frac{v_r}{D_m}$	Schmidt number with representative viscosity	—
$Sh_B = \frac{k_Ld_{eq}}{D_m}$	bubble Sherwood number	—
$Sh_N = \frac{k_Ld_N}{D_m}$	impeller Sherwood number	—
$St = k_La\tau_M$	Stanton number	—
$St_G = k_La(V/Q_G)$	modified Stanton number	—
T	temperature	θ
t	time	T
t_c	contact time	T
t_{cr}	characteristic time of the liquid (Eq. (57))	T
t_{01}, t_{02}	initial relaxation time (Eq. (11))	T
X	longitudinal distance from the gas distributor	L
X_n	constant defined by Eq. (33)	—
x	length	L
U	single-bubble velocity	LT^{-1}
U_{slug}	slug velocity	LT^{-1}

u	linear liquid velocity	LT^{-1}
u_o	gas velocity through the orifice	LT^{-1}
V	volume of the bubbling layer	L^3
V_B	bubble volume	L^3
V_L	volume of the bubble free liquid	L^3
v	velocity	LT^{-1}
$vvm = Q_G/V$	aeration rate	T^{-1}
$dv/dx = D$	shear velocity	T^{-1}
w	velocity	
w_{app}	apparent wake velocity	LT^{-1}
$w_G = \frac{w_{SG}}{E_G}$	gas velocity	LT^{-1}
$w_L = \frac{w_{SL}}{1-E_G}$	liquid velocity	LT^{-1}
w_S	relative gas velocity (bubble-swarm velocity) with regard to liquid (Eq. (23))	LT^{-1}
w_{SL}	superficial liquid velocity	LT^{-1}
w_{SG}	superficial gas velocity	LT^{-1}

Greek letters

α_c	coil heat transfer coefficient	$MT^{-3}\theta^{-1}$
α_j	jacket heat transfer coefficient	$MT^{-3}\theta^{-1}$
β	constant	—
γ	constant	—
δ	film thickness	L
$\varepsilon = E_G$	mean relative gas hold-up	—
θ	mixing time	T
θN	dimensionless mixing time	—
λ	heat conductivity	$MLT^{-3}\theta^{-1}$
$\lambda_1\lambda_2$	exponents in Eq. (14)	—
η	dynamic viscosity	$ML^{-1}T^{-1}$
η_r	representative dynamic viscosity	$ML^{-1}T^{-1}$
$\nu = \frac{\eta}{\varrho_L}$	kinematic viscosity of liquid	L^2T^{-1}
$\nu_r = \eta_r/\varrho_L$	representative kinematic viscosity	L^2T^{-1}
ϱ_L	liquid density	ML^{-3}
σ	surface tension	MT^{-2}
σ_1	normal stress Eq. (56)	$ML^{-1}T^{-2}$
σ_I, σ_{II}	normal stress	$ML^{-1}T^{-2}$
τ	shear stress	$ML^{-1}T^{-2}$
τ_M	mean liquid residence time	θ
τ_T	dynamical pressure of turbulence	$ML^{-1}T^{-2}$
Ω	angular speed of the rotational viscosimeter	T^{-1}
ω	angular speed of the impeller	T^{-1}

Indices

c	coil
e	equilibrium
eq	equivalent
g	gas
j	jacket
L	liquid
p	primary
r	representative
S	Sauter

Polymers employed by different research groups

CMC	carboxymethylcellulose (e.g. Tylose C 300, Hoechst Co., RK 5000 Wolf/Walsrode)
CMC Na	sodium carboxymethylcellulose (e.g. medium viscosity type 4 MH Hercules Powder Co.)
Carbopol	Carboxypolymethylene (e.g. 934, Goodrich Co.)
Carbowax	polyethylene glycol
Cyanamer	Polyacrylamide (e.g. P-250 American Cyanamide Co.)
ET-947	polyacrylamide
J-100	polyacrylamide
WSR-301	polyethylene oxide
Lanogen	polyethylene glycol
Macrogol	polyethylene glycol
PAA	polyacrylamide (e.g. Separan AP 30 Dow Chem. Co.)
PEG	polyethylene glycol
PEO	polyethylene oxide
Polyox	polyethylene oxide (e.g. WSR-301 blend A-3254)

11 References

1. Margaritis, A., Zajic, J. E.: Biotech. Bioeng. *20*, 939 (1978)
2. Deindoerfer, F. H., Gaden, E. L. Jr.: Appl. Microbiol. *3*, 253 (1955)
3. Bongenaar, J. J. T. et al.: Biotech. Bioeng. *15*, 201 (1973)
4. Charles, M.: Adv. Biochem. Eng. *8*, 1 (1978)
5. Metz, B., Kossen, N. W. F., van Suijdam, I. C.: Adv. Biochem. Eng. *11*, 103 (1979)
6. Costes, J., Alran, C.: Int. J. Multiphase Flow *4*, 535 (1978)
7. Buchholz, R., Adler, I., Schügerl, K.: Europ. J. Appl. Microbiol. Biotech. *7*, 135, 241 (1979)
8. Crabtree, J. R., Bridgewater, J.: Chem. Eng. Sci. *26*, 839 (1971)
9. de Nevers, N., Wu, Jen-Liang: A.I.Ch.E. Journal *17*, 182 (1971)
10. Otake, T., Tone, S., Nakato, K., Mitsuhashi, Y.: Chem. Eng. Sci. *32*, 377 (1977)
11. Franz, K., Buchholz, R., Schügerl, K.: Part. I Chem. Eng. Commun. *5*, 165 (1980)
12. Danckwerts, P. V.: Gas-liquid reactions, New York: McGraw-Hill Book Co. 1970
13. Meskat, W.: In: Hengstenberg, J., Sturm, B., Winkler, O. (eds.). Viskosimetrie in Messen und Regeln. Berlin, Heidelberg, New York: Springer Verlag 1957
14. Jain, M. K.: Z. Angew. Mech. *35*, 12 (1955)
15. Mooney, M.: J. Rheol. *2*, 210 (1931)
16. Krieger, J. M., Maron, S. H.: J. Appl. Phys. *23*, 147 (1952)
17. Krieger, J. M., Elrod, H.: J. Appl. Phys. *24*, 134 (1953)
18. Krieger, J. M., Maron, S. H.: J. Appl. Phys. *25*, 72 (1954)

19. Pawlowski, J.: Kolloid. Z. *130*, 129 (1953)
20. Schulz-Grunow, F., Weyman, H.: Kolloid. Z. *131*, 61 (1953)
21. Tillmann, W.: Kolloid. Z. *131*, 66 (1953)
22. Finke, A., Heinz, W.: Kolloid. Z. *154*, 167 (1957)
23. von Brachel, H., Schümmer, P.: Rheol. Acta *12*, 35 (1973)
24. Chmiel, H., Schümmer, P.: Colloid Polym. Sci. *252*, 886 (1974)
25. Schümmer, P., R. H. Worthoff: Rheol. Acta *15*, 1 (1976)
26. Kuo, Y., Tanner, R. I.: Rheol. Acta *13*, 443 (1974)
27. Schumpe, A., Deckwer, W. D.: Biotech. Bioeng. *21*, 1079 (1979)
28. Schumpe, A., Deckwer, W. D.: Oxygen solubilities in synthetic fermentation media. M/3 p. 154. First European Congress on Biotechnology, Interlaken 1978
29. Koch, B.: Diplomarbeit, Univers. Hannover 1979
30. Quicker, G.: Diplomarbeit, Univers. Hannover 1980
31. Popovic, M., Niebeschütz, H., Reuß, M.: Europ. J. Appl. Microbiol. Biotech. *8*, 1 (1979)
32. Reid, R. C., Prausnitz, J. M., Sherwood, T. K.: The properties of gases and liquids. New York: McGraw Hill Book Co. 1977
33. Hayduk, W., Cheng, S. C.: Chem. Eng. Sci. *26*, 635 (1971)
34. Lohse, M.: Diplomarbeit, University of Hannover 1979
35a. Astarita, G.: Ind. Eng. Chem. Fundam. *4*, 236 (1965)
35b. Calderbank, H.: Trans. Inst. Chem. Eng. *37*, 173 (1959)
36. Voigt, J., Hecht, V., Schügerl, K.: Chem. Eng. Sci. *35*, 1317 (1980)
37. Hecht, V., Voigt, J., Schügerl, K.: Chem. Eng. Sci. *35*, 1325 (1980)
38. Harkins, W. D., Jordan, H. F.: J. Am. Chem. Soc. *52*, 1751 (1930)
39. Kalischewski, K., Bumbullis, W., Schügerl, K.: Europ. J. Appl. Microbiol. Biotech. *7*, 21 (1979)
40. Lucassen-Reynders, E. H., Lucassen, J.: Adv. Coll. and Interface Science, Elsevier *2*, 347 (1969)
41. Lucassen, J., van den Tempel, M.: J. Colloid, Interface Sci. *41*, 491 (1972)
42. Calderbank, P. H.: Mass transfer in fermentation equipment. In: Biochemical and biological engineering science. Blakebrough, N. (ed.), p. 102. New York: Academic Press 1967
43. Buchholz, R., Schügerl, K.: Europ. J. Appl. Microbiol. Biotech. *6*, 301 and 315 (1979)
44. Serizawa, A., Kataoka, K., Michioshi, I.: Int. J. Multiphase Flow *2*, 221 (1975)
45. Burgess, J. M., Calderbank, P. H.: Chem. Eng. Sci. *30*, 743 (1975)
46. Pilhofer, T., Miller, H. D.: Chem. Ing. Techn. *44*, 295 (1972)
47. Brentrup, L. et al.: Bioreactoren. Keune, H., Scheunemann, R., eds., p. 137. DFVLR 1977
48. Brentrup, L.: Dissertation, Univers. Dortmund 1979
49. Buchholz, R., Zakrzewski, W., Schügerl, K.: Chem. Ing. Techn. *51*, 568 (1979)
50. Franz, K.: Diplomarbeit, Univers. Hannover 1979
51. Peschke, G.: Diplomarbeit, Univers. Hannover 1980
52. Wernau, W. C., Wilke, C. R.: Biotech. Bioeng. *15*, 571 (1973)
53. Linek, V., Vacek, V.: Biotech. Bioeng. *18*, 1537 (1976)
54. Linek, V., Benes, P.: Biotech. Bioeng. *20*, 903 (1978)
55. Shioya, S., Dunn, I. J.: Chem. Eng. Sci. *33*, 1529 (1978)
56. Dang, N. D. P., Karrer, D., Dunn, J.: Biotech. Bioeng. *19*, 853 (1977)
57. Joosten, G. E. H., Schilder, J. G. M., Janssen, J. J.: Chem. Eng. Sci. *32*, 563 (1977)
58. Deckwer, W. D., Burghardt, R., Zoll, G.: Chem. Eng. Sci. *29*, 2177 (1974)
59. Burghardt, R., Deckwer, W. D.: Verfahrenstechnik *10*, 429 (1976)
60. Deckwer, W. D., Adler, I., Zaidi, A.: Can J. Chem. Eng. *56*, 43 (1978)
61. Oels, U., Schügerl, K., Todt, J.: Chem. Ing. Techn. *48*, 73 (1976)
62. Oels, U., Lücke, J., Schügerl, K.: Chem. Ing. Techn. *49*, 59 (1977)
63. Voigt, J., Schügerl, K.: Chem. Eng. Sci. *34*, 1221 (1979)
64. König, B., Lippert, J., Schügerl, K.: Ger. Chem. Eng. *2*, 371 (1979)
65. Levenspiel, O., Godfrey, J. H.: Chem. Eng. Sci. *29*, 1723 (1974)
66. Kalischewski, K., Schügerl, K.: Ger. Chem. Eng. *1*, 140 (1978)
67. Elstner, F.: Dissertation, Univers. Dortmund 1978
68. Calderbank, P. H., Johnson, D. S. L., London, J.: Chem. Eng. Sci. *25*, 235 (1970)
69. Koide, K., Orito, Y., Hara, Y.: Chem. Eng. Sci. *29*, 417 (1974)
70. Hallensleben, J.: Dissertation, Univers. Hannover 1980

71. Todt, J. et al.: Chem. Eng. Sci. *32*, 369 (1977)
72. König, B. et al.: Ger. Chem. Eng. *1*, 199 (1978)
73. Niebeschütz, H.: Diplomarbeit, Univers. Hannover 1977
74. Davidson, J. F., Schüler, B. O. G.: Trans. Inst. Chem. Eng. *58*, 144 (1960)
75. Hallensleben, J. et al.: Chem. Ing. Techn. *49*, 663 (1977)
76. Hill, J. H.: Trans. Inst. Chem. Eng. *52*, 1 (1974)
77. Acharya, A., Ulbrecht, J. J.: A.I.Ch.E. Journal *24*, 348 (1978)
78. Astarita, G., Apuzzo, G.: A.I.Ch.E. Journal *11*, 815 (1965)
79. Davies, R. M., Taylor, G. I.: Proc. Roy. Soc. *A200*, 375 (1950)
80. Uno, S., Kintner, R. C.: A.I.Ch.E. Journal *2*, 420 (1956)
81. Dimitrescu, D. T.: Z. Angew. Mech. *23*, 139 (1943)
82. Hirose, T., Moo-Young, M.: Can. J. Chem. Eng. *47*, 265 (1969)
83. Bhavaraju, M., Mashelkar, R. A., Blanch, H.: 69th Annual AIChE Meeting Chicago 1976 and in [77]
84. Buchholz, H. et al.: Chem. Eng. Sci. *33*, 1061 (1978)
85. Hill, H. J., Darton, R. C.: Trans. Instn. Chem. Eng. *54*, 258 (1976)
86. Metzner, A. B. et al.: A.I.Ch.E. Journal *7*, 3 (1961)
87. Mendelson, H. D. A.: A.I.Ch.E. Journal *13*, 250 (1967)
88. Levich, V. G.: Physicochemical Hydrodynamics, chapter 8. New York: Prentice Hall 1962
89. Boussinesq, J.: J. Math. *6*, 285 (1905)
90. Zana, E., Leal, L. G.: Int. J. Multiphase Flow *4*, 237 (1978)
91. van't Riet, K., Bruin, W., Smith, J. M.: Chem. Eng. Sci. *31*, 407 (1976)
92. Höcker, H., Langner, G.: Rheol. Acta *16*, 400 (1977)
93. Höcker, H.: Dissertation, Univers. Dortmund 1979
94. Kiepke, K.: In: Bioreaktoren BMFT-Statusseminar "Bioverfahrenstechnik". Braunschweig-Stöckheim 1977. Keune, H., Scheunemann, R. (eds.). p. 187. DFVLR 1977
95. Keitel, G.: Dissertation, Univers. Dortmund 1978
96. Hiby, J. W.: In: Verfahrenstechnische Fortschritte beim Mischen, Dispergieren und bei der Wärmeübertragung in Flüssigkeiten, p. 7. VDI-GVC Düsseldorf 1978; Fortschr. Verfahrenstech. *17*, 137 (1979)
97. Swallow, B., Finn, R. K., Einsele, A.: Design of aerated fermentors for non-Newtonian liquids, p. 21. First Europ. Cong. Biotechnology, Interlaken 1978
98. Metzner, A. B., Otto, R. E.: A.I.Ch.E. Journal *3*, 3 (1957)
99. Calderbank, P. H., Moon-Young, M. B.: Trans. Instn. Chem. Eng. London *37*, 26 (1959)
100. Taguchi, H., Miyamato, S.: Biotech. Bioeng. *8*, 43 (1966)
101. Ranade, V. R., Ulbrecht, J.: Paper No. F6. Sec. Europ. Conf. on Mixing 1977
102. Weihrauch, W.: Verfahrenstechnik *2*, 243 (1966)
103. Franz, K., Buchholz, R., Schügerl, K.: Chem. Eng. Commun. *5*, 178 (1980)
104. Kalischewski, K., Schügerl, K.: Colloid Polym. Sci. *257*, 1099 (1979)
105. White, W. D., McEligot, D. M.: Trans. ASME, J. Basic Eng. *92*, 411 (1970)
106. Friebe, H. W.: Rheol. Acta *15*, 329 (1976)
107. Lee, T. S.: Proc. of the 5th Intern. Congress on Rheology. Vol. 4, p. 125. Tokyo: University Press 1970
108. Kipke, K.: (personal communication)
109. Perez, J. F., Sandall, O. C.: A.I.Ch.E. Journal *20*, 770 (1974)
110. Páca, J., Ěttler, P., Grégr, V.: J. Appl. Chem. Biotech. *26*, 309 (1976)
111. Loucaides, R., Manamey, W. J.: Chem. Eng. Sci. *28*, 2161 (1973)
112. Yagi, H., Yoshida, F.: Ind. Eng. Chem. Process Des. Dev. *14*, 488 (1975)
113. Ranade, V. R., Ulbrecht, J. J.: A.I.Ch.E. Journal *24*, 796 (1978)
114. Henzler, H. J.: In: Verfahrenstechnische Fortschritte beim Mischen, Dispergieren und bei der Wärmeübertragung in Flüssigkeiten, p. 91. VDI-GVC Düsseldorf 1978
115. Blakebrough, N., Sambamurthy, K.: Biotech. Bioeng. *8*, 25 (1966)
116. Prest, W. M., Porter, R. S.: J. Appl. Polym. Sci. *14*, 2697 (1970)
117. Zlokarnik, M.: Adv. Biochem. Eng. *8*, 133 (1978)
118. Steiff, A.: Dissertation, Univers. Dortmund 1976
119. Edney, H. G. S., Edwards, M. F.: Trans. Instn. Chem. Eng. *54*, 160 (1976)
120. Nagata, S. et al.: Heat Transfer Japanese Res. *5*, 75 (1976)

121. Kipke, K.: Bioreaktoren BMFT-Statusseminar „Bioverfahrenstechnik". March, 2 to 4, 1979, p. 327. KFA-Jülich: Red. J. Gartzen 1979
122. Kahilainen, H., Kurki-Suomio, I., Laine, J.: Chem. Ing. Techn. *51*, 1143 (1979)
123. Steiff, A.: In: Verfahrenstechnische Fortschritte beim Mischen, Dispergieren und bei der Wärmeübertragung in Flüssigkeiten, p. 195. VDI-GVC Düsseldorf, 1978
124. Suryanarayanan, S., Mujanar, B. A., Rao, M. Raja: Ind. Eng. Chem. Process Des. Dev. *15*, 564 (1976)
125. Skelland, A. H. P., Dimmic, G. R.: Ind. Eng. Chem. Process Des. Dev. *8*, 267 (1969)
126. Nooruddin M. Raja Rao: Ind. J. Technol. *4*, 131 (1966)
127. Schügerl, K., Lücke, J., Oels, U.: Adv. Biochem. Eng. *7*, 1 (1977)
128. Schügerl, K. et al.: Adv. Biochem. Eng. *8*, 63 (1978)
129. Gerstenberg, H.: Fortschr. Verfahrenstech. *16*, Abt. D. (1978)
130. Gerstenberg, H.: Chem. Ing. Techn. *51*, 208 (1979)
131. Schlingmann, H. et al.: Proc. 12th Symp. on Computer Application in Chemical Engineering. Montreaux: 8.—11. 4. 1979
132. Wippern, D.: Dissertation, Univers. Hannover 1979
133. Buchholz, H. et al.: Europ. J. Appl. Microbiol. Biotechnol. *6*, 115 (1978)
134. Levenspiel, O.: Chemical reaction engineering. New York: John Wiley and Sons 1964
135. Moo-Young, M., Chan, K. W.: Can. J. Chem. Eng. *49*, 187 (1971)
136. Paquet, M., Cholette, A.: Can. J. Chem. Eng. *50*, 348 (1972)
137. Williams, C. N., Hubbard, D. W.: A.I.Ch.E.J. *24*, 154 (1978)
138. Sinclair, C. G., Brown, D. E.: Biotech. Bioeng. *12*, 1001 (1970)
139. Riquarts, H. P., Pilhofer, Th.: Verfahrenstechnik *12*, 77 (1978)
140. Jekat, H.: Dissertation, TU München 1975
141. Shah, Y. T., Stiegel, G. J., Sharma, M. M.: A.I.Ch.E.J. *24*, 369 (1978)
142. Alexander, B. F., Shah, Y. T.: Chem. Eng. J. *11*, 153 (1976)
143. Aoyama, Y. et al.: J. Chem. Eng. Japan *1*, 158 (1968)
144. Cova, D. R.: Ind. Eng. Chem. Process Des. Develop *13*, 292 (1974)
145. Hikita, H., Kikukawa, H.: Chem. Eng. J. *8*, 191 (1974)
146. Eissa, S., Schügerl, K.: Chem. Eng. Sci. *30*, 1251 (1975)
147. Bach, H. F., Pilhofer, Th.: Partikel Technologie Nürnberg 1977, Brauer, H., Moleurs, O. (eds.), p F49
148. Bach, H. F., Pilhofer, Th.: Ger. Chem. Eng. *1*, 270 (1978)
149. Buchholz, H. et al.: Chem. Ing. Techn. *50*, 227 (1978)
150. Buchholz, H.: Diplomarbeit TU Hannover 1976
151. Buchholz, R., Schügerl, K.: (unpublished)
152. Acharya, A., Mashelkar, R. A., Ulbrecht, J.: Chem. Eng. Sci. *32*, 863 (1977)
153. Acharya, A., Mashelkar, R. A., Ulbrecht, J.: Ind. Eng. Chem. Fundam. *17*, 231 (1978)
154. Alran, C., Angelino, H.: Chem. Eng. Sci. *27*, 593 (1972)
155. Swope, R. D.: Can. J. Chem. Eng. *49*, 169 (1971)
156. Bhavaraju, S. M., Russel, T. W. F., Blanch, H. W.: A.I.Ch.E.J. *24*, 454 (1978)
157. Schügerl, K., Lücke, J.: Dechema-Monographien. Biotechnologie Vol. 81, Nr. 1670—1692, p. 59. Weinheim: Verlag Chemie 1977
158. Scheele, G. F., Meister, B. J.: A.I.Ch.E.J. *14*, 9 (1968)
159. Akita, K., Yoshida, F.: Ind. Eng. Chem. Process Des. Develop. *12*, 76 (1973)
160. Akita, K., Yoshida, F.: Ind. Eng. Chem. Process Des. Develop. *13*, 84 (1974)
161. Baykara, Z. S., Ulbrecht, J.: Biotech. Bioeng. *20*, 287 (1978)
162. Schügerl, K.: Journées d'études gaz-liquide transfert de matière agitation et mélange, Resp. Roques, M. M., Roustan, M., p. IV.3. INSA-Toulouse September 1979
163. Voigt, J., Schügerl, K.: Chem. Eng. Sci. *34*, 1221 (1979)
164. Hecht, V.: Diplomarbeit, TU Hannover 1978
165. Voigt, J., Dissertation, Univers. Hannover 1980
166. Henzler, H. J.: Chem. Ing. Techn. *52*, 643 (1980)
167. Marquart, R.: Dissertation, Univers. Stuttgart 1977
168. Marquart, R., Blenke, H.: Verfahrenstechnik *12*, 721 (1978)
169. Marquart, R.: Verfahrenstechnik *13*, 527 (1979)
170. Wasan, D. T. et al.: A.I.Ch.E. Journal *18*, 928 (1972)

171. Schulz, N.: Chem. Ing. Techn. *51*, 693 (1979)
172. Chavan, V. V., Arumugam, M., Ulbrecht, J.: A.I.Ch.E. Journal *21*, 613 (1975)
173. Beckner, J. L., Smith, J. M.: Trans. Instn. Chem. Engrs. *44*, T224 (1966)
174. Chavan, V. V., Ulbrecht, J.: Ind. Eng. Chem. Proc. Des. Dev. *12*, 472 (1973)
175. Bumbullis, W.: Dissertation, University of Hannover 1980
176. Gisekus, H., Langer, G.: Rheol. Acta *16*, 1 (1977)
177. Smith, J. M., Riet, K. v.: Paper F4, Second Europ. Conf. on Mixing. Mons/Belgium 1978
178. Höcker, H., Langer, G., Werner, U.: Chem. Ing. Techn. *52*, 752 (1980)
179. Höcker, H., Langer, G., Werner, U.: Chem. Ing. Techn. *52*, 916 (1980)

Mechanisms and Occurrence of Microbial Oxidation of Long-Chain Alkanes

H. J. Rehm, I. Reiff
Institut für Mikrobiologie, Universität Münster
D-4400 Münster, Federal Republic of Germany

The different primary oxidation steps of long-chain aliphatic hydrocarbons by microorganisms and the pathways of degradation are reviewed.

Furthermore, the occurrence of the different degradation pathways in microorganisms is described. Some relations with regard to taxonomy of microorganisms can be observed.

In the concluding remarks lacks of our knowledge of microbial alkane oxidation are described. Especially, data are missing on the primary oxidation steps, on enzymes of the different pathways and on the regulation of alkane oxidation in most microorganisms.

1 Introduction

The degradation of aliphatic hydrocarbons by microorganisms has been previously reviewed especially by Klug and Markovetz[86] and by Einsele and Fiechter[30]. These authors provide a summary of the literature concerning mechanisms and the occurrence of microbial oxidation of long-chain alkanes up to 1970. Later reviews have focussed on other main points, e.g. product formation[2], alkane uptake[108], transformation[143] or they were not written in English (e.g. Tanaka and Fukui[160], Neryng[110], Rehm[132] or published only as short communications[33, 133].

Since in the meantime a number of papers on the various mechanisms of degradation and on the occurrence of these mechanisms in different microorganisms have appeared, a new review may be helpful. For reviews of microbial alkane oxidation focussed on other main points, see Fukui and Tanaka[41].

Alkanes occur not only in petroleum but also in different organisms, e.g. in bacteria[169] fungi, (e.g.[105, 176] and other microorganism[12], green algae[118], plants, and animals (e.g.[76, 99, 16]). With regard to the extensive paraffin formation in nature, different degradation mechanisms have been developed.

In this paper only mechanisms of microbial oxidation of long-chain *n*-alkanes will be discussed. Most research in this field has been conducted with C_{10}–C_{20} *n*-alkanes, but some microorganisms are also able to oxidize alkanes up to C_{44}[51]. It can be assumed that the degradation mechanisms for these very long-chain alkanes are identical with or similar to those of C_{10}–C_{20} *n*-alkanes. For a review of the oxidation of methane and other short-chain alkanes and related substances see Sahm[136], for cyclic hydrocarbons see Perry[123], for branched alkanes see Jones[69], Pirnik[128], for halogenated hydrocarbons see Jones and Howe[72], Bourguin and Gibson[15], for product formation see Fukui and Tanaka[41], for microbial and enzymatic denitrification of nitroalkanes see Soda et al.[151] for microbial cooxidations involving hydrocarbons see Perry[124].

The plasmid-determined alkane oxidation in *Pseudomonas* was recently reviewed by Shapiro et al.[146]; the genetics of *Saccharomycopsis lipolytica* with emphasis on the genetics of hydrocarbon utilization has been reviewed by Bassel and Ogrydziak[7].

2 Primary Oxidation Step of the Alkane Molecule

In *Micrococcus cereficans* growing on alkanes Stewart et al.[155] found an incorporation of ^{18}O into the corresponding fatty acids. These results were confirmed by Imada et al.[67]; thus, it can be concluded that in many cases the initial reaction is catalyzed by an oxygenase. The following four mechanisms for the initial step in alkane oxidation can be assumed especially for terminal oxidations.

2.1 A Mixed Functional Oxidation System without Involvement of Cytochrome P-450

This system was found by Peterson et al.[125] in cell-free extracts of *Pseudomonas oleovorans*. The authors partly isolated three components: rubredoxin, a NADH-dependent reductase and an alkane-1-hydroxylase (Fig. 1).

Peterson et al.[126] described an ω-oxidation of fatty acids, e.g. of laurate according to the system:

$$\text{NADH} + \text{H}^+ + \text{laurate} + \text{O}_2 \rightarrow \text{NAD}^+ + \text{H}_2\text{O} + \omega\text{-hydroxylaurate}$$

The hydroxylase system was confirmed by Kusunose et al.[88] who detected an increased hydroxylating activity in the extracts from *P. desmolytica* resulting from the addition of flavinadenine dinucleotid. With mutants of *P. putida* (syn. *P. oleovorans*) Nieder and Shapiro[111] found an inducible hydroxylase system for the monoterminal oxidation of long-chain *n*-alkanes coded by a plasmid.

In *P. aeruginosa* a system without cytochrome P-450 also seems to be present[130]. Cell-free extracts of this organism could oxidize different alkanes (*n*-pentane to *n*-

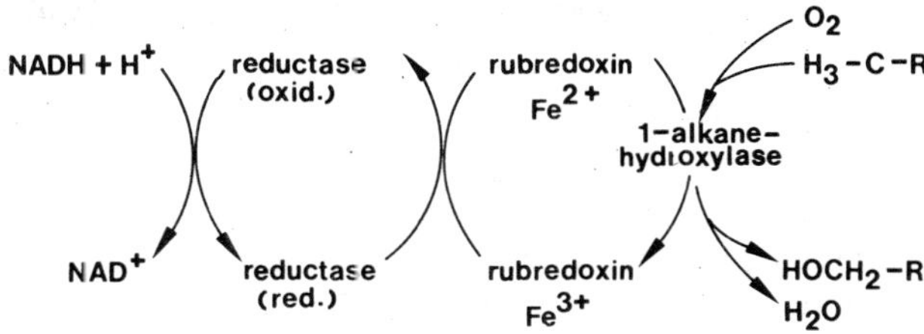

Fig. 1 Mixed functional alkane oxidation system without cytochrome P-450

decane), alkylbenzoenes and (alkyl)cycloalkanes. Straight-chain alkanes were more readily oxidized than branched ones. These results led the authors to the hypothesis that only molecules which can assume a more or less planar conformation have access to the active centre of the hydroxylase.

Hammer and Liemann[52] describe the activiation of a non-cytochrome P-450 dependent *n*-decane hydroxylase system of a marine *Pseudomonas* sp. by 1 mM Mg^{2+} in combination with 1 mM Fe^{2+} resulting in more than 300% enzyme activity. The enzyme complex was inhibited by 2—5 mM KCN and was assumed to be closely associated with the corresponding alcohol dehydrogenase.

2.2 A Mixed Functional Oxidation System with Involvement of Cytochrome P-450

This system is probably similar to the microsomal hydroxylase[46,48] from rat liver or from rabbit liver. The function of this microsomal hydroxylase from rabbit liver is supposed to be a peroxidase-like mechanism[50,113]; see also Sato and Omura[137]. Cardini and Jurtshuk[20] found spectral characteristics of cytochrome P-450 in cell-free extracts of *Corynebacterium* sp. which were able to oxidize *n*-octane. Furthermore, a specific requirement for NADH + H^+, molecular O_2 and flavoprotein was stated. The reaction was sensitive to CO.

In *Candida tropicalis*[28] and in *Cunninghamella bainieri*[32] the cytochrome P-450 hydroxylase system was dependent on NADPH + H^+ (Fig. 2).

The formation of P-450 in *Candida tropicalis* was shown to be inducible by long-chain alkanes, alkenes, secondary alcohols, and ketones[43]. Hexadecane as the sole carbon source derepressed the alkane oxidizing enzymes at least 150 times compared with the specific activity of glucose-grown cells of *C. tropicalis*. No

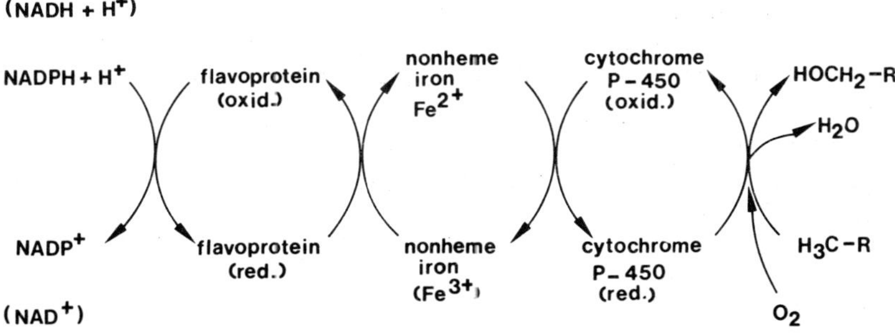

Fig. 2 Mixed functional alkane oxidation system with cytochrome P-450

effect could be found on the specific activity of alcohol and aldehyde dehydrogenase[46]. Together with P-450, a microsomal NADPH-cytochrome c-reductase is increased[42]. Tetradecane induced the formation of cytochrome P-450 up to 0.31 μmol/mg protein in the microsomal fraction[44]. The cytochrome P-450 concentration increased linearly with increasing specific hexadecane uptake rate, especially when the continuously cultivated cells grew under oxygen limitation (below 2.54 kPa). As a consequence, cytochrome P-450 seems to be the rate-limiting step of alkane uptake and alkane oxidation[46].

The enzymes of *C. tropicalis* oxidize *n*-tetradecane, drugs and especially laurate. An oxidizing enzyme system of *Cunninghamella bainieri* was partly purified; it was inhibited by CO, but not by KCN. Cytochrome P-450 was also found in microsomes of anaerobically grown *Saccharomyces cerevisiae*[97, 182]. Normally, *S. cerevisiae* does not oxidize alkanes so that the function of cytochrome P-450 in this microorganism for the alkane oxidation is still unknown. There is an influence of oxygen on the formation of P-450 in relation to the cytochrome a content of *S. cerevisiae* cells grown on glucose as the substrate (for details see Schunck et al.[141]).

Furthermore, *Endomycopsis lipolytica*[25] and *Candida guilliermondii*[168] contain cytochrome P-450, and its involvement in the activation of oxygen in the alkane oxidation was shown by Schunck et al.[140]. These authors stated the existence of a NADPH-dependent cytochrome P-450 alkane hydroxylase system in *C. guilliermondii*. Probably, a P-450-dependent system of *Pseudomonas putida*[183] cannot oxidize alkanes but only camphor.

Gmünder[46] suggests that cytochrome P-450, which is located in the microsomes, catalyzes the monoterminal hydroxylation of hexadecane in the endoplasmic reticulum.

2.3 Formation of Hydroperoxides via Free-Radical Intermediates and Reduction to Primary and Secondary Alcohols

As reviewed before[106] Leadbetter and Foster[89] suggested the formation of alkyl hydroxyperoxides arising via a free-radical mechanism in the microbial oxidation

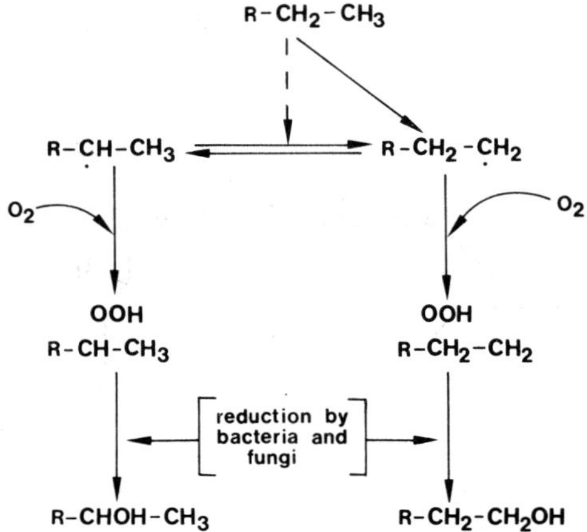

Fig. 3 Formation of hydroperoxides via free-radical intermediates[89]

of *n*-alkanes. Subsequently, a reduction to alcohols and oxidation to ketones takes place (Fig. 3).

Hydroperoxides were suggested as intermediates in the microbial *n*-alkane oxidation by Imelik[68]. The results of Stewart et al.[155] concerning the incorporation of $^{18}O_2$ into alkanes were consistent with the formation of hydroperoxides. Updegraff and Bovey[171] detected the reduction of hydroperoxides by bacteria and fungi.

Some time later Finnerty et al.[34] detected an oxidation of 1-dodecyl, 1-tetradecyl, 1-hexadecyl and 1-octadecyl hydroperoxides by *Micrococcus cereficans*, and suggested on the basis of these results that the formation of alkyl hydroperoxides — beneath the direct introduction of an OH group into the alkane molecule by a mixed function oxidase — merits consideration as a mechanism in involved in the bacterial oxidation of alkanes.

A possible role of free-radicals in the oxidation of methane by *Methylococcus capsulatus* was proposed by Hutchinson et al.[62]. For long-chain alkanes a metabolic pathway as described in their paper has not been observed.

2.4 Dehydrogenation to Alkenes by an NAD^+-Dependent Dehydrogenase

According to this mechanism an alkene will first be formed which can react to the alcohol in two ways (Fig. 4):

a) formation of an epoxide
b) addition of water

An anaerobic dehydrogenation of *n*-heptane by crude extracts from *Pseudomonas aeruginosa* in the presence of NAD^+ was described by Senez and Azoulay[145] and Chouteau et al.[23]. This mechanism of *n*-alkane dehydrogenation leading to the corresponding 1-alkene has also been found by Wagner et al.[172], especially with *Nocardia* in a culture grown on *n*-hexadecane, and by Abbott and Casida[1] using resting cells of *Nocardia salmonicolor* grown on glucose. Iizuka et al.[65,66] suggested the formation of 1-decene from *n*-decane by resting cells and cell-free extracts of *Candida rugosa* under aerobic and later[63] under anaerobic conditions. Further analytical proof of the formation of 1-decene was provided by Lebeault and Azoulay[90] and Gallo et al.[43] with *C. tropicalis*. In these experiments, no 1-decene could be isolated.

Using labeled 1-bromo-heptadecane, Jones[70] detected that in *Torulopsis gropen-*

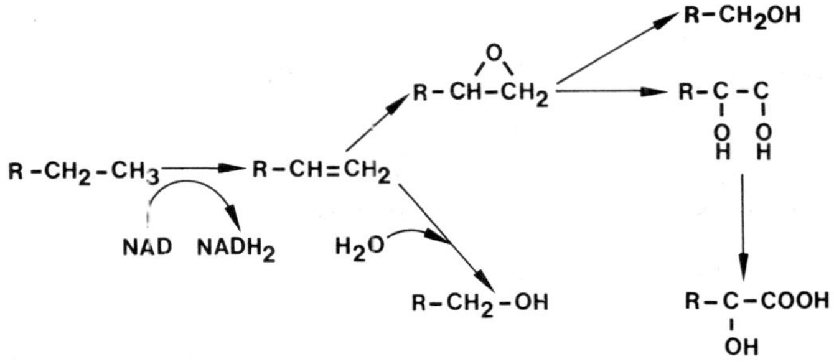

$R{-}CH_2{-}CH_3$ → $R{-}CH{=}CH_2$ (NAD → $NADH_2$)
$R{-}CH{=}CH_2$ → $R{-}CH(O)CH_2$ (epoxide) → $R{-}CH_2OH$
$R{-}CH(O)CH_2$ → $R{-}C(OH){-}C(OH)$ → $R{-}C(OH){-}COOH$
$R{-}CH{=}CH_2$ + H_2O → $R{-}CH_2{-}OH$

Fig. 4 Dehydrogenation of alkanes to alkenes and subsequent reactions

giesseri 1-alkanes are not intermediates of the terminal oxidation of alkanes by yeast.

Due to a methodical investigation with *Candida parapsilosis*, it has been stated[154] that growing cells formed 1-tetradecene from tetradecane during 0.5 to 7 h of incubation. Starved cells formed 1-tetradecene in high amounts during the first 0.5 h of incubation. 1-Tetradecene was produced in high amounts (0.17 mg/ml) by cell-free extracts in the presence of NAD^+ and ATP.

In *Pseudomonas oleovorans* 1-alkenes (C_6—C_{12}) are oxidized either in traces or in small amounts to 1,2-epoxyalkanes[3]. In the same microorganism 1-octene was oxidized to 1,2-epoxyoctane by growing cells[142].

3 Metabolic Pathways in the Oxidation of Long-Chain Alkanes

3.1 Terminal Oxidation Pathways

In the terminal oxidation pathways one or both of the terminal methyl groups are oxidized. A monoterminal oxidation pathway with the formation of the corresponding fatty acids in bacteria, yeasts, and some fungi has been reviewed up to 1971 by Klug and Markovetz[86] and Einsele and Fiechter[30] (Fig. 5):

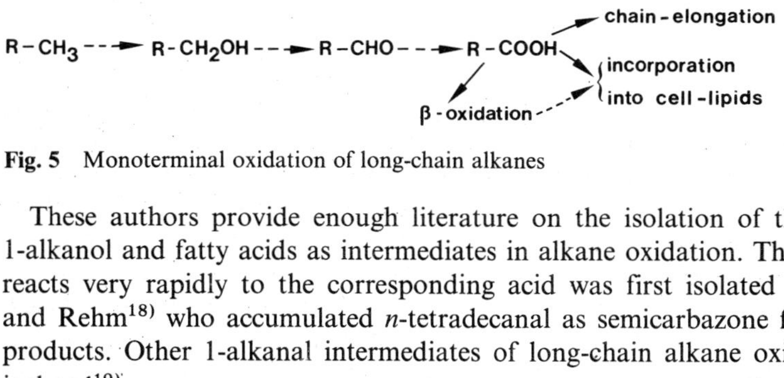

$$R-CH_3 \dashrightarrow R-CH_2OH \dashrightarrow R-CHO \dashrightarrow R-COOH$$

R-COOH → chain-elongation; → incorporation into cell-lipids; → β-oxidation ⇢ incorporation into cell-lipids

Fig. 5 Monoterminal oxidation of long-chain alkanes

These authors provide enough literature on the isolation of the corresponding 1-alkanol and fatty acids as intermediates in alkane oxidation. The aldehyde which reacts very rapidly to the corresponding acid was first isolated by Büning-Pfaue and Rehm[18] who accumulated *n*-tetradecanal as semicarbazone from the reaction products. Other 1-alkanal intermediates of long-chain alkane oxidation were also isolated[19].

Pseudomonas aeruginosa contains a soluble NADP-linked dehydrogenase displaying activity toward primary alcohols. Further it possesses two NAD(P)-independent primary alcohol dehydrogenases which can be induced by primary alcohols, α-, ω-diols or *n*-alkanes[96].

In a marine *Pseudomonas* sp., an alcohol dehydrogenase seems to be closely associated with the oxygenase system by a kind of multienzyme complex[52]. Prior to these results Tassin et al.[162] purified a membrane-bound alcohol dehydrogenase from *P. aeruginosa* which does not use pyridine nucleotides (NAD or NADP) as coenzymes. No spectral evidence for the involvement of a flavin as a prosthetic group has been found by the authors. The dehydrogenase shows a high affinity for long-chain primary alcohols (K_m for 1-tetradecanol = 4,5 μM).

These results are in agreement with the findings of Benson and Shapiro[10] who found in *P. putida* a plasmid-coded octanol dehydrogenase with an activity independent of NAD.

Table 1. Substrate utilization tests of alkane-negative mutants of *Saccharomycopsis lipolytica*

Phenotypic designation	Growth on carbon source:						No. of mutants
	Hydro-carbon	Alcohol	Aldehyde	Fatty acid	Acetate	Glucose	
A	—	+	+	+	+	+	6
B	—	—	+	+	+	+	0
C	—	—	—	+	+	+	0
D	—	—	—	—	+	+	14
E	—	—	—	—	—	+	8
						Total	28

+ = growth; — = no growth

In cell-free extracts of *Candida tropicalis* an NAD^+-dependent alcohol dehydrogenase was detected revealing a slight substrate specificity in the range from C_{11}–C_{16} alcohols[91]. The same authors decribed an aldehyde dehydrogenase which was also NAD^+-dependent.

Pseudomonas aeruginosa contains a $NAD(P)^+$-dependent aldehyde dehydrogenase involved in the oxidation of alkanes[53,11]. One of two soluble aldehyde dehydrogenases purified from *P. aeruginosa* by Guerillot and Vandescasteele[49] was NAD-dependent with a main activity for aldehydes of short- and middle-chain length; the second was NADP-dependent with a good activity for aldehydes of higher chain length. When the bacteria were grown on *n*-paraffin, a new membrane-bound NAD-dependent aldehyde dehydrogenase activity was produced.

Furthermore, *Acinetobacter calcoaceticus* possesses a $NADP^+$-dependent aldehyde dehydrogenase for alkane degradation[152].

The influence of carbon and nitrogen sources on the levels of several $NADP^+$- and NAD^+-linked dehydrogenases in relation to alkane oxidation in *Candida tropicalis* and *C. lipolytica* was investigated by Hirai et al.[54].

Recently, Bassel and Ogrydziak[7] succeeded in obtaining mutants of *Saccharomycopsis lipolytica* which were unable to utilize *n*-decane as a carbon source. Table 1 summarizes the substrate utilization studies of these authors designed to identify tentatively the genetic blocks associated with these mutants.

These results confirm the biochemical findings on the monoterminal degradation pathways via alkanols, alkanals and alkanoic acids and perhaps those of the C_2-units (see Fig. 5).

The corresponding fatty acids formed from the alkanes can be oxidized by β-oxidation so that C_2-units are prepared in high amounts. As far as is known, all microorganisms which are growing on aliphatic hydrocarbons as the sole carbon source, incorporate into their lipids a large number of the fatty acids formed as intermediates. The fatty acids are partially elongated by C_2-units, or C_2-units are split via β-oxidation.

As a rule, microorganisms cultivated on odd-carbon alkanes contain high amounts of the corresponding odd-carbon fatty acids in their lipids, whereas

Table 2. Cellular fatty acid composition from bacteria, degrading long-chain n-alkanes (C_{10}—C_{18})

Organism	Substrate	Relative proportion of fatty acids (% w/w)																	Ref.	Remarks
		10:0	11:0	12:0	13:0	14:0	14:1 +14:2	15:0	15:1	16:0	16:1	17:0	17:1	18:0	18:1	18:2	18:3	19:0 +19:1 +20:0		
Pseudomonas aeruginosa	C_{16}	triglyceride fraction				4.4	0.2	0.2	—	22.3	6.7	tr	—	3.4	19.7	35.2		2.2	[135]	4 d, 31 °C, reactor
—	C_{16}	polar fraction				14.3	6.8	1.9	3.9	62.6	12.6	0.6	6.3	51.6	51.9	—		7.4	[135]	4 d, 31 °C, reactor; 10.9% $C_{22:1}$ fatty acid
—	C_{13}				tr	2.56		2.37		32.62	1.52	2.34		44.04	7.37			tr	[29]	4 d, 37 °C, static conditions
Pseudomonas sp.	C_{10}	18.3	—	1.5	—	2.0		—	—	23.4	29.8	—	—	—	25.0				[80]	
—	C_{11}	0.7	2.6	2.4	1.6	1.4		6.6	3.6	15.3	15.3	16.5	13.8	—	18.4				[80]	
—	C_{14}	1.4	—	4.8	—	8.3		—	—	37.7	19.8	—	—	—	28.0				[80]	
—	C_{15}	1.0	1.8	1.6	1.1	0.4		15.8	9.1	5.1	18.3	12.0	15.4	—	18.1				[80]	
—	C_{17}	2.8	3.3	2.4	1.3	0.9		6.2	3.1	7.8	12.2	17.6	21.6	—	20.8				[80]	
—	C_{18}	2.0	—	5.5	—	7.0		—	—	35.6	15.9	—	—	7.5	26.5				[80]	
Micrococcus cerificans	C_{10}	3.78	—	7.55	—	4.36		—	—	39.50	14.52	—	—	6.68	23.61				[101]	
—	C_{11}	0.74	0.93	2.61	tr	1.68		1.86	tr	18.65	14.18	11.19	15.64	6.34	26.12				[101]	
—	C_{12}	0.46	—	6.65	—	0.84		—	—	23.26	14.30	—	—	6.63	47.52				[101]	
—	C_{13}	1.59	1.68	7.23	7.70	2.16		1.20	tr	17.32	11.55	7.70	7.70	6.74	27.43				[101]	
—	C_{14}	1.70	—	7.25	—	28.43		—	—	10.88	16.93	—	—	11.60	23.21				[101]	
—	C_{15}	1.12	1.21	1.30	1.12	0.24		32.60	36.22	tr	4.53	3.32	3.77	11.32	3.26				[101]	
—	C_{16}	0.22	—	7.14	—	2.04		—	—	29.58	43.34	—	—	7.14	2.04				[101]	
—	C_{17}	tr	2.00	1.40	1.76	0.36		6.58	3.59	2.70	1.60	20.95	49.89	1.60	7.48				[101]	
—	C_{18}	tr	—	8.92	—	1.93		—	—	16.34	20.80	—	—	10.40	41.60				[101]	
—	C_{14}	12.05	—	15.03	—	23.51	6.80	—	—	4.68	14.28	—	—	—	18.19				[102]	
—	C_{15}	1.21	2.83	5.45	6.15	0.78	—	44.67	26.63	1.53	1.68	0.68	2.78	1.14	2.6				[102]	
—	C_{16}	2.96	—	3.65	—	1.75	0.28	—	36.05	52.55	—	—	—	0.86	0.44				[102]	
—	C_{17}	0.88	2.06	1.97	1.03	8.10	0.30	3.25	2.89	2.53	16.84	52.20	—	4.98	0.56				[102]	
Acinetobacter sp.	C_{15}					4.12		3.48	11.53	11.87	16.20	17.45	17.11	—	18.45				[103]	[101, 102]
—	C_{16}					4.00		—		79.00	17.00	—	—	—	—				[103]	[101, 102]
—	C_{17}					—		14.92		1.98	2.23	49.97	26.74	—	4.16				[103]	[101, 102]

Bacillus stearo-thermophilus	C_{13}		—	1.7	3.2		59.8		11.8	—	23.0		tr	—			[75]	5 d, shaking flask culture with cooxidation substrate
—	C_{14}		—	2.4	2.5		61.7		8.0	—	25.4		tr	—			[75]	5 d, shaking flask culture with cooxidation substrate
Mycobacterium sp. 7 EIC	C_{13}	1.5	1.2	28.7	1.4		2.8	0.2	22.5	4.5	1.5	11.8	1.6	15.5			[27]	see [26]
Corynebacterium sp.	C_{17}	—	0.8	1.2	0.4		23.7	1.0	1.2	—	43.5	27.0	—	1.2			[27]	see [26]
Arthrobacter simplex	C_{13}			5.1	1.1	—	5.8	—	20.4	7.7	2.9	17.4	2.0	37.7	—	—	[181]	30 °C, shaking flasks, 48 h
-	C_{14}			0.5	6.9		1.9		19.9	11.2	0.4	2.9	1.7	54.5			[181]	30 °C, shaking flasks, 48 h
—	C_{15}			1.1	1.5	—	33.4	—	10.0	2.3	2.3	29.1	1.4	18.8	—	—	[181]	30 °C, shaking flasks, 48 h
—	C_{16}			tr	1.5	—	2.5	—	38.9	53.0	tr	1.1	tr	3.0	—	—	[181]	30 °C, shaking flasks, 48 h
—	C_{17}			tr	tr	—	9.7	—	1.8	0.2	22.6	46.4	tr	19.2	—	—	[181]	30 °C, shaking flasks, 48 h
—	C_{18}			tr	1.8	—	0.9	—	30.3	6.8	0.9	tr	5.0	54.3	—	—	[181]	30 °C, shaking flasks, 48 h
—	C_{19}			tr	tr	—	9.6	—	2.0	tr	21.2	3.4	0.7	13.1	—	39.9	[181]	30 °C, shaking flasks, 48 h
Brevibacterium sp. JOB5	C_{13}	2.9	0.2	19.6	3.5		5.2	0.5	24.1	6.3	4.1	2.7	3.0	18.0			[27]	see [26]
—	C_{17}	—	0.6	0.9	0.4		35.2	2.7	3.4	1.4	19.7	30.1	—	3.3			[27]	see [26]
Mycobacterium sp. (strain OFS)	C_{13}	1.4	0.9	21.4	1.8		6.6	1.7	15.3	14.3	3.6	9.8		9.6			[26]	shaking flasks, 72 h
—	C_{14}	1.0	2.9	0.9	39.0		—	—	7.2	4.0	—	-		19.4			[26]	shaking flasks, 72 h
—	C_{15}	0.8	0.5	6.9	0.9		74.0	8.9	—	—	0.8	5.7		—			[26]	shaking flasks, 72 h
—	C_{16}	0.5	1.0	0.2	5.9		1.5	—	50.0	28.2	—	—		5.0			[26]	shaking flasks, 72 h
—	C_{17}	2.5	0.3	2.3	0.5		18.1	2.7	3.8	2.5	19.4	27.2		19.5		0.7	[26]	shaking flasks, 72 h
—	C_{11}	2.1	0.4	1.1	3.3		2.9	0.4	30.6	13.8	1.2	2.0	0.8	20.7			[27]	see [26]
—	C_{12}	5.1	4.1	tr	7.0		1.4	0.1	27.5	18.1	0.5	0.6	—	17.5			[27]	see [26]
—	C_{18}	—	—	—	2.9		1.3	0.4	28.3	15.5	0.4	0.5	1.9	26.3			[27]	see [26]
—	C_{20}	—	—	—	3.0		tr	tr	22.0	18.0	—	tr	tr	30.0			[27]	see [26]

Table 2 (continued)

Organism	Sub-strate	Relative proportion of fatty acids (% w/w)																	Ref.	Remarks
		10:0	11:0	12:0	13:0	14:0	14:1 +14:2	15:0	15:1	16:0	16:1	17:0	17:1	18:0	18:1	18:2	18:3	19:0 +19:1 +20:0		
—	C_{24}		—	0.1	0.3	3.0		10.4	2.5	21.8	13.8	—	—	—	21.6			0.3	27)	see 26)
—	C_{28}		0.4	0.8	3.1	6.0		21.1	3.7	15.8	10.6	3.0	8.5	—	13.8				27)	see 26)
Mycobacterium vaccae JOB5	C_{14}			2.0	tr	54.5		tr	—	5.7	3.2	tr	—	4.2	19.1			—	81)	3–5 d, shaking flask culture, 26 °C
—	C_{15}			3.6	4.7	1.4		51.0	14.4	4.5	2.4	tr	tr	1.8	9.1			1.2	81)	3–5 d, shaking flask culture, 26 °C
—	C_{16}			1.9	tr	0.9		2.3	—	45.5	7.4	1.5	5.0	tr	13.1			—	81)	3–5 d, shaking flask culture, 26 °C
—	C_{17}			5.3	11.9	4.4		17.9	—	3.3	2.8	22.6	19.1	2.9	5.2			tr	81)	3–5 d, shaking flask culture, 26 °C
—	C_{18}			5.3	1.7	8.0		1.8	—	26.5	17.5	tr	—	5.3	15.2			—	81)	3–5 d, shaking flask culture, 26 °C
Nocardia salmonicolor PSU-N-18	C_{16}	1.7		—		5.5				46.0	14.5			1.3	14.5				1)	resting cells (glucose grown); 1.6% $C_{8:0}$ fatty acid
Cellular fatty acid composition from yeasts, degrading long-chain n-alkanes (C_{10}—C_{18})																				
Candida lipolytica	C_{14}		tr	tr	tr	2.98	—	4.48	—	13.45	7.45	tr	tr	44.80	10.10	14.90			85)	
—	C_{15}		—	tr	tr	0.8	—	20.80	1.73	2.60	4.30	0.8	25.40	12.30	8.60	19.73			85)	
—	C_{16}		—	tr	5.0	5.0	—	5.00	—	30.00	20.00	tr	tr	10.00	25.00	—			85)	
—	C_{17}		1.30	—	1.96	tr	—	3.90	3.90	5.20	2.60	14.30	36.60	19.60	3.90	5.20			85)	
—	C_{18}		—	2.6	4.0	tr	—	tr	—	21.40	6.70	tr	5.30	6.70	16.70	34.80			85)	
—	C_{12}	tr		11.7	—	5.0	0.6	tr	—	10.4	17.4	—	0.9	3.1	43.1	7.9			107)	30 °C, shaking flasks, 44.5 h
—	C_{14}	—		tr	tr	19.5	2.6	tr	tr	8.5	24.3	tr	1.0	0.8	29.6	13.5			107)	30 °C, shaking flasks, 44 h
—	C_{16}	—		—	—	0.1	—	0.6	—	30.2	27.7	—	1.0	2.1	30.0	8.3			107)	30 °C, shaking flasks, 44 h
—	C_{18}	1.0		—	—	0.8	—	—	—	16.8	12.3	—	3.4	7.9	41.2	16.7			107)	30 °C, shaking flasks, 44.5 h

—	C_{11}		1.1		0.3	tr		2.5	—	13.4	14.6	tr	5.7	2.4	47.9	12.1			[107]	30 °C, shaking flasks, 44.5 h
—	C_{13}		tr		8.5	0.3		16.6	1.8	2.0	6.1	1.5	31.6	0.6	20.6	10.5			[107]	30 °C, shaking flasks, 44 h
—	C_{15}		—		—	0.3		20.0	2.9	0.3	1.4	1.5	56.5	—	14.3	7.2 (18:1 + 17:2)			[107]	30 °C, shaking flasks, 44 h
—	C_{17}		—		0.2	0.4		2.1	0.8	0.7	0.6	4.6	77.9	—		0.5			[107]	30 °C, shaking flasks, 44.5 h; 12.4% $C_{17:2}$ fatty acid
—	C_{16}			1.68	2.93	1.96	2.97	9.99	3.66	26.70	9.03	2.88		5.55	17.00	6.28			[73]	30 °C, shaking flask culture, 72 h
C. parapsilosis	C_{15}							32.8	3.0	2.6	1.6	5.5	29.9	tr	8.4	7.8	1.0		[153],	2 d, shaking flasks; 7.1% $C_{17:2}$ fatty acid
—	C_{15}							32.3	2.1	2.0	1.2	3.8	30.7	0.4	6.1	10.6	0.4		[153]	3 d, reactor; 10.1% $C_{17:2}$ fatty acid
C. tropicalis	C_{12}	tr	—	24.4	tr	1.9	tr	tr	—	10.4		tr	2.0	12.5	33.5	8.4		tr	[61]	reactor
—	C_{13}	tr	0.5	0.4	28.4	0.8	tr	7.2	0.4	7.3		1.3	5.3	14.1	24.7	4.5		tr	[61]	reactor
—	C_{14}	0.8	—	2.4	tr	33.3	5.1	tr	—	9.6		tr	5.5	5.3	19.1	7.0			[61]	reactor
—	C_{15}	1.0	1.2	0.9	0.5	0.5	—	22.8	4.5	2.6		2.0	15.6	18.6	16.9	4.6		5.4	[61]	reactor
—	C_{16}	tr	—	1.0	tr	0.4	tr	0.1	tr	28.9		tr	5.5	8.3	21.2	7.6			[61]	reactor
—	C_{17}	tr	1.0	1.3	1.2	2.5	1.4	1.9	0.5	8.1		17.1	25.3	24.8	3.4	0.8		5.5	[61]	reactor
—	C_{12}			1.4		3.8	0.3	0.4		17.2	12.5	0.5	2.4	2.9	40.5	16.2	1.8		[107]	30 °C, shaking flasks, 44 h
—	C_{14}			tr		10.6	0.6	tr		15.1	24.9	tr	2.1	0.6	28.1	14.0	3.9		[107]	30 °C, shaking flasks, 34 h
—	C_{16}			—		0.8	—	—		33.8	33.7	—	1.1	0.6	23.0	6.0	1.0		[107]	30 °C, shaking flasks, 34 h
—	C_{18}			—		—	—	—		7.2	1.3	—	0.3	5.6	67.0	15.8	2.7		[107]	30 °C, shaking flasks, 46 h
—	C_{11}				—	—		1.9	—	5.8	2.1	8.6	30.0	1.3	27.0	20.3	2.9		[107]	30 °C, shaking flasks, 44.5 h
—	C_{13}				2.7	—		11.5	0.4	2.0	2.3	6.0	49.2	tr	17.9	7.1	0.8		[107]	30 °C, shaking flasks, 47 h
—	C_{15}				—	0.2		18.3	0.9	0.8	1.4	3.4	50.9	—	18.3	4.5	1.2		[107]	30 °C, shaking flasks, 35 h
—	C_{17}				—	tr		0.5	—	0.5	0.3	17.8	65.6	—	12.5	1.8	0.8		[107]	30 °C, shaking flasks, 35 h

Table 2 (continued)

Organism	Sub-strate	Relative proportion of fatty acids (% w/w) 10:0	11:0	12:0	13:0	14:0	14:1 +14:2	15:0	15:1	16:0	16:1	17:0	17:1	18:0	18:1	18:2	18:3	19:0 +19:1 +20:0	Ref.	Remarks
C. petrophillum	C_{13}							9.3	0.6	6.1	6.9		28.8		31.9	16.3			[109]	3 d, shaking flask culture, 30 °C
—	C_{16}									19.2	23.1		1.4		50.0	7.7			[109]	3 d, shaking flask culture, 30 °C
Mycotorula japonica	C_{10}	1.9	2.8	2.2	2.0	5.3		2.7	2.1	27.1	2.9	1.3	tr	13.0	12.3	7.8			[178]	30 °C, shaking flask culture, 48 h; neutral lipids
—	C_{11}	1.5	2.8	2.6	3.6	3.9		2.6	1.3	19.0	2.4	14.7	7.0	10.4	5.7	6.0			[178]	30 °C, shaking flask culture, 48 h; neutral lipids
—	C_{12}	1.7	2.3	6.8	2.7	4.6		2.6	2.3	22.7	2.9	5.6	3.7	6.7	8.6	6.0			[178]	30 °C, shaking flask culture, 48 h; neutral lipids
—	C_{13}	0.9	0.9	2.2	9.9	5.2		7.8	tr	16.3	3.6	9.2	6.0	6.7	7.0	8.3			[178]	30 °C, shaking flask culture, 48 h; neutral lipids
—	C_{14}	0.7	1.3	1.2	1.6	15.8		3.4	tr	27.4	10.1	3.5	1.9	4.3	4.4	13.4			[178]	30 °C, shaking flask culture, 48 h; neutral lipids
—	C_{15}	0.7	0.9	1.6	2.6	5.1		28.7	5.1	10.5	0.9	9.0	9.4	7.3	2.6	3.1			[178]	30 °C, shaking flask culture, 48 h; neutral lipids
—	C_{16}	0.6	1.3	1.1	tr	0.9		1.0	tr	37.9	6.9	1.3	0.6	3.3	8.1	19.2			[178]	30 °C, shaking flask culture, 48 h; neutral lipids
Cellular fatty acid composition from molds, degrading long-chain n-alkanes (C_{10}—C_{18})																				
Absidia spinosa	C_{12}		—	8.6	—	3.9		0.4		25.3	4.2	0.3	0.1	1.7	18.8	11.9	24.6	0.1	[58]	8 d, surface
—	C_{13}		0.2	0.1	10.1	0.6		23.2		6.3	1.1	8.2	14.1	0.6	4.6	4.9	12.3	0.2	[58]	8 d, surface; 7.8% $C_{17:2}$ 5.8% $C_{17:3}$ fatty acids
Cunninghamella echinulata	C_{12}	0.2	0.2	20.1	0.3	2.2		2.5		33.1	2.2	—	—	2.3	21.0	2.6	0.1	tr	[122]	3 d, surface
—	C_{13}	0.4	2.7	0.1	32.3	0.5		6.3		9.4	2.5	3.4	3.0	1.9	11.3	6.1	5.3	tr	[122]	3 d, surface
—	C_{14}	9.4	—	1.4	0.2	26.5		0.5		21.0	3.1	tr	—	2.7	22.3	4.5	3.8	tr	[122]	3 d, surface
—	C_{16}	0.5	—	1.2	—	3.1		0.2		56.2	2.9	—	tr	2.1	21.4	6.7	3.4	tr	[122]	3 d, surface

—	C_{12}		—	10.0	—	2.9		0.3		24.4	2.8	0.2	0.2	3.9	27.0	16.2	11.8	0.2	58)	8 d, surface
—	C_{13}		0.4	0.1	24.0	0.4		17.0		5.6	0.8	8.5	16.6	0.9	9.0	8.8	1.4	2.8	58)	8 d, surface; 5.7% $C_{17:2}$ 0.8% $C_{17:3}$ fatty acids
C. elegans	C_{13}				13.1	tr		6.2		17.3		3.8	1.3	2.9	9.7	26.1	12.1	tr	21)	25 °C, 10 d, stationary culture; further $C_{20:2}$ to $C_{22:2}$ and methyl branched fatty acids
—	C_{14}				tr	23.2		tr		21.6		tr	—	3.5	3.8	22.1	10.2	1.7	21)	25 °C, 10 d, stationary culture; further $C_{20:2}$ to $C_{22:2}$ and methyl branched fatty acids
—	C_{15}				1.2	1.1		23.2		4.0		21.8	6.0	1.3	12.3	22.8	1.5	tr	21)	25 °C, 10 d, stationary culture; further $C_{20:2}$ to $C_{22:2}$ and methyl branched fatty acids
—	C_{11}			—	—	4.54	—	3.06	—	76.97	—	2.39	—	13.03	—	—	—	—	45)	10 d, shaking flask culture, 27 °C
—	C_{12}			2.65	—	2.05	0.93	—	—	25.53	5.16	—	—	4.89	43.52	8.66	6.55	—	45)	10 d, shaking flask culture, 27 °C
—	C_{13}			—	9.57	1.23	—	11.07	tr	9.39	1.32	11.40	11.78	6.59	21.60	9.04	6.93	tr	45)	10 d, shaking flask culture, 27 °C
—	C_{14}			9.26	—	15.27	1.88	—	—	23.22	6.01	—	—	3.43	31.28	7.53	2.05	—	45)	10 d, shaking flask culture, 27 °C
—	C_{15}			1.82	—	0.29	—	26.57	tr	6.75	0.62	8.84	13.34	7.62	14.78	8.63	10.94	—	45)	10 d, shaking flask culture, 27 °C
—	C_{16}			—	—	2.19	—	—	—	51.74	6.25	—	—	2.78	15.23	10.67	11.13	—	45)	10 d, shaking flask culture, 27 °C
—	C_{17}			—	—	—	13.02	—	—	5.09	—	54.64	25.06	tr	8.66	—	—	—	45)	10 d, shaking flask culture, 27 °C
*Mortierella isabellina**	C_{12}		—	8.6	0.1	3.0		1.0		23.8	6.1	0.6	0.7	1.1	24.4	12.9	16.5	tr	58)	8 d, surface
—	C_{13}		tr	tr	14.1	0.9		12.8		11.1	3.1	4.3	11.9	0.9	12.5	8.8	11.8	0.9	58)	8 d, surface; 5.2% $C_{17:2}$ 1.5% $C_{17:3}$ fatty acids
*Aspergillus flavus**	C_{12}	2.1	—	0.5	—	0.6		1.1		17.5	1.1	0.4	1.6	1.1	6.4	38.1	2.3	0.3	122)	6 d, surface
—	C_{13}	2.1	tr	tr	0.3	tr		0.5		21.0	0.7	0.2	1.9	0.9	13.3	41.4	0.3	0.3	122)	6 d, surface
—	C_{14}	1.1	tr	0.7	tr	3.2		tr		11.7	0.8	0.2	1.7	1.0	5.1	45.0	13.1	1.2	122)	6 d, surface

Table 2 (continued)

Organism	Substrate	10:0	11:0	12:0	13:0	14:0	14:1 + 14:2	15:0	15:1	16:0	16:1	17:0	17:1	18:0	18:1	18:2	18:3	19:0 + 19:1 + 20:0	Ref.	Remarks
		Relative proportion of fatty acids (% w/w)																		
A. versicolor	C_{10}	tr	—	6.8	3.4	8.5	—	3.4	—	42.4	3.4	5.1	3.4	—	20.3	—			[95]	10 d, stationary flasks, 30 °C
—	C_{10}	tr	tr	tr	3.0	4.0	5.5	1.0	5.0	12.5	55.0	10.0	—	—	3.0	—			[95]	17 d, stationary flasks, 30 °C
—	C_{11}	4.4	tr	—	—	8.8	—	—	—	29.4	17.6	8.8	tr	4.4	13.2	10.3			[95]	8 d, stationary flasks, 30 °C
—	C_{11}	—	6.0	0.5	2.7	0.5	3.3	0.8	3.8	10.4	32.9	8.2	1.1	—	3.3	24.7			[95]	17 d, stationary flasks, 30 °C; 1.1% $C_{8:0}$ fatty acid
—	C_{16}	tr	2.9	2.9	—	2.9	—	—	—	47.1	23.5	—	—	tr	8.8	5.9			[40]	7 d, stationary flasks, 30 °C
—	C_{16}	70.1	—	—	—	—	—	—	—	10.4	2.6	—	—	—	3.9	10.4			[95]	17 d, stationary flasks, 30 °C
—	C_{17}	tr	tr	tr	—	—	—	—	—	71.4	—	—	—	—	14.3	—			[95]	8 d, stationary flasks, 30 °C
—	C_{17}	tr	3.4	tr	—	—	—	—	—	69.0	—	—	—	—	13.8	6.9			[95]	16 d, stationary flasks, 30 °C
Penicillium lilacinum	C_{10}	5.5	3.4	tr	2.1	2.1	—	2.8	—	48.3	5.5	—	—	—	27.6	—			[95]	8 d, stationary flasks, 30 °C
—	C_{10}	3.0	2.1	tr	1.4	1.5	—	1.1	2.5	20.7	22.1	4.1	0.7	0.8	10.3	27.3			[95]	17 d, stationary flasks, 30 °C; 1.9% $C_{8:0}$ fatty acid
—	C_{11}	—	6.7	1.5	—	3.7	—	3.0	5.9	40.7	18.5	—	—	—	11.9	—			[95]	8 d, stationary flasks, 30 °C; 3.7% $C_{8:0}$ fatty acid
—	C_{11}	—	5.9	tr	—	tr	—	tr	tr	23.5	22.1	8.8	—	—	29.4	—			[95]	17 d, stationary flasks, 30 °C; 5.9% $C_{8:0}$ fatty acid
—	C_{16}	88.7	—	—	0.5	0.7	—	0.7	—	6.1	1.0	—	—	—	—	—			[95]	5 d, stationary flasks, 30 °C

—	C_{16}	49.8	—	—	0.4	0.4	—	0.5	—	21.4	3.6	2.1	—	—	3.6	17.8			[95]	17 d, stationary flasks, 30 °C
—	C_{17}	tr	66.7	tr	1.7	2.2	—	2.2	—	4.4	tr	3.3	3.9	13.3	tr	1.1			[95]	5 d, stationary flasks, 30 °C
—	C_{17}	tr	90.5	—	—	—	—	—	—	tr	—	2.1	2.1	3.1	tr	tr			[95]	17 d, stationary flasks, 30 °C
P. zonatum	C_{13}				16.2	tr		7.4		15.7		4.2	1.3	2.5	8.5	22.1	15.2	tr	[21]	25 °C, 10 d, stationary culture; further $C_{20:2}$ to $C_{22:2}$ and methyl branched fatty acids
—	C_{14}				tr	24.5		tr		21.9		tr	—	4.2	3.2	24.1	8.4	2.1	[21]	25 °C, 10 d, stationary culture; further $C_{20:2}$ to $C_{22:2}$ and methyl branched fatty acids
—	C_{15}				2.2	tr		23.4		3.0		23.2	5.8	1.1	11.7	23.0	2.3	tr	[21]	25 °C, 10 d, stationary culture; further $C_{20:2}$ to $C_{22:2}$ and methyl branched fatty acids
Cladosporium resinae	C_{10}		tr			tr		tr		16.72	0.80	1.12	0.68	5.84	43.82	29.61	0.96	tr	[24]	25 °C reactor
	C_{11}		tr	tr	tr	tr		0.54		31.17	2.43	0.68	1.93	3.63	34.08	22.93	1.52		[24]	25 °C reactor
—	C_{12}		tr	tr	—	tr	tr	tr		24.52	1.33	1.05	0.89	3.88	51.94	14.13	1.18		[24]	25 °C reactor
—	C_{13}				1.21	0.90		2.17		21.63	0.93	1.60	3.42	3.57	37.52	24.78	1.40		[24]	25 °C reactor
—	C_{14}				—	6.45		tr	tr	15.20	2.60	tr	tr	1.95	34.15	36.87	0.99		[24]	25 °C reactor
Cladosporium sp.	C_{10}	86.6	—	1.4	—	—	—	—	—	4.6	1.8	—	—	—	1.4	3.7	—		[94]	13 d, stationary flasks, 30 °C
—	C_{11}	0.7	48.3	—	—	1.1		0.4		17.8	8.9	6.7	—	—	4.5	9.3	—		[94]	11 d, stationary flasks, 30 °C; 1.5% $C_{9:0}$ fatty acid
—	C_{16}	88.9	—	tr	—	0.4		—		7.4	1.9	—	—	—	0.6	0.8	—		[94]	16 d, stationary flasks, 30 °C
—	C_{17}	—	83.0	—	—	—		0.4		1.4	—	1.7	9.0	—	—	2.1	—		[94]	16 d, stationary flasks, 30 °C
—	C_{10}	32.4	1.4	tr	—	—		—		29.7	2.1	tr	—	tr	10.8	22.7	0.5		[94]	18 d, stationary flasks, 30 °C
—	C_{11}	—	4.6	1.7	—	tr		0.4		42.3	—	—	—	tr	9.2	39.4	2.1		[94]	17 d, stationary flasks, 30 °C

Table 2 (continued)

Organism	Substrate	Relative proportion of fatty acids (% w/w)																	Ref.	Remarks
		10:0	11:0	12:0	13:0	14:0	14:1 +14:2	15:0	15:1	16:0	16:1	17:0	17:1	18:0	18:1	18:2	18:3	19:0 +19:1 +20:0		
—	C_{16}	23.5	—	tr	—	tr		—		36.3	—	—	—	tr	19.0	20.8	tr		[94]	22 d, stationary flasks, 30 °C
—	C_{17}	—	96.8	—	—	—		tr		0.2	—	—	2.4	—	—	tr	—		[94]	24 d, stationary flasks, 30 °C
Hormodendrum hordei	C_{10}	15.5	—	4.1	—	—		—		33.1	5.0	3.1	—	—	5.8	33.1	—		[94]	12 d, stationary flasks, 30 °C
—	C_{11}	0.6	16.2	13.0	—	tr		tr		26.1	14.9	tr	—	—	2.6	20.3	1.7		[94]	15 d, stationary flasks, 30 °C; 4.3% $C_{9:0}$ fatty acid
—	C_{16}	89.8	2.8	—	—	—		—		3.4	2.8	—	—	—	0.6	0.3	—		[94]	16 d, stationary flasks, 30 °C
—	C_{17}	6.8	73.3	tr	tr	0.3		0.3		6.3	4.7	0.3	4.2	—	—	1.3	—		[94]	15 d, stationary flasks, 30 °C
—	C_{10}	16.6	—	0.7	—	—		—		41.3	—	—	—	tr	7.4	33.6	—		[94]	18 d, stationary flasks, 30 °C
—	C_{11}	4.5	6.3	2.1	—	tr		—		21.1	11.4	0.3	—	tr	11.0	40.5	1.7		[94]	20 d, stationary flasks, 30 °C; 0.6% $C_{9:0}$ fatty acid
—	C_{16}	17.4	tr	5.4	—	0.4		—		34.7	10.4	—	—	—	6.9	12.5	—		[94]	23 d, stationary flasks, 30 °C; 6.9% $C_{9:0}$ fatty acid
—	C_{17}	1.3	92.9	—	—	—		tr		1.9	1.3	tr	1.3	—	—	tr	—		[94]	22 d, stationary flasks, 30 °C
Cellular fatty acid composition from green algae, degrading long-chain n-alkanes (C_{10}—C_{18})																				
Chlorella vulgaris	C_{12}									35.6					11.5	41.38		22.99	[139]	3 d, bubble column, cooxidation (photosynthesis)

—	C_{13}	1.7	2.9		32.4		17.78	23.2	21.97	[139]	3 d, bubble column, cooxidation (photo-synthesis)
—	C_{15}		5.6	8.9	44.2	16.3	16.7	7.0	1.4	[139]	3 d, bubble column, cooxidation (photo-synthesis)
—	C_{16}				58.9		10.5	15.3	15.3	[139]	3 d, bubble column, cooxidation (photo-synthesis)

Table 3. Extracellular fatty acids from different microorganisms

Organism	Substrate	Relative proportion of fatty acids (% w/w)														Ref.	Remarks	Calculation
		11:0	12:0	13:0	14:0	15:0	15:1	16:0	16:1	17:0	17:1	18:0	18:1	18:2	19:0			
Pseudomonas aeruginosa	C_{16}				tr	0.6	tr	33.5	tr	6.6		43.0	8.6	1.1		[135]	4 d, 31 °C, reactor	g per 100 g fatty acids
Acinetobacter sp.	C_{16}				tr			69.0	31							[103]	as described in 101	g per 100 g fatty acids
Bacillus macerans	C_{13}				2.33	59.70		11.10								[74]	shaking flask culture, 30 °C with cooxidation substrate	g per 100 g fatty acids
—	C_{14}				2.41	66.75		7.25								[74]	shaking flask culture, 30 °C with cooxidation substrate	g per 100 g fatty acids
Candida lipolytica	C_{14}	tr	tr	tr	2.98	4.48	—	13.45	7.45	tr	tr	44.80	10.10	14.9(		[85]	60 h, 30 °C, reactor	g per 100 g fatty acids
—	C_{15}	—	tr	tr	0.8	20.80	1.73	2.60	4.30	0.8	25.40	12.30	8.60	19.73		[85]	60 h, reactor; 2.5 mg $C_{13:1}$	g per 100 g fatty acids
—	C_{16}	—	tr	5.0	5.0	5.00	—	30.00	20.00	tr	tr	10.00	25.00	—		[85]	60 h, 30 °C, reactor	g per 100 g fatty acids
—	C_{17}	1.3	—	1.96	tr	3.90	3.90	5.20	2.60	14.30	36.60	19.60	3.90	5.20		[85]	60 h, 30 °C, reactor	g per 100 g fatty acids
—	C_{18}	—	2.6	4.00	tr	tr	—	21.40	6.70	tr	5.30	6.70	16.70	34.80		[85]	60 h, 30 °C, reactor	g per 100 g fatty acids
C. parapsilosis	C_{15}		2.0	0.6	0.8	52.0	2.1	7.6	3.6	8.2	41.0	0.3	13.7	22.5		[153]	after 2 d shaking culture; 1.0 mg $C_{14:1}$, 20.6 mg $C_{17:2}$, 3.0 mg $C_{18:3}$	mg per 10 g alkane
—	C_{15}		1.0	0.3	2.0	18.0	1.2	2.8	0.6	3.1	11.0	0.6	3.1	1.8		[153]	3 days, reactor; 0.1 mg $C_{14:1}$, 1.1 mg $C_{17:2}$	mg per 100 g alkane
Mycotorula japonica	C_{10}	3.1	3.1	tr	2.8	tr	tr	14.6	tr	tr	tr	2.8	16.0	tr		[178]	48 h, 30 °C, shaking flask culture	
—	C_{11}	8.6	1.3	tr	1.3	0.8	tr	16.5	tr	7.3	tr	18.1	10.0	4.3		[178]	48 h, 30 °C, shaking flask culture	
—	C_{12}	2.2	14.3	1.5	tr	tr	tr	8.9	tr	tr	tr	5.2	tr	2.4		[178]	48 h, 30 °C, shaking flask culture	
—	C_{13}	2.4	1.3	1.2	tr	0.3	tr	0.8	tr	tr	tr	1.2	tr	2.0		[178]	48 h, 30 °C, shaking flask culture	

—	C_{14}	0.0	0.4	tr	8.1	tr	tr	7.0	tr	tr	tr	6.1	tr	5.3		[178]	48 h, 30 °C, shaking flask culture	
—	C_{15}	2.7	1.2	tr	tr	2.7	tr	1.0	tr	tr	tr	tr	tr	tr		[178]	48 h, 30 °C, shaking flask culture	
—	C_{16}	2.3	2.5	3.9	3.9	7.8	6.9	14.0	16.1	tr	tr	tr	tr	tr		[178]	48 h, 30 °C, shaking flask culture	
Cladosporium resinae	C_{12}		69.2		21.6			6.7				2.6				[150]	26 °C, shaking flask culture (stationary phase)	% total fatty acids (neutral lipid fraction)
—	C_{16}		70.2		11.4			14.7				3.4				[150]	26 °C, shaking flask culture (stationary phase)	% total fatty acids (neutral lipid fraction)
Chlorella vulgaris	C_{12}							35.63					11.49	41.38	22.99	[139]	3 d, cooxidation	% total fatty acids of cell lipids
—	C_{13}	1.67			2.93			32.43					17.78	23.22	21.97	[139]	3 d, cooxidation	% total fatty acids of cell lipids
—	C_{15}				5.58	8.84		44.19				16.28	16.74	6.98	1.40	[139]	3 d, cooxidation	% total fatty acids of cell lipids
—	C_{16}							58.87					10.48	15.32	15.32	[139]	3 d, cooxidation	% total fatty acids of cell lipids

microorganisms cultivated on even-carbon alkanes possess high amounts of the even-carbon fatty acids (see Table 2).

Table 2 summarizes some of the fatty acid patterns of lipids of different microorganisms grown on alkanes.

In Table 3 are compiled extracellular fatty acids excreted by microorganisms due to a mainly terminal alkane degradation pathway.

In long-time experiments of up to 24 days Lin et al.[94,95] analyzed the extracellular fatty acids from *Cladosporium* sp., *Hormodendrum hordei*, *Penicillium lilacinum* and *Aspergillus versicolor*. They determined especially C_7, C_{10}, C_{11} and partly C_9 and C_{12}–C_{15} fatty acids in addition to citric acid as the main product. Because of the very long duration of these experiments no indications on the metabolic pathways were given.

In *Torulopsis gropengiesseri* a metabolic pathway occurs which hinders a further degradation of the alkanoic acid by forming a glycolipid[71]. This pathway is illustrated by the following scheme (Fig. 6).

In many cases, the β-oxidation described above, α-oxidation or α-, β- and ω-oxidation of the fatty acids takes place. In *Arthrobacter simplex* grown on various *n*-alkanes, it was proposed that both the α- and β-oxidation of fatty acids operated equally[181]. Normally, the C_2-units are metabolized via a citric acid cycle or, in addition, via a glyoxylic acid cycle, e.g. by a citric acid forming *Candida parapsilosis*[117].

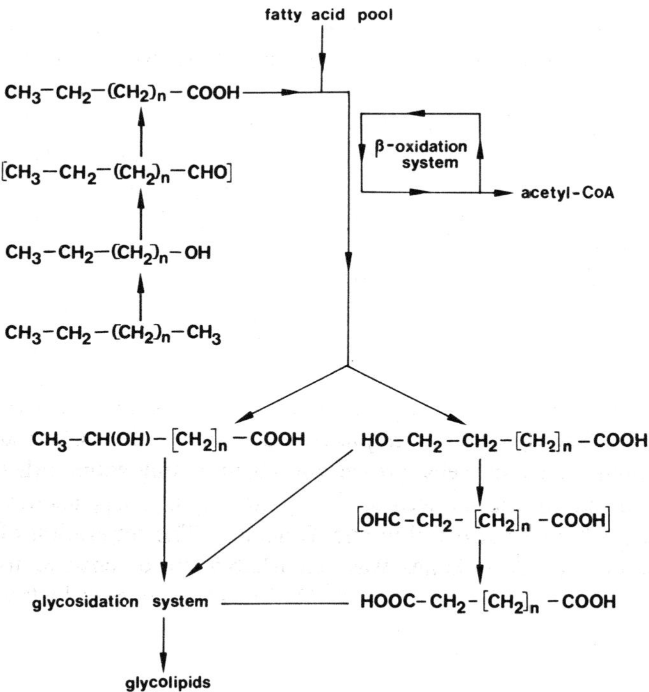

Fig. 6 Alkane oxidation and formation of a glycolipid in *Torulopsis gropengiesseri*[71]

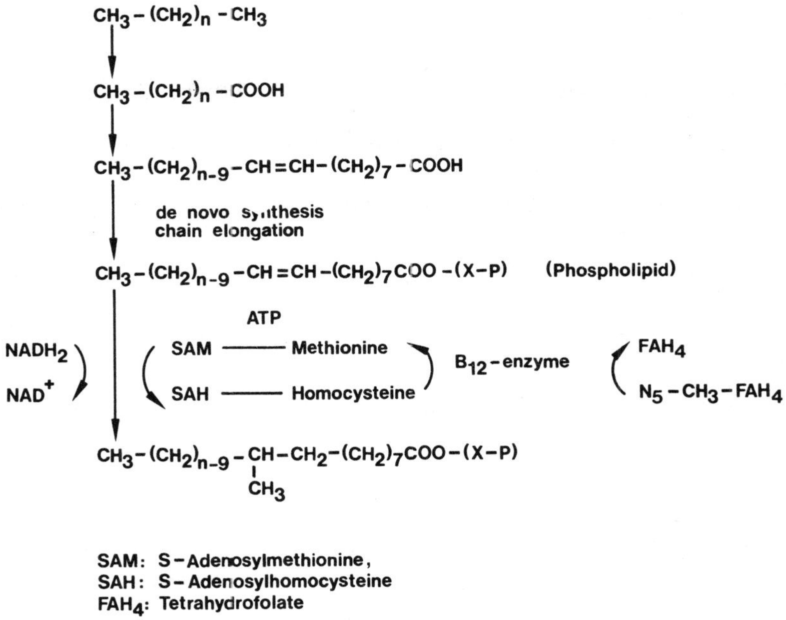

$CH_3-(CH_2)_n-CH_3$

↓

$CH_3-(CH_2)_n-COOH$

↓

$CH_3-(CH_2)_{n-9}-CH=CH-(CH_2)_7-COOH$

↓ de novo synthesis, chain elongation

$CH_3-(CH_2)_{n-9}-CH=CH-(CH_2)_7COO-(X-P)$ (Phospholipid)

↓ $NADH_2$ / NAD^+; ATP; SAM — Methionine; SAH — Homocysteine; B_{12}-enzyme; FAH_4 / $N_5-CH_3-FAH_4$

$CH_3-(CH_2)_{n-9}-CH(CH_3)-CH_2-(CH_2)_7COO-(X-P)$

SAM: S-Adenosylmethionine,
SAH: S-Adenosylhomocysteine
FAH_4: Tetrahydrofolate

Fig. 7 Synthesis of branched-chain fatty acids from C_{14}—C_{16} n-alkanes in *Corynebacterium simplex*[180)]

The following pathway of methyl-branched fatty acid synthesis in *Corynebacterium simplex* grown on C_{14}—C_{16} *n*-alkanes was proposed by Yanagawa et al.[180)] (Fig. 7).

In *Candida lipolytica*, *C. brumptii*, *Pichia farinosa* and perhaps some other yeasts Tabuchi and Hara[157)] and Tabuchi and Serizawa[158)] found methyl isocitrate and proposed a methyl citric acid cycle occuring in these microorganisms, especially when odd-carbon *n*-alkanes are degraded. A mutant of *Candida lipolytica* accumulates great amounts of methyl isocitric acid and other C_7-acids from odd-carbon *n*-alkanes, due to the absence of a methyl isocitrate cleaving enzyme.

In the methyl citric acid cycle propionyl-CoA is oxidized to pyruvate via C_7-tricarboxylic acids[159)] as follows (Fig. 8).

It is not our intention to review every paper dealing with the influence of long-chain alkanes on the terminal metabolism and other peripheric metabolic activities. The catalase activities of alkane-grown *C. tropicalis* and some other *Candida* spp. are much higher than those of cells grown on glucose and some other substrates[163)]. This catalase activity was induced in *C. tropicalis* by C_{10}–C_{13}–alkanes and is located in microbodies[164)] (see also Fukui and Tanaka[41)]. The respiration of *C. tropicalis* cells grown on C_{10}–C_{13}–alkanes was remarkably more sensitive to various respiratory inhibitors than glucose-grown cells[165)]. The Q_{O_2}-values of hydrocarbon-grown cells obtained from *n*-alkanes were noticeably high as compared with those from glucose[166)].

Some authors suggest that there probably exists a monoterminal oxidation pathway involving dehydrogenation with the formation of a 1-alkene. This

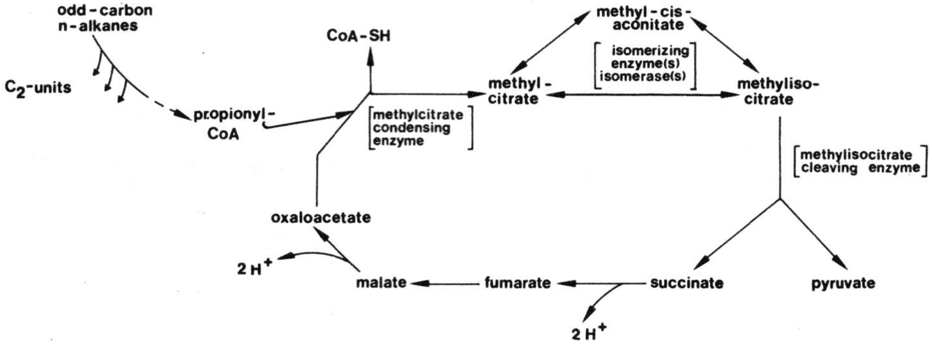

Fig. 8 Methylcitric acid cycle in a mutant of *Candida lipolytica*[159]

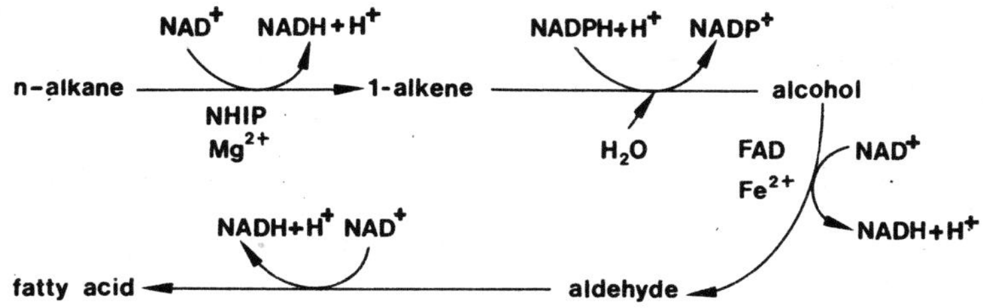

Fig. 9 Anaerobic alkane degradation pathways in *Pseudomonas* sp.[119]

pathway was postulated by Iizuka et al.[65,66] to occur in *C. rugosa*. No information is available about the fatty acid patterns of microorganisms which oxidize alkanes mainly according to this pathway. There are still many doubts about this pathway; microorganisms in which alkenes were determined are listed in Table 7.

Recently, in a *Pseudomonas* sp., an anaerobic *n*-alkane degradation pathway to fatty acids was observed due to characterization of seven enzymes which may correspond to this route[119] (Fig. 9).

Diterminal oxidation can be observed in some bacteria and yeasts. Investigations about this pathway have also been reviewed by Einsele and Fiechter[30] and Klug and Markovetz[86] until 1971 (Fig. 10).

It has been postulated that in the first step one methyl group is oxidized to the corresponding fatty acid. The second step involves oxidation of the ω-methyl group. The formation of an 1,ω-alkanediol, alkanedial or ω-hydroxyalkanal has not been observed.

Table 4 shows some extracellular dioic acids excreted by microorganisms which are often active through a predominantly monoterminal, but also through an important diterminal degradation pathway. There are many industrial patents for the production of dioic acids from long-chain alkanes (see[41]).

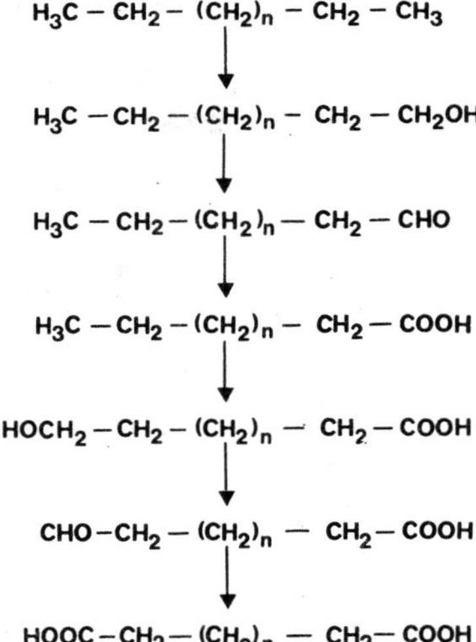

Fig. 10 Diterminal oxidation of long-chain alkanes

3.2 Subterminal Oxidation Pathways

The first step of a subterminal oxidation involves the formation of a secondary alcohol and the second step the generation of the corresponding ketone. Nothing is known about the alcohol oxidase which catalyses this step. In *Fusarium lini* dehydrogenation of two methylene groups could be excluded to occur[59].

Until 1971, mainly the oxidation at C_2 of long-chain alkanes was described[86]. The further degradation of the ketones in the 2-position, e.g. of 2-tridecanone in *Pseudomonas aeruginosa*, is assumed to be a Baeyer-Villiger type reaction yielding undecyl acetate which is split into 1-undecanol and acetate[39,36,37,38].

A reaction similar to the Baeyer-Villiger rearrangement in microorganisms (*Cladosporium resinae*) was first reported for the progesteron oxidation[35]. Rahim and Sih[129] described an enzymatic cleavage of the pregnane side chain. In both cases, an acetate moiety was split off from the steroid.

An undecyl acetate esterase was formed by *Pseudomonas cepacia* grown on 2-tridecanone. The K_m value for undecyl acetate was 2.3×10^{-2} M. Two esterases with the same molecular weight (approximately 34.500) were identified[148]. Undecyl acetate esterase was strongly inhibited by organophosphates and other esterase inhibitors. The enzyme was inducible when 2-tridecanone, 2-tridecanol, undecyl acetate and, to a lesser extent, 1-undecanol were the growth substrates[148]. An acetyl esterase from *Nocardia* sp. involved in 2-butanone degradation was purified and characterized[31].

From *Cladosporium cladosporioides* an acetate-ester synthesizing enzyme was characterized[179]. This mold forms also ketones in 2-position[22].

The Baeyer-Villiger type reaction takes place by incorporation of molecular oxygen in the presence of NADPH. This is an oxygenase type reaction[17]. The oxygenase was characterized by Markovetz[104] from *Pseudomonas*. Dodecyl acetate was also isolated from *Mortierella isabellina* and identified by Hoffmann et al.[57].

The formation of alcohols and ketones at C-3, C-4, C-5 and higher positions

Table 4. Extracellular dioic fatty acids from yeasts and models

Organism	Substrate	Relative proportion of dioic acids (% w/w)									Ref.	Remarks	Calculation
		5	6	7	8	9	10	11	12	13			
Candida albicans	C_{12}		28	—	8	—	—	—	—	—	[147]	3 d, shaking flask culture, 30 °C	mg l^{-1} medium
C. cloacae	C_9	96	—	187	—	546	—	—	—	—	[147]	3 d, shaking flask culture, 30 °C	mg l^{-1} medium
—	C_{10}	—	119	—	24	—	427	—	—	—	[147]	3 d, shaking flask culture, 30 °C	mg l^{-1} medium
—	C_{11}	—	—	78	—	22	—	232	—	—	[147]	3 d, shaking flask culture, 30 °C	mg l^{-1} medium
—	C_{12}	—	98	—	113	—	—	—	610	—	[147]	3 d, shaking flask culture, 30 °C	mg l^{-1} medium
—	C_{13}	—	—	65	—	21	—	—	—	192	[147]	3 d, shaking flask culture, 30 °C	mg l^{-1} medium
—	C_{14}	—	28	—	31	—	—	—	—	—	[147]	3 d, shaking flask culture, 30 °C	mg l^{-1} medium
—	C_{15}	12	—	186	—	28	—	—	—	—	[147]	3 d, shaking flask culture, 30 °C	mg l^{-1} medium
—	C_{16}	—	207	—	71	—	—	—	—	—	[147]	3 d, shaking flask culture, 30 °C	mg l^{-1} medium
—	C_{17}	6	—	176	—	23	—	—	—	—	[147]	3 d, shaking flask culture, 30 °C	mg l^{-1} medium
—	C_{18}	—	39	—	118	—	5	—	—	—	[147]	3 d, shaking flask culture, 30 °C	mg l^{-1} medium
—	C_{12}		247	—	—	—	46	—	880	—	[147]	2 d, resting cells	mg l^{-1} medium
—	C_{13}		—	167	—	87	—	—	—	427	[147]	2 d, resting cells	mg l^{-1} medium
—	C_{14}		67	—	6	—	—	—	11	—	[147]	2 d, resting cells; 147 mg l^{-1} C_{14} dioic acid	mg l^{-1} medium
—	C_{15}		—	182	—	16	—	—	—	2	[147]	2 d, resting cells; 9 mg l^{-1} C_{15} dioic acid	mg l^{-1} medium
—	C_{16}		107	—	13	—	6	—	26	—	[147]	2 d, resting cells; 4 mg l^{-1} C_{16} dioic acid	mg l^{-1} medium
—	C_{17}		—	126	—	24	—	—	—	11	[147]	2 d, resting cells	mg l^{-1} medium
—	C_{18}		46	—	—	—	—	—	—	—	[147]	2 d, resting cells	mg l^{-1} medium
C. guilliermondii	C_{13}	—	—	—	—	9	—	—	—	13	[147]	3 d, shaking flask culture, 30 °C	mg l^{-1} medium
C. haplophyla	C_{12}	—	—	—	9	—	—	—	21	—	[147]	3 d, shaking flask culture, 30 °C	mg l^{-1} medium
—	C_{13}	—	—	21	—	15	—	—	—	10	[147]	3 d, shaking flask culture, 30 °C	mg l^{-1} medium
C. intermedia	C_{12}	—	237	—	27	—	14	—	—	—	[147]	3 d, shaking flask culture, 30 °C	mg l^{-1} medium
C. lipolytica	C_{12}	—	17	—	14	—	—	—	—	—	[147]	3 d, shaking flask culture, 30 °C	mg l^{-1} medium
—	C_{13}	6	—	57	—	12	—	—	—	—	[147]	3 d, shaking flask culture, 30 °C	mg l^{-1} medium
—	C_{12}	—	4	—	50	—	5	—	13	—	[147]	3 d, shaking flask culture, 30 °C	mg l^{-1} medium
C. maltosa	C_{12}	—	65	—	19	—	10	—	—	—	[147]	3 d, shaking flask culture, 30 °C	mg l^{-1} medium
—	C_{13}	—	—	42	—	—	—	—	—	8	[147]	3 d, shaking flask culture, 30 °C	mg l^{-1} medium
C. parapsilosis	C_{12}	—	85	—	57	—	56	—	526	—	[147]	3 d, shaking flask culture, 30 °C	mg l^{-1} medium
—	C_{13}	28	—	80	—	10	—	—	—	53	[147]	3 d, shaking flask culture, 30 °C	mg l^{-1} medium

—	C_{15}	256.5	—	39.5	—	2.2	—	0.1	—	84.3	[153]	3 d, shaking flask culture, 30 °C	mg g^{-1} alkane
—	C_{15}	82.7	1.1	23.1	0.6	0.1	tr	4.0	1.8	3.1	[153]	3 d; 71.3 mg l^{-1} C_3 dioic acid, 0.3 mg l^{-1} C_{14} dioic acid	mg g^{-1} alkane
C. rugosa	C_{10}						16.7				[64]	4 d, shaking flask culture, 30 °C	mg l^{-1} medium
C. zeylanoides	C_{12}	—	6	—	4	—	—	—	—	—	[147]	3 d, shaking flask culture, 30 °C	mg l^{-1} medium
Aspergillus versicolor	C_{10}	—	—	—	—	—	tr	—	—		[95]	stationary flask culture, 30 °C, 10 d	% total fatty acid present
—	C_{10}	—	—	—	—	—	tr	—	—		[95]	stationary flask culture, 30 °C, 17 d	% total fatty acid present
—	C_{11}	—	—	—	—		—	30.7	—		[95]	stationary flask culture, 30 °C, 8 d	% total fatty acid present
—	C_{11}	—	—	3.2	—		—	37.2	—		[95]	stationary flask culture, 30 °C, 17 d	% total fatty acid present
—	C_{16}	—	—	—	—		—		—		[95]	stationary flask culture, 30 °C, 7 d	% total fatty acid present
—	C_{16}	—	1.0	—	—		—	—	—		[95]	stationary flask culture, 30 °C, 17 d	% total fatty acid present
—	C_{17}	—	—	—	—		—	—	—		[95]	stationary flask culture, 30 °C, 8 d	% total fatty acid present
—	C_{17}	—	—	—	—		—	—	—		[95]	stationary flask culture, 30 °C, 16 d	% total fatty acid present
Penicillium lilacinum	C_{10}	—	4.8	3.2	0.6		—	4.8	—		[95]	stationary flask culture, 30 °C, 8 d	% total fatty acid present
—	C_{10}	—	12.7	8.3	—		1.3	3.2	11.5		[95]	stationary flask culture, 30 °C, 17 d	% total fatty acid present
—	C_{11}	—	11.1	31.1	—		—	—	—		[95]	stationary flask culture, 30 °C, 8 d	% total fatty acid present
—	C_{11}	—	4.8	19.3	—		—	—	—		[95]	stationary flask culture, 30 °C, 17 d	% total fatty acid present
—	C_{16}	—	10.5	4.2	—		—	5.6	—		[95]	stationary flask culture, 30 °C, 5 d	% total fatty acid present
—	C_{16}	—	12.8	2.6	—		—	—	2.6		[95]	stationary flask culture, 30 °C, 17 d	% total fatty acid present
—	C_{17}	—	—	1.0	—		—	—	—		[95]	stationary flask culture, 30 °C, 5 d	% total fatty acid present
—	C_{17}	—	—	2.1	—		—	—	—		[95]	stationary flask culture, 30 °C, 17 d	% total fatty acid present

Table 4 (continued)

Organism	Sub-strate	Relative proportion of dioic acids (% w/w)									Ref.	Remarks	Calculation
		5	6	7	8	9	10	11	12	13			
Cladosporium sp.	C_{10}						—	1.2	3.1		[94]	stationary flask culture, 30 °C, 13 d	% total fatty acid present
—	C_{10}						—	1.5	1.0		[94]	stationary flask culture, 30 °C, 18 d	% total fatty acid present
—	C_{11}						—	2.9	0.9		[94]	stationary flask culture, 30 °C, 11 d	% total fatty acid present
—	C_{11}						—	1.2	—		[94]	stationary flask culture, 30 °C, 17 d	% total fatty acid present
—	C_{16}						—	—	—		[94]	stationary flask culture, 30 °C, 16 d	% total fatty acid present
—	C_{16}						—	—	2.1		[94]	stationary flask culture, 30 °C, 22 d	% total fatty acid present
—	C_{17}						—	—	—		[94]	stationary flask culture, 30 °C, 16 d	% total fatty acid present
—	C_{17}						—	—	1.1		[94]	stationary flask culture, 30 °C, 24 d	% total fatty acid present
Hormodendrum hordei	C_{10}						—	0.5	0.5		[94]	stationary flask culture, 30 °C, 12 d	% total fatty acid present
—	C_{10}						—	0.9	4.7		[94]	stationary flask culture, 30 °C, 18 d	% total fatty acid present
—	C_{11}						—	2.3	2.4		[94]	stationary flask culture, 30 °C, 15 d	% total fatty acid present
—	C_{11}						13.2	3.2	—		[94]	stationary flask culture, 30 °C, 20 d	% total fatty acid present
—	C_{16}						—	7.7	6.9		[94]	stationary flask culture, 30 °C, 16 d	% total fatty acid present
—	C_{16}						—	—	0.2		[94]	stationary flask culture, 30 °C, 23 d	% total fatty acid present
—	C_{17}						—	1.0	3.1		[94]	stationary flask culture, 30 °C, 15 d	% total fatty acid present
—	C_{17}						—	0.4	1.1		[94]	stationary flask culture, 30 °C, 22 d	% total fatty acid present

has been reported. *Pseudomonas aeruginosa* formed 1-, 2-, 3- and/or 4- and/or 5-decanol from decane and the corresponding ketones[40]. Another strain of *P. aeruginosa*, *P. fluorescens* and *Acetobacter peroxidans* formed 2-, 3- and partly 4-tetradecanone from tetradecane[138]. An *Arthrobacter* sp. was also capable of transforming *n*-hexadecane to 2-, 3- and 4-hexadecanones thus furnishing evidence of the accumulation of 2- and 3-hexadecanols as oxidative intermediates when yeast extract or peptone was used as an additional growth substrate. The addition of glucose promoted the transformation of *n*-hexadecane to alcoholic and ketonic oxidation products by resting cells of this bacterium[82].

Aspergillus flavus, *A. ochraceus*, *Penicillium javanicum* and other Penicillia and *Verticillium* sp. formed 2-, 3-, 4- and/or 5- and/or 6-ketones from dodecane and *n*-tridecane[120,121,122], *Cladosporium cladosporioides* produced 2-, 3- and 4-ketones from *n*-tridecane[22].

With *Mortierella isabellina* the subterminal degradation pathway of long-chain alkanes could be ensured by the isolation of the intermediates, especially of the possible esters, by use glass capillar gas chromatography[55,56,58,57].

These results were confirmed with *Bacillus* spp.[75] and *Fusarium lini*[167]. Cell-free extracts of *F. lini* revealed an esterase activity giving rise to cleavage of all isomeric esters containing 12 carbon atoms, synthesized by chemical methods, the corresponding alcohols and acids. When the chain length of the alcohol component in the ester increased from one to six carbon atoms, the esterase activity decreased. Minimal esterase activity was reached when both the alcohol and the acid component had a chain length of six carbon atoms. The esterase resembled lipases[167].

These results with *Mortierella isabellina*, bacilli and *Fusarium lini* demonstrate that also a cleavage of the long-chain carbon molecule by a Baeyer-Villiger type reaction in positions other than C-2 can be observed in microorganisms.

Some microorganisms are not able to oxidize long-chain alkanes as the sole carbon source. They can only oxidize these alkanes when another carbon source, e.g. glucose is present. Such a coupled oxidation which may probably be a cooxidation as defined by Horvath[60] and by Raymond et al.[131] has been stated to occur in bacilli in the mesophilic temperature range by Kachholz and Rehm[74,75]. In these bacilli, different types of subterminal oxidation of *n*-tridecane and *n*-tetradecane were observed (see Table 7). One type (e.g. *Bacillus coagulans*) preferably underwent oxidation in 2- or 3-position, another type (e.g. *B. stearothermophilus*) preferably in 4-, 5- or 6-position[74].

Some *Streptomyces* spp., e.g. *S. griseus* and *S. violaceoruber*, also degrade *n*-tetradecane only by an oxidation coupled with glucose metabolism. The pathway was a subterminal one[48]. *S. eurythermus* may oxidize alkanes as the sole carbon source. This strain revealed a subterminal degradation pathway perhaps beneath a not very active monoterminal pathway[48].

Furthermore, *Chlorella vulgaris* can degrade *n*-tridecane and — perhaps in minute amounts — also *n*-dodecane, *n*-tetradecane, *n*-pentadecane and *n*-hexadecane in the presence of CO_2 and light or partly in the presence of glucose[139]. The degradation pathway is also a subterminal one.

In bacilli, subterminal diols, ketols and diketones — corresponding to the chain length of the alkanes — could be found in cooxidation culture so that a disubterminal

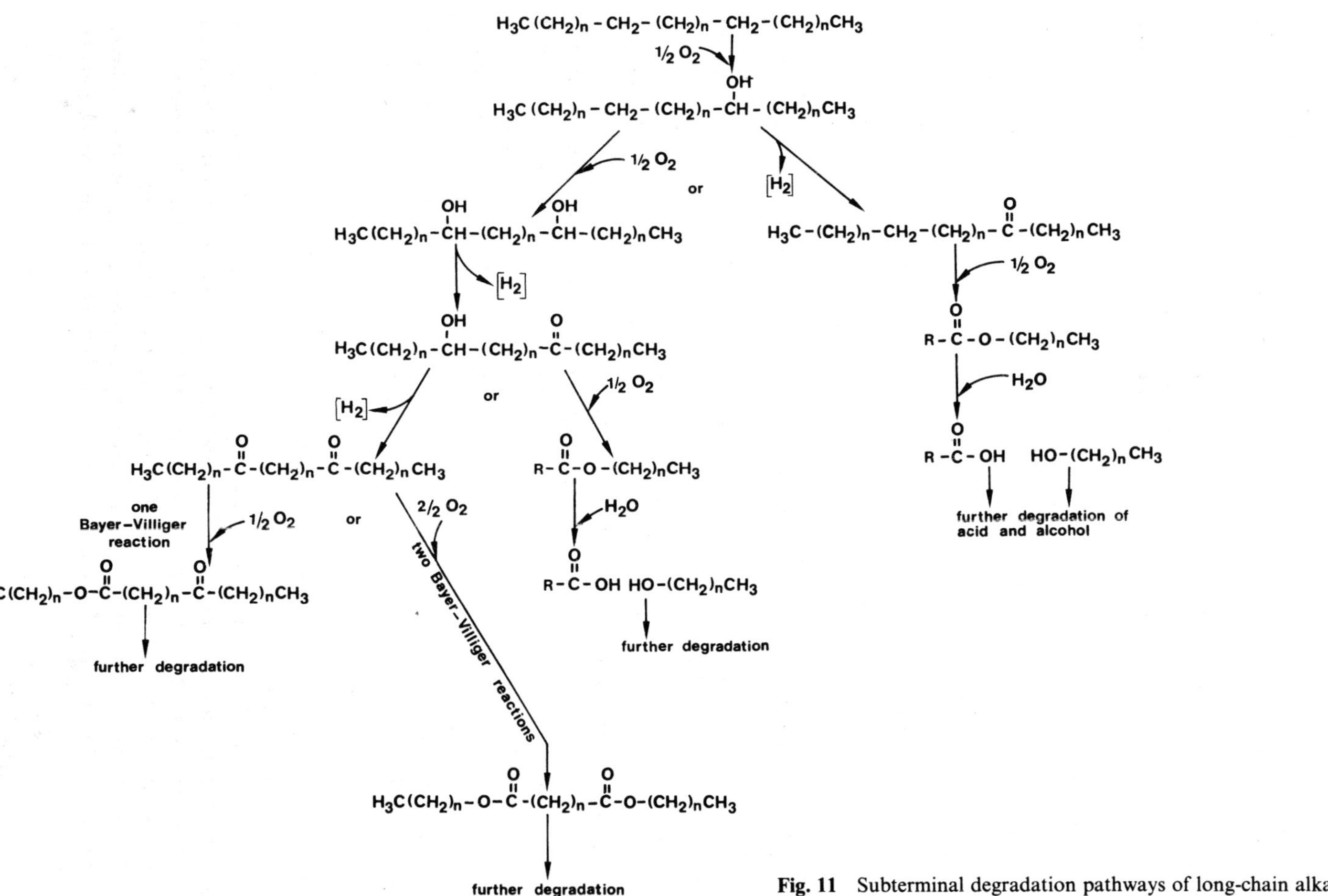

Fig. 11 Subterminal degradation pathways of long-chain alkanes

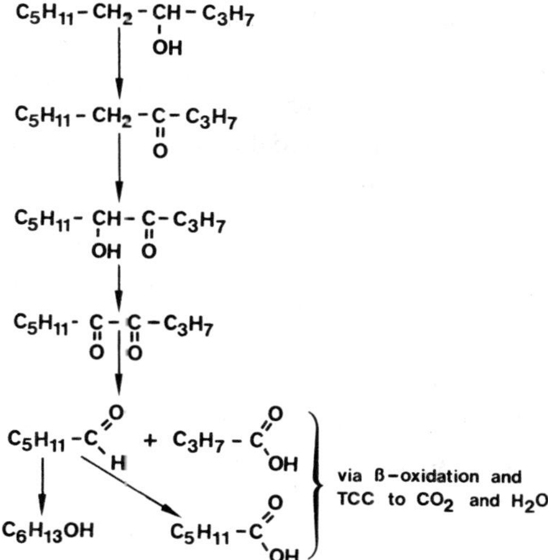

Fig. 12 Oxidation of 4-decanol via ketone-hydroxyketone-dione[93)]

degradation pathway may exist. The two functional groups were located at the secondary and primary C-atom at the opposite ends of the molecule. These substances have hitherto not been found in microorganisms which degrade long-chain hydrocarbons as the sole carbon source. It has been suggested that in this case non-specific enzyme reactions take place[75)].

Further information about cooxidation of hydrocarbons is given by Perry[124)].

The following diagram demonstrates the degradation of long-chain alkanes following different subterminal pathways (Fig. 11).

In *Pseudomonas* sp. an oxidation pathway of secondary alcohols via ketone-hydroxyketone-dione followed by cleavage into aldehyde and acid has been described[93)]. The authors have proposed a model for the oxidation of 4-decanol (Fig. 12).

No further data are available about this pathway concerning the degradation of other long-chain alkanes. We did not succeed in uncovering this pathway in Mucorales, Moniliales and Streptomycetes[134)] and bacilli[75)]. In all cases, we isolated the esters corresponding to the monoketones.

4 Degradation Pathways for Long-Chain Alkanes in Microorganisms

Nearly 20 years of research on alkane degradation have passed and many results on the degradation of long-chain *n*-alkanes have been published. For this reason, it may be useful to find out correlations between the systematic position of microorganisms and the preference of alkane degradation pathways.

Table 7 demonstrates the intermediates of long-chain alkane degradation pathways formed by different microorganisms. We do not claim completeness concerning the microorganisms. With great precaution, it may be possible to draw some qualitative conclusions and with much more precaution it may also be possible to make some quantitative statements about the pathways and their intensity in metabolism (see Table 2 to 4).

4.1 Criteria for a Metabolic Alkane Degradation Pathway

What are the criteria for a monoterminal pathway?

a) An important criterion for a monoterminal oxidation is the isolation of extracellular degradation products, especially primary alcohols, aldehydes and the corresponding acids. In Table 3 are compiled degradation products.
   It is evident that these intermediates are only products not metabolized in a further metabolic pathway. Therefore, it is possible that at a high oxidation rate only small amounts of intermediates are excreted. Thus from the amounts of excreted intermediates only qualitative rather than quantitative conclusions can be drawn.
b) A second criterion for a monoterminal oxidation is the determination of the enzymes catalyzing the primary oxidation steps. These enzymes need not be specific for this pathway. There are no meaningful results available on the specificity of these enzymes.
c) Most authors who have determined fatty acids in cell lipids of alkane-grown microorganisms found in active alkane-degrading cells fatty acid patterns corresponding to the *n*-alkane chain length in monoterminal oxidizing microorganisms. The data listed in Table 5 clearly illustrate these relations, e.g. in yeasts, so that it can be stated that most yeasts generally reveal a monoterminal degradation pathways. The alkanes are oxidized to the corresponding fatty acids which are incorporated — either directly or after C_2-elongation or C_2-splitting — into the lipids due to β-oxidation.

Table 5. Ratios of odd-chain fatty acids and C_{17} acids to total cellular fatty acids in *Candida lipolytica* cells grown on n-alkanes and glucose

Growth substrate	Odd-chain acids (%)	C_{17} acid (%)
Glucose	2.9	1.6
n-Undecane	9.6	5.7
n-Dodecane	0.9	0.9
n-Tridecane	60.0	33.1
n-Tetradecane	1.0	1.0
n-Pentadecane	80.9	58.0
n-Hexadecane	1.6	1.0
n-Heptadecane	97.8	94.9
n-Octadecane	3.4	3.4

The yeast was harvested during the late exponential growth phase

These relations are illustrated by the results of Tanaka et al.[161)] with *Candida lipolytica*.

What are the criteria for a diterminal pathway?

a) The most important criterion — perhaps the only one until now — is the isolation of diterminal degradation products, especially dioic acids excreted into the substrate.

Table 4 shows that some *Candida* spp. and perhaps some Moniliales are active through a diterminal degradation pathway. It must be stated that for Moniliales only such products could hitherto be isolated in long-time experiments[94,95)]. In

Table 6. Dicarboxylic acid (DC) production from n-alkanes by *Candida cloacae* 310

n-Alkane	DC produced (mg l^{-1})									
	DC-5	DC-6	DC-7	DC-8	DC-9	DC-10	DC-11	DC-12	DC-13	DC-14
n-C_9	96		187		546					
n-C_{10}		119		24		427				
n-C_{11}			78		22		232			
n-C_{12}		98		113				610		
n-C_{13}			65		21				192	
n-C_{14}		28		31						79
n-C_{15}	12		186		28					
n-C_{16}		207		71						
n-C_{17}	6		176		23					
n-C_{18}		39		118		5				

Medium containing 10% (v/v) n-alkane as the sole carbon source. Shaking culture at 30 °C for 72 h. The pH value was adjusted to 7.0 with 1 N KOH twice a day

this case, a de novo synthesis of dioic acids cannot be excluded so that the hints on this pathway are very vague. We did not succeed in obtaining dioic acids in Mucorales[57,58].

Table 6 shows a good agreement between chain length of *n*-alkanes and the production of the corresponding dioic acids in *Candida cloacae*.

b) No results are available on the isolation of specific enzymes involved in diterminal degradation pathways. Perhaps, it can be assumed that the monoterminal oxidation systems acts at both methyl groups of the alkane chain.

c) In addition, the cell lipids give no indications regarding this pathway.

What are the criteria for a subterminal pathway?

a) The oxidation products, e.g. secondary alcohols, ketones and especially esters and the corresponding ester splitting products prove the existence of this pathway, as described recently[57].
   Concerning quantitative statements, the same reservations as for the monoterminal pathway should be made.

b) Regarding enzymes especially esterase have been described. They do not display a high specificity[148,149,167].

c) The fatty acid patterns of subterminal degrading microorganisms show no direct relation to the chain length of the *n*-alkane substrate.
   In *Cladosporium resinae* a correlation between the chain length of *n*-alkane substrates and the cellular fatty acids could not be deduced (Table 2)[24] but monoterminal products were isolated so that a monoterminal degradation pathway was stated for the same organism[173] (Table 2).

Also, during cooxidation, the same criteria are evident. In this case, it must be stated that products which are presumed to be specific of alkane oxidation are not formed from the other carbon source, e.g. glucose[75].

4.2 Degradation Pathways in Different Microorganism Species

A large number of papers deal with the uptake and utilization of long-chain alkanes and their primary degradation products. Most of them do not provide any

Table 7. Intermediates formed by different bacteria, degrading long-chain n-alkanes (C_{10}—C_{18})

Microorganisms	Terminal pathways			Subterminal pathways mono-subterminal			Ref.	Remarks
	mono-terminal	diterminal	1,2-alkenes	only 2-position	2-, 3-, 4-position	4-, 5-, 6-position		
Pseudomonas aeruginosa			+++				145)	cell-free extracts, anaerobic, C
—	++		+				23)	resting cells; C
—	+				+		40)	C_{10}
—	+++						130)	enzyme extracts; C_5—C_{10}
—	++				+		138)	C_{14}
P. cepacia				++			17)	cell-free extracts; C_{13}
P. fluorescens	++			+	tr		138)	C_{14}
P. oleovorans	++						125)	cell-free extracts; C
P. putida	+++						111)	C
Pseudomonas sp.	++	+			+		79)	C_{11}
Acetobacter peroxidans	++			+	tr		138)	C_{14}
Acinetobacter calcoaceticus	++						152)	enzyme extracts; C
Acinetobacter sp.	+++						103, 155)	C_{16}
Achromobacter sp.			+++				77, 78)	C_{10}—C_{14}
Micrococcus cereficans	+		+				172)	C_{16}
—	++						34)	C
Bacillus coagulans	+				++	+	74)	C_{13}; cooxidation
—					++	+	74)	C_{14}; cooxidation
B. lentus	+				+	+	74)	C_{13}; cooxidation
—					+	++	74)	C_{14}; cooxidation
*B. macerans**)	+				+	+++	74)	C_{13}, C_{14}; cooxidation
*B. stearothermophilus**)**)	+				+	+++	75)	C_{13}, C_{14}; cooxidation (mesophilic)
Corynebacterium simplex	+++						180)	C_{18}; (C_{14}—C_{17}) mixture
—	+++	++					77, 78)	C_{10}—C_{14}
—	++	++					6)	cell-free extracts; C_{10}
Arthrobacter paraffineus	+++						156)	C_{12}—C_{14}
—					+		83)	C_{15}, C_{16}; cooxidation
Brevibacterium erythrogenes	+++						127)	C_{15}, C_{16}
Mycobacterium phlei	+		+				172)	C_{16}

*) disubterminal intermediates isolated; **) intermediates in 7-, 8-, 9-position isolated

M. rhodochrous	+						40)	C_{10}
M. smegmatis	+	++		+			100)	C_{11}
M. vaccae	++						81)	C_{14}—C_{18}
Nocardia hydrocarbonoxydans n. sp.	++						112)	C_{10}, C_{11}
N. petroleophila n. sp.	++						144)	C_{10}, C_{11}
N. salmonicolor			+				1)	resting cells, C_{16}
Nocardia sp.	+		+				172)	C_{16}
—				++			17)	cell-free extracts, C_{13}
Streptomyces eurythermus	+				+	++	48)	C_{14}; cooxidation
S. griseus	+				+	++	48)	C_{14}; cooxidation
S. violaceoruber					+	+	48)	C_{14}; cooxidation
Anacystis nidulans	—	—	—	—	—	—	94)	C_{10}—C_{14}
Intermediates formed by different yeasts, degrading long-chain n-alkanes (C_{10}—C_{18})								
Pichia sp.	++	++					115)	C
P. haplophyla		+					147)	C_{12}
—		++					147)	C_{13}
Candida albicans		++					147)	C_{12}
C. cloacae		++					170, 147)	C_9—C_{18}
C. guilliermondii		+					147)	C_{13}
—		++					87)	C_{11}—C_{16}
—	+++						140)	cell-free extracts; C
C. intermedia		+++					147)	C_{12}
C. lipolytica				+			85)	C_{14}, C_{15}, C_{16}, C_{17}, C_{18}
—		++					147)	C_{13}
—	+++						158)	C; methylcitrate
C. maltosa		++					147)	C_{12}, C_{13}
C. parapsilosis		+++					147)	C_{12}
—		++					147)	C_{13}
—	+++			+			153)	C_{12}, C_{13}, C_{14}
—	+++			++			153)	C_{15}
—			+				154)	cell-free extracts; C_{14}
C. rugosa		++					64)	C
—			+++				65, 63)	cell-free extracts + NAD(P)H; C
C. tropicalis	+++	+	+				91)	cell-free extracts; C
—		+					116)	C
—	+++						28)	cell-free extracts; C
C. zeylanoides		+					147)	C_{12}

Table 7 (continued)

Microorganisms	Terminal pathways			Subterminal pathways mono-subterminal			Ref.	Remarks
	mono-terminal	diterminal	1,2-alkenes	only 2-position	2-, 3-, 4-position	4-, 5-, 6-position		
Rhodotorula sp.	+		+				172)	C_{16}
Torulopsis gropengiesseri	+++		excluded	+			71, 70)	cell-free extracts; C
Intermediates formed by different molds, degrading long-chain n-alkanes (C_{10}—C_{18})								
Absidia spinosa	++	—	—		++	++	55)	C_{12}, C_{13}
Cunninghamella bainieri	+++						32)	cell-free extracts; C
C. blakesleeana (– strain)	+++						4)	C_{14}
C. echinulata	+	—	—		++	+	55)	C_{12}, C_{13}
Mortierella isabellina	+	—	—		++	+++	55, 57)	C_{12}, C_{13}
Rhizopus nigricans	+	—	—		tr	tr	55)	C_{12}, C_{13}
Aspergillus flavus (2 strains)					+	+++	122)	C_{12}, C_{13}
A. ochraceus					+++	++	122)	C_{12}, C_{13}
A. niger					++	++	122)	C_{12}, C_{13}
A. versicolor	++	++					94)	C_{11}
Penicillium javanicum					++	+++	122)	C_{12}, C_{13}
P. lilacinum	++	++					94)	C_{10}, C_{11}, C_{16}, C_{17}
Penicillium sp.					+		4)	C
Penicillium sp. (2 strains)					++	+++	122)	C_{12}, C_{13}
Botrytis sp.	++	++					177)	(C_9—C_{18} mixture)
Cladosporium resinae	+++						173, 174	growing cells, C_{12}, C_{16}
Verticillium sp.					+	+++	122)	C_{12}, C_{13}
Fusarium lini					++	+++	167)	C_{12}
Intermediates formed by *Chlorella vulgaris*, degrading long-chain n-alkanes (C_{10}—C_{18})								
Chlorella vulgaris					+	+	139)	C_{15}; cooxidation

useful information about the degradation pathway. From Tables 2—4, the following degradation pathways for some microorganism species can be deduced.

4.2.1 Degradation Pathways of Bacteria

Among the large number of bacteria, mostly gram-negative aerobe rods and cocci and Actinomycetes and related organisms have been examined concerning the degradation pathways of long-chain alkanes (Table 7). Most species act via a monoterminal pathway, some of them producing alkenes. It may be assumed — if experiments have been carried out — that the formation of *n*-alkenes can also be observed with many other species.

Especially bacilli and *Streptomyces* spp., mostly in cooxidation, generate subterminal products. One type forms oxidation products in 2-, 3-, 4-position, another type preferably at 4-, 5-, 6-position. This can lead to the conclusion that there exist four oxidation types: a monoterminal type, a diterminal type, a subterminal type with an oxigenase preferably oxidizing at the position 2-, 3- or 4- and a second subterminal type preferably oxidizing at 4-, 5- or 6-positions. Furthermore, a disubterminal type was found in bacilli[75]. No data on the differences and the specificity of the enzymes involved are available.

4.2.2 Degradation Pathways of Yeasts

The degradation pathway in yeasts is predominantly a monoterminal one and only partly a diterminal one. While 1-alkenes could be detected, no reliable data on the involvment of 1-alkenes in an important degradation pathway are available.

In general, the fatty acid patterns show a relation to the number of carbon atoms in the given *n*-alkanes (Table 2). These patterns are sometimes blurred by an additional diterminal pathway and perhaps by the oxidation products arising from the secondary alcohol. A subterminal oxidation other than that occurring at the 2-position could not be detected.

4.2.3 Degradation Pathways of Molds

Among the Zygomycetes the investigated Mucorales always give rise to the formation of monoterminal products but no diterminal intermediates are produced (Table 7). 1-Alkenes have been formed in all molds. It is evident that subterminal intermediates generated by oxidation both at C-2, C-3, or C-4 and C-4, C-5 or C-6 of the alkanes have been isolated in large amounts. This demonstrates that molds can oxidize the methylene group whereas yeasts preferably attack the methyl group.

The fatty acid patterns of *Cunninghamella echinulata*, *C. elegans*, *Absidia spinosa* and partly *Mortierella isabellina* are characteristic of a monoterminal degradation pathway (Table 2). For this reason, they preferentially undergo a monoterminal oxidation in spite of the relatively high amounts of the subterminal intermediates produced.

Moniliales, included *Cladosporium resinae*, mostly show subterminal fatty acid patterns. Therefore, the subterminal pathway may be preferred here irrespective of many monoterminal intermediates formed by *C. resinae*, *Penicillium lilacinum* and *Aspergillus versicolor* (long-time experiments[94]). Concerning the last two molds it may be that no attempts to detect subterminal intermediates have been made.

4.2.4 Degradation Pathways of *Chlorella vulgaris*

This organism only forms subterminal oxidation products in small amounts by cooxidation, especially from pentadecane. Since the fatty acid patterns are subterminal a subterminal degradation pathway may be assumed when a de novo formation of cellular fatty acids can be excluded. This has not yet been confirmed experimentally.

4.3 Long-Chain n-Alkane Degradation with Regard to Systematics of Microorganisms

The taxonomic value of the property of alkane degradation has hitherto been investigated only with regard to assimilation. Partly alkane degrading microorganisms only were listed in ecological studies, e.g. bacteria[5,92], Mucorales[55], Moniliales[120], and many other molds[114,9] and *Prototheca zopfii*[175].

Studies on 220 strains of *Arthrobacter*, *Brevibacterium*, *Mycobacterium*, *Corynebacterium*, *Cellulomonas* and *Nocardia* revealed that the ability for alkane degradation was rather a specificity for strains nil for species[47]. Also with 500 phytopathogenic bacteria a relationship between alkane degradation and taxonomic position could not be established[8].

Nyns et al.[114] concluded for hydrocarbon assimilation through molds that this property is mainly found in the case of Mucorales and Moniliales but not necessarily with related species, nor is it typical of one species but rather the property of individual strains. These authors argue that "the property of assimilation of hydrocarbons appears to lack taxonomic value".

Bos and De Bruyn[14] and Bos[13] discussed the significance of hydrocarbon assimilation in yeast identification. They found that, in general, all strains of a species share either the ability to assimilate hydrocarbons or the failure to do so. In some cases, the simple hydrocarbon assimilation test may be useful. Also, in subgrouping the genera *Candida* and *Torulopsis*, the test may be of value because some perfect genera like *Hansenula*, *Kluyveromyces* and *Saccharomyces* lack hydrocarbon-assimilating representatives.

Table 7 shows that intermediates resulting from different degradation pathways have partly some relations to taxonomy, e.g.

- bacilli especially form intermediates form subterminal pathways. Most species can only cause cooxidization,
- ability for degradation among Streptomycetes is not widely distributed. Mostly they can only cooxidize long-chain alkanes,
- *Pseudomonas*, *Acetobacter*, *Micrococcus*, *Acinetobacter* and the group of *Corynebacterium*, *Arthrobacter*, *Mycobacterium* can cause rapid oxidation and their main pathway involves a monoterminal oxidation. Other pathways may occur, but it seems that they are not very important,
- yeasts, especially *Candida* (partly *Saccharomycopsis* and *Endomycopsis*), *Pichia* and *Torulopsis* can only give rise to terminal degradation infrequently form intermediates in the 2-position. Another subterminal pathway does not exist. Most yeasts oxidize alkanes very quickly,
- in molds both pathways, a terminal and a subterminal one, can occur. In Phycomycetes, diterminal degradation products could not be detected.

These results show that the common remark "alkane degradation" cannot be used for taxonomic purposes but that the remark „pathway for alkane degradation" or "intermediates which give hints at pathways for alkane degradation" may be useful for some taxonomic purposes.

It seems to be certain that microorganisms rapidly oxidizing alkanes preferably act through terminal pathways and that all microorganisms which oxidize alkanes less rapidly preferably react via subterminal pathways.

5 Concluding Remarks

Alkane degradation has been extensively investigated during the last 20 to 30 years. Investigations started from two viewpoints, from an ecological and from a metabolic one. Insufficient data are available on enzymes involved in different pathways, on the primary oxidation steps, especially of the subterminal pathway. Furthermore, results are missing about the regulation of primary steps of alkane oxidation and of the steps involved in the subsequent pathways of nearly all microorganisms, especially eukaryotic microorganisms.

Inspite of the decreasing availability of petroleum it is of interest to know the degradation pathways of microbial alkane oxidation, not only with respect to basic research but also with regard to the production of primary and secondary metabolites.

6 Acknowledgement

We are very grateful to Mrs. Bötticher, Mrs. Niemann und Miss Steckel for their technical assistance.

7 References

1. Abbott, B. J., Casida, L. E.: J. Bacteriol. *96*, 925 (1968)
2. Abbott, B. J., Gledhill, W. E.: Adv. Appl. Microbiol. *14*, 249 (1971)
3. Abbott, B. J., Hou, C. T.: Appl. Microbiol. *26*, 86 (1973)
4. Allen, J. E., Markovetz, A. J.: J. Bacteriol. *103*, 426 (1970)
5. Austin, B. et al.: Appl. Environ. Microbiol. *34*, 60 (1977)
6. Bacchin, P., Robertiello, A., Viglia, A.: Appl. Microbiol. *28*, 737 (1974)
7. Bassel, J., Ogrydziak, D. M.: In: Genetics of Industrial Microorganisms (Sebek, O. K., Laskin, A. I., eds.), p. 160. Washington, Amer. Soc. Microbiology 1979
8. Bel'tyukova, K. I. et al.: Fitopatog. Bakt. (Mater. Uses Konf. Bakt. Bolezn. Rast. J.) *1972*, 123 (Publ. 1975)
9. Bemmann, W., Tröger, R.: Zentralbl. Bakteriol. Abt. II *129*, 742 (1975)
10. Benson, S., Shapiro, J.: J. Bacteriol. *126*, 794 (1976)
11. Bertrand, J. C., Gallo, M., Azoulay, E.: Biochimie *55*, 343 (1973)
12. Bird, C. W., Lynch, J. M.: Chem. Soc. Rev. *3*, 309 (1974)
13. Bos, P.: Some Aspects of Hydrocarbon Assimilation by Yeasts. Thesis, Techn. Hogeschool Delft 1975
14. Bos, P., De Bruyn, J. C.: Antonie van Leeuwenhoek, J. Microbiol. Serol. *39*, 99 (1973)
15. Bourguin, A. W., Gibson, D. T.: Water Chlorination: Environ. Impact Health Eff., Proc. Conf. 1977 (Pub. 1978) *2*, 253 (1978)
16. Brieskorn, C. H.: In: „Unsichtbare" Fette und Lipoide in Lebensmitteln und ihre Bedeutung für die menschliche Ernährung (Zöllner, N., ed.), Darmstadt: Steinkopff 1973

17. Britton, L. N., Brand, J. M., Markovetz, A. J.: Biochim. Biophys. Acta *369*, 45 (1974)
18. Büning-Pfaue, H., Rehm, H. J.: Arch. Mikrobiol. *86*, 213 (1972)
19. Büning-Pfaue, H., Rehm, H. J.: Arch. Mikrobiol. *86*, 231 (1972a)
20. Cardini, G., Jurtshuk, P.: J. Biol. Chem. *243*, 6070 (1968)
21. Cerniglia, C. E., Perry, J. J.: J. Bacteriol. *118*, 844 (1974)
22. Chen, D. C. T., Markovetz, A. J.: Z. Allg. Mikrobiol. Morphol. Physiol. Ökol. Mikroorg. *14*, 525 (1974)
23. Chouteau, J., Azoulay, E., Senez, J. C.: Nature *194*, 576 (1962)
24. Cooney, J. J., Proby, C. M.: J. Bacteriol. *108*, 777 (1971)
25. Delaissé, J. M., Nyns, E. J.: Arch. Intern. Physiol. Biochim. *82*, 179 (1974)
26. Dunlap, K. R., Perry, J. J.: J. Bacteriol. *94*, 1919 (1967)
27. Dunlap, K. R., Perry, J. J.: J. Bacteriol. *96*, 318 (1968)
28. Duppel, W., Lebeault, J. M., Coon, M. J.: Eur. J. Biochem. *36*, 583 (1973)
29. Edmonds, P., Cooney, J. J.: J. Bacteriol. *98*, 16 (1969)
30. Einsele, A., Fiechter, A.: Adv. Biochem. Engng. *1*, 170 (1971)
31. Eubanks, E. F., Forney, F. W., Larson, A. D.: J. Bacteriol. *120*, 1133 (1974)
32. Ferris, J. P. et al.: Arch. Biochem. Biophys. *175*, 443 (1976)
33. Finnerty, W. R.: Abstracts XII Intern. Congr. Microbiol., p. 18, München 1978
34. Finnerty, W. R., Hawtrey, E., Kallio, R. E.: Z. Allg. Mikrobiol. *2*, 169 (1962)
35. Fonken, G. S., Murray, H. C., Reinecke, L. M.: J. Amer. Chem. Soc. *82*, 5507 (1960)
36. Forney, F. W., Markovetz, A. J.: J. Bacteriol. *96*, 1055 (1968)
37. Forney, F. W., Markovetz, A. J.: Biochem. Biophys. Res. Commun. *37*, 31 (1969)
38. Forney, F. W., Markovetz, A. J.: J. Bacteriol. *102*, 281 (1970)
39. Forney, F. W., Markovetz, A. J., Kallio, R. E.: J. Bacteriol. *93*, 649 (1967)
40. Fredricks, K.: Antonie van Leeuwenhoek *33*, 41 (1967)
41. Fukui, S., Tanaka, A.: Adv. Biochem. Engng. *17*, 1 (1980)
42. Gallo, M., Bertrand, J. C., Azoulay, E.: FEBS Lett. *19*, 45 (1971)
43. Gallo, M. et al.: Biochim. Biophys. Acta *296*, 624 (1973)
44. Gallo, M., Roche, B., Azoulay, E.: Biochim. Biophys. Acta *419*, 425 (1976)
45. Gerasimova, N. M., Lin, L. Z.: Mikrobiologiya *44*, 460 (1975)
46. Gmünder, F. K.: Die Assimilation von Hexadecan durch *Candida tropicalis*. Dissertation ETH Zürich 1979
47. Golovlev, E. L., Skryabin, G. K., Golovleva, L. A.: Atyp. Mycobacteria, Proc. Symp. *1971*, 91 (Publ. 1973)
48. Großebüter, W., Reiff, I., Rehm, H. J.: Eur. J. Appl. Microbiol. Biotechnol. *8*, 139 (1979)
49. Guerillot, L., Vandescasteele, J. P.: Eur. J. Biochem. *81*, 185 (1977)
50. Gunsalus, I. C., Pederson, T. C., Sligar, S. G.: Ann. Rev. Biochem. *44*, 377 (1975)
51. Haines, J. R., Alexander, M.: Appl. Microbiol. *28*, 1084 (1974)
52. Hammer, K. D., Liemann, F.: Zbl. Bakt. Hyg., I. Abt. Orig. B *162*, 169 (1976).
53. Heydeman, M. T., Azoulay, E.: Biochim. Biophys. Acta *77*, 545 (1963)
54. Hirai, M. et al.: Agr. Biol. Chem. *40*, 1819 (1976)
55. Hoffmann, B., Rehm, H. J.: Eur. J. Appl. Microbiol. *3*, 19 (1976)
56. Hoffmann, B., Rehm, H. J.: ibid. *3*, 31 (1976a)
57. Hoffmann, B., Schulte, E., Rehm, H. J.: ibid. *3*, 273 (1977)
58. Hoffmann, B., Rehm, H. J.: ibid. *5*, 189 (1978)
59. Hortmann, L., Rehm, H. J.: Manuscript in preparation for publ. in Eur. J. Appl. Microbiol. Biotechnol. (1981)
60. Horvath, R. S.: Bacteriol. Rev. *36*, 146 (1972)
61. Hug, H., Fiechter, A.: Arch. Mikrobiol. *88*, 87 (1973)
62. Hutchinson, D. W., Whittenbury, R., Dalton, H.: J. Theor. Biol. *58*, 325 (1976)
63. Iida, M., Iizuka, H.: Z. Allg. Mikrobiol. *10*, 245 (1970)
64. Iizuka, H., Iida, M., Toyoda, S.: Z. Allg. Mikrobiol. *6*, 335 (1966)
65. Iizuka, H., Iida, M., Unami, Y., Hoshino, Y.: Z. Allg. Mikrobiol. *8*, 145 (1968)
66. Iizuka, H., Iida, M., Fujita, S.: Z. Allg. Mikrobiol. *9*, 223 (1969)
67. Imada, Y. et al.: Biotech. Bioeng. *9*, 45 (1967)
68. Imelik, B.: C. R. Acad. Sci. *226*, 2082 (1948)
69. Jones, D. F.: J. Chem. Soc. (C), 2809 (1968)

70. Jones, D. F.: ibid. 2827 (1968a)
71. Jones, D. F., Howe, R.: ibid. 2801 (1968)
72. Jones, D. F., Howe, R.: ibid. 2816 (1968a)
73. Jwanny, E. W.: Z. Allg. Mikrobiol. *15*, 423 (1975)
74. Kachholz, T., Rehm, H. J.: Eur. J. Appl. Microbiol. *4*, 101 (1977)
75. Kachholz, T., Rehm, H. J.: ibid. *6*, 39 (1978)
76. Kaufmann, H. P.: Analyse der Fette und Fettprodukte, Bd. I. Berlin–Göttingen–Heidelberg: Springer 1958
77. Kester, A. S., Foster, J. W.: Bacteriol. Proc. 168 (1960)
78. Kester, A. S., Foster, J. W.: J. Bacteriol. *85*, 859 (1963)
79. Killinger, A.: Arch. Mikrobiol. *73*, 153 (1970)
80. Killinger, A.: ibid. *73*, 160 (1970a)
81. King, D. H., Perry, J. J.: Can. J. Microbiol. *21*, 85 (1975)
82. Klein, D. A., Henning, F. A.: Appl. Microbiol. *17*, 676 (1969)
83. Klein, D. A., Davis, J. A., Casida, L. E. jr.: Antonie van Leeuwenhoek *34*, 495 (1968)
84. Kleine Borgmann, B., Rehm, H. J.: Manuscript in preparation for publ. in Eur. J. Appl. Microbiol. Biotechnol. (1981)
85. Klug, M. J., Markovetz, A. J.: J. Bacteriol. *93*, 1847 (1967)
86. Klug, M. J., Markovetz, A. J.: Adv. Microbial Physiol. *5*, 1 (1971)
87. Krauel, H., Weide, H.: Z. Allg. Mikrobiol. *18*, 47 (1978)
88. Kusunose, M. et al.: Agr. Biol. Chem. *31*, 990 (1967)
89. Leadbetter, E. R., Foster, J. W.: Arch. Mikrobiol. *35*, 92 (1960)
90. Lebeault, J. M., Azoulay, E.: Lipids *6*, 444 (1971)
91. Lebeault, J. M. et al.: Arch. Mikrobiol. *72*, 140 (1970)
92. Le Petit, J. et al.: Ann. Microbiol. *126* A, 367 (1975)
93. Lijmbach, G. W. M., Brinkhuis, E.: Antonie van Leeuwenhoek *39*, 415 (1973)
94. Lin, H. T., Iida, M., Iizuka, H.: J. Ferment. Technol. *49*, 206 (1971)
95. Lin, H. T., Iida, M., Iizuka, H.: ibid. *49*, 771 (1971a)
96. van der Linden, A. C., Huybregtse, R.: Antonie van Leeuwenhoek *35*, 344 (1969)
97. Lindenmayer, A., Smith, L.: Biochim. Biophys. Acta *93*, 445 (1964)
98. Lu, A. Y. H., Coon, M. J.: J. Biol. Chem. *243*, 1331 (1968)
99. Lüdecke, C.: In: Analyse der Fette und Fettprodukte, Bd. I (Kaufmann, H. P., ed.), Berlin–Göttingen–Heidelberg: Springer 1958
100. Lukins, H. B., Foster, J. W.: J. Bacteriol. *85*, 1174 (1963)
101. Makula, R. A., Finnerty, W. R.: ibid. *95*, 2102 (1968)
102. Makula, R. A., Finnerty, W. R.: ibid. *112*, 398 (1972)
103. Makula, R. A., Lookwood, P. J., Finnerty, W. R.: ibid. *121*, 250 (1975)
104. Markovetz, A. J.: Abstr. XII Intern. Congr. Microbiology, p. 18, München, 3–8 Sept. 1978
105. Mazzola, E. B., Golik, S. I.: Rev. Fac. Ing. Qca. U.N.L. *39*, 169 (1970)
106. McKenna, E. J., Kallio, R. E.: Ann. Rev. Microbiol. *19*, 183 (1965)
107. Mishina, M. et al.: Agr. Biol. Chem. *37*, 863 (1973)
108. Miura, Y.: Adv. Biochem. Eng. *9*, 31 (1978)
109. Mizuno, M., Shimojima, Y., Iguchi, T., Takeda, I., Senoh, S.: Agr. Biol. Chem. *30*, 506 (1966)
110. Neryng, A.: Postepy Mikrobiol. *15*, 57 (1976)
111. Nieder, M., Shapiro, J.: J. Bacteriol. *122*, 93 (1975)
112. Nolof, G., Hirsch, P.: Arch. Mikrobiol. *44*, 266 (1962)
113. Nordblom, G. D., White, R. E., Coon, M. J.: Arch. Biochem. Biophys. *175*, 524 (1976)
114. Nyns, E. J., Auquière, J. P., Wiaux, A. L.: Antonie von Leeuwenhoek *34*, 441 (1968)
115. Ogino, S., Yano, K., Tamura, G., Arima, K.: Agr. Biol. Chem. *29*, 1009 (1965)
116. Okuhara, M., Kubochi, Y., Harada, T.: Agr. Biol. Chem. *35*, 1367 (1971)
117. Omar, S. H., Rehm, H. J.: Eur. J. Appl. Microbiol. Biotechnol. *11*, 35 and 42 (1980)
118. Paoletti, C. et al.: Lipids *11*, 258 (1976)
119. Parekh, V. R., Traxler, R. W., Sobek, J. M.: Appl. Environm. Microbiol. *33*, 881 (1977)
120. Pelz, B. F., Rehm, H. J.: Arch. Mikrobiol. *84*, 20 (1972)
121. Pelz, B. F., Rehm, H. J.: Naturwissenschaften *59*, 513 (1972a)

122. Pelz, B. F., Rehm, H. J.: Arch. Mikrobiol. *92*, 153 (1973)
123. Perry, J. J., CRC Crit. Rev. Microbiol. *5*, 387 (1977)
124. Perry, J. J.: Microbiol. Rev. *43*, 59 (1979)
125. Peterson, J. A. et al.: J. Biol. Chem. *242*, 4334 (1967)
126. Peterson, J. A. et al.: Arch. Biochem. Biophys. *131*, 245 (1969)
127. Pirnik, M. P., Atlas, R. M., Bartha, R.: J. Bacteriol. *119*, 868 (1974)
128. Pirnik, M. P.: CRC Crit. Rev. Microbiol. *5*, 413 (1977)
129. Rahim, M. A., Sih, C. J.: J. Biol. Chem. *241*, 3615 (1966)
130. van Ravenswaay Claasen, J. C., van der Linden, A. C.: Antonie van Leeuwenhoek *37*, 339 (1971)
131. Raymond, R. L., Jamison, V. W., Hudson, J. O.: Lipids *6*, 453 (1971)
132. Rehm, H. J.: In: Dechema-Monographien Nr. 1670–1692, Biotechnologie. Rehm, H. J. (ed.), Bd. 81, p. 145. Weinheim–New York, Verlag Chemie 1977
132. Rehm, H. J.: In: Dechema-Monographien Nr. 1670–1692, p. 145. Biotechnologie. Rehm, H. J. (ed.), Bd. 81, Weinheim–New York, Verlag Chemie 1977
133. Rehm, H. J.: XII Intern. Congr. Microbiology, p. 18, München, 3–8 Sept. 1978, Abstr., Symp. No. 13.2 (1978)
134. Rehm, H. J., Reiff, I.: Unpublished (1979)
135. Romero, E. M., Brenner, R. R.: J. Bacteriol. *91*, 183 (1966)
136. Sahm, H.: Adv. Biochem. Engng. *6*, 77 (1977)
137. Sato, R., Omura, T. (eds.): Cytochrome P-450. Tokyo, Kodansha Ltd. and New York–San Francisco–London: Academic Press 1978
138. Schnabl, H., Rehm, H. J.: Naturwissenschaften *58*, 55 (1971)
139. Schröder, E., Rehm, H. J.: Eur. J. Appl. Microbiol. Biotechnol. *11* (1981) in press
140. Schunck, W. H. et al.: Acta Biol. Med. Germ. *37*, K3 (1978)
141. Schunck, W. H., Riege, P., Kuhl, R.: Pharmazie *33*, 412 (1978a)
142. Schwartz, R. D., McCoy, C. J.: Appl. Environm. Microbiol. *31*, 78 (1976)
143. Sebek, O. K., Kieslich, K.: In: Ann. Rep. Fermentation Proc. (Perlman, D., ed.) *1*, 267 (1977)
144. Seeler, G.: Arch. Mikrobiol. *43*, 213 (1962)
145. Senez, J. C., Azoulay, E.: Biochim. Biophys. Acta *47*, 306 (1961)
146. Shapiro, J., Fennewald, M., Benson, S.: In: Genetics of Industrial Microorganisms (Sebek, O. K., Laskin, A. I., eds.), p. 147. Proc. 3rd Intern. Symp. on Genetics of Ind. Microorganisms, Wisconsin 1978. Washington: Amer. So. Microbiol. 1979
147. Shiio, I., Uchio, R.: Agr. Biol. Chem. *35*, 2033 (1971)
148. Shum, A. C., Markovetz, A. J.: J. Bacteriol. *118*, 880 (1974)
149. Shum, A. C., Markovetz, A. J.: ibid. *118*, 890 (1974a)
150. Siporin, C., Cooney, J. J.: Appl. Microbiol. *29*, 604 (1975)
151. Soda, K., Misono, H., Hirasawa, T., Moriguchi, M.: In: Microbiology fór Environment Cleaning (Arima, K., ed.), p. 487. Ministry of Education; Grant No. 212204-1977 (1978)
152. Sorger, H., Aurich, H.: Wiss. Z. Univers. Leipzig, Math.-Naturwiss. R. *27*, 35 (1978)
153. Souw, P., Luftmann, H., Rehm, H. J.: Eur. J. Appl. Microbiol. *3*, 289 (1977)
154. Souw, P., Schulte, E., Rehm, H. J.: XII Intern. Congr. Microbiology, München, 3—8 Sept. 1978, Abstr., No. F. 8, p. 145 (1978)
155. Stewart, J. E. et al.: J. Bacteriol. *78*, 441 (1959)
156. Suzuki, T., Ogawa, K.: Agr. Biol. Chem. *36*, 457 (1972)
157. Tabuchi, T., Hara, S.: ibid. *38*, 1105 (1974)
158. Tabuchi, T., Serizawa, N.: ibid. *39*, 1055 (1975)
159. Tabuchi, T., Uchiyama, H.: ibid. *39*, 2035 (1975)
160. Tanaka, A., Fukui, S.: Kagaku (Kyoto) *30*, 686 (1975)
161. Tanaka, A. et al.: Eur. J. Appl. Microbiol. *3*, 115 (1976)
162. Tassin, J. P., Celier, C., Vandecasteele, J. P.: Biochim. Biophys. Acta *315*, 220 (1973)
163. Teranishi, Y. et al.: Agr. Biol. Chem. *38*, 1213 (1974)
164. Teranishi, Y. et al.: ibid. *38*, 1221 (1974a)
165. Teranishi, Y., Tanaka, A., Fukui, S.: ibid. *38*, 1779 (1974b)
166. Teranishi, Y. et al.: ibid. *38*, 1581 (1974c)
167. Thiele, H., Rehm, H. J.: Eur. J. Appl. Microbiol. Biotechnol. *6*, 361 (1979)
168. Tittelbach, M., Rohde, H. G., Weide, H.: Z. Allg. Mikrobiol. *16*, 155 (1976)

169. Tornabene, T. G.: Microb. Energy Convers., Proc. Semin. 1976, p. 281 (1977)
170. Uchio, R., Shiio, I.: Agr. Biol. Chem. *36*, 1389 (1972)
171. Updegraff, D. M., Bovey, F. A.: Nature *181*, 890 (1958)
172. Wagner, F., Zahn, W., Bühring, U.: Angew. Chem. *79*, 314 (1967)
173. Walker, J. D., Cooney, J. J.: Can. J. Microbiol. *19*, 1325 (1973)
174. Walker, J. D., Cooney, J. J.: J. Bacteriol. *115*, 635 (1973a)
175. Walker, J. D., Scott Pore, R.: Appl. Environm. Microbiol. *35*, 694 (1978)
176. Weete, J.: Phytochem. *11*, 1201 (1972)
177. Yamada, K., Torigoe, Y.: J. Agr. Chem. Soc. *40*, 364 (1966)
178. Yamaguchi, K., Kurosawa, M.: Agr. Biol. Chem. *40*, 719 (1976)
179. Yamakawa, Y., Goto, S., Yokotsuka, J.: Agr. Biol. Chem. *42*, 269 (1978)
180. Yanagawa, S. et al.: Agr. Biol. Chem. *36*, 2123 (1972)
181. Yano, I., Furukawa, Y., Kusunose, M.: Eur. J. Biochem. *23*, 220 (1971)
182. Yoshida, Y. et al.: Biochem. Biophys. Res. Comm. *78*, 1005 (1977)
183. Yu, C. A., Gunsalus, I. C.: J. Biol. Chem. *249*, 107 (1974)

Metabolism of Alkanes by Yeasts

S. Fukui, A. Tanaka

Laboratory of Industrial Biochemistry, Department of Industrial Chemistry, Faculty of Engineering, Kyoto University, Yoshida, Sakyo-ku, Kyoto 606, Japan

One of the specific features of alkane-utilizing yeasts is the conspicuous appearance of peroxisomes. This review describes the metabolism of alkanes in yeasts with special emphasis on the physiological function of peroxisomes. The subtle diversity in alkane utilization pathway in yeasts is mediated by subcellular localization of enzymes.

In microsomes, alkanes are hydroxylated to the corresponding fatty alcohols which are further oxidized to fatty acids *via* aldehydes in microsomes, mitochondria and peroxisomes, respectively. Degradation of fatty acids to acetyl-CoA *via* the β-oxidation pathways is carried out exclusively in peroxisomes while fatty acids formed in microsomes and mitochondria are incorporated into cellular lipids, each after being activated to acyl-CoAs. Acetyl-CoA produced in peroxisomes is converted to C_4-compounds by the cooperative action of peroxisomes and mitochondria. Some regulation of enzymes in alkane-assimilating yeasts is also discussed.

1 Introduction

Despite extensive studies on the application of alkane-utilizing microorganisms for the production of a variety of useful substances[1,2], fundamental knowledge is presumably insufficient to understand the metabolic and physiological features of microorganisms which can utilize such unconventional carbon sources.

An important characteristic of alkane assimilation by microorganisms is the flow of carbon from alkane substrates to cellular carbohydrates *via* fatty acids, which is quite different from the case of conventional substrates, i.e. carbohydrates (Table 1). Another characteristic is that the main test organisms in alkane assimilation are *Candida* yeasts, whose metabolic features are fairly different from those of *Saccharomyces* yeasts being utilized for the metabolic and physiological studies as typical eukaryotic microorganisms.

Alkane assimilation by microorganisms is divided into several steps (Table 2):

Table 1. Metabolic features of different carbon sources

Substrate	Cellular components	
	Fatty acids	Carbohydrates
Carbohydrates	Synthesis	Degradation
C_2-, C_3-Compounds	Synthesis	Synthesis
Alkanes, fatty acids	Degradation	Synthesis

Table 2. Metabolism of alkanes

Uptake of alkanes
Oxidation of alkanes to the corresponding fatty acids
Degradation of fatty acids to acetyl-CoA
Incorporation of fatty acids into cellular lipids
Synthesis of TCA cycle intermediates
Synthesis of carbohydrates
Syntheses of other cellular components

(1) Incorporation of alkanes into cells; (2) oxidation of alkanes to the corresponding fatty acids; (3) activation of fatty acids to their CoA esters; (4) subsequent metabolism of fatty acyl-CoAs — degradation to acetyl-CoA or incorporation of fatty acyl moieties into cellular lipids; (5) synthesis of TCA cycle intermediates from acetyl-CoA *via* the glyoxylate cycle; (6) gluconeogenesis; (7) syntheses of amino acids, nucleic acids, and others including a portion of cellular fatty acids.

Steps 1 and 2 are the most specific processes of microbial alkane utilization. Steps 3, 4 and 5 are common to those in the metabolism of fatty acids, and step 6 to that of gluconeogenic substrates such as ethanol and acetate, respectively. Step 7 is common to that of various other carbon sources including carbohydrates.

This review summarizes the available information on alkane-utilizing yeasts according to the above-mentioned classification with special emphasis on the metabolic function of peroxisomes which appear conspicuously in the cells.

2 Uptake of Alkanes

The first step of alkane assimilation by microorganisms is the uptake of exogenous alkanes by cells and further transport to the site where alkanes undergo the initial oxidation. Studies on the incorporation phenomena of alkanes have been mainly carried out in the field of biochemical engineering. The results accumulated hitherto are summarized in Table 3. In the meantime, several microbial products are known to stimulate the growth of microorganisms on alkanes, due to their strong activity to emulsify alkanes. An "emulsifying factor" produced by a strain of *Candida* consists of peptide and fatty acids[3]. In contrast to these works dealing with the effect of external circumstances, far lesser works have been carried out on the cytological and cytochemical aspects of the microbial alkane uptake.

Table 3. Uptake mechanisms of alkanes

1. Uptake through direct contact between alkane droplets and microbial cells
2. Uptake *via* accommodated alkane phase
3. Uptake of (pseudo)solubilized alkanes

Meissel et al.[4,5] have observed electron microscopically special channels in the cell wall of alkane-grown cells of *C. tropicalis*. Similar channels accompanied by slime-like outgrowths (Fig. 1) have also been detected by the authors[6] together with protrusions on the cell surface of alkane-grown *C. tropicalis* (Fig. 2) at an early phase of growth. As seen in Fig. 1, the slime-like outgrowths reach the cell membrane through the electron-dense channels, and endoplasmic reticulum is arranged regularly beneath each channel. This observation has led to the hypothesis that alkanes attached to the protrusions (or slime-like outgrowths) may migrate through the channels to the endoplasmic reticulum, the site of alkane hydroxylation.

The protrusions (Fig. 2) or outgrowths (Fig. 1) do not seem to be alkanes or lipids which are simply attached to the cell wall, because these constituents cannot be removed by washing the cells with organic solvents or detergents. Fiechter et al.[7–9] have demonstrated that alkane microemulsion adheres by a non-enzymatic mechanism to the cell wall of *C. tropicalis* and that a lipopolysaccharide located at the cell wall is responsible for this affinity. The lipopolysaccharide whose formation is induced by alkanes is isolated and characterized as mannan containing approximately 4% covalently linked fatty acids[10].

These results shed a faint light on the mechanism of microbial alkane uptake. However, biochemical information on the transport process is still insufficient at present.

3 Initial Oxidation of Alkanes

Alkanes uptaken by cells are subjected to initial oxidation followed by further metabolism. At present, three different mechanisms are reported for the oxidation:
1) Hydroxylation of alkanes by mono-oxygenase (mixed-function oxidase);
2) dehydrogenation of alkanes to the corresponding alkenes;
3) peroxidation of alkanes.

Alkane molecules are susceptible to such oxidations at either terminal (mono-terminal) and, in some cases, at both terminal (di-terminal), subterminal or internal positions of the carbon chain (Fig. 3). However, the mono-terminal oxidation is considered to be the main pathway of alkane utilization by yeasts, although the di-terminal oxidation in yeasts has been recognized in some strains and applied to the production of long-chain dicarboxylic acids.

The most well-known mono-oxygenase system is that of *Pseudomonas*, consisting of rubredoxin, NADH-rubredoxin reductase and ω-hydroxylase (Fig. 4)[11]. On the other hand, *Corynebacterium* sp. contains another hydroxylation system composed of cytochrome P-450 and NADH-cytochrome *c* reductase[12]. Several lines of evidence strongly indicate that the yeast initial alkane oxidation system involves cytochrome P-450.

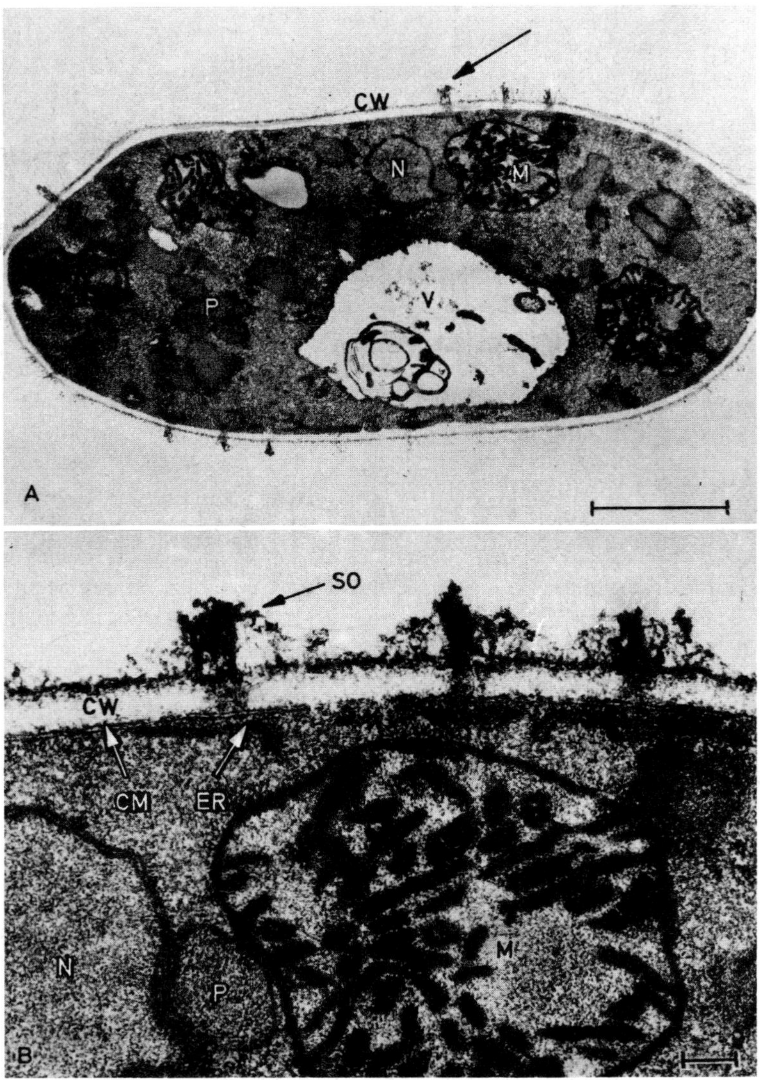

Fig. 1 Thin-section of *C. tropicalis* grown on alkanes **(A)** and an enlarged photograph **(B)**[6]. Development of slime-like outgrowths and conspicuous appearance of peroxisomes are observed. Abbreviations: CM = cell membrane; CW = cell wall; ER = endoplasmic reticulum; M = mitochondrion; N = nucleus; P = peroxisome; SO = slime-like outgrowth; V = vacuole. Bar, 1 μm **(A)** or 100 nm **(B)**

Liu and Johnson[13] have reported an inducible, particulate oxidation system in *C. intermedia* requiring molecular oxygen for activity. Participation of cytochrome P-450 and NADPH-cytochrome c reductase in the *C. tropicalis* and *C. guilliermondii* systems has been reported by Coon's group[14] and Azoulay's group[15], and by Müller et al.[16]. Azoulay's group has intensively studied the subcellular localization of the hydroxylation system, regarding it as microsomal one[17-20]. They have partially purified cytochrome P-450[21] and NADPH-cytochrome c reductase[22] from *C. tropicalis* and reconstituted the system with these two enzymes[21]. On the other hand, Coon's group has resolved the system of *C. tropicalis* into three components — cytochrome P-450, NADPH-cytochrome c

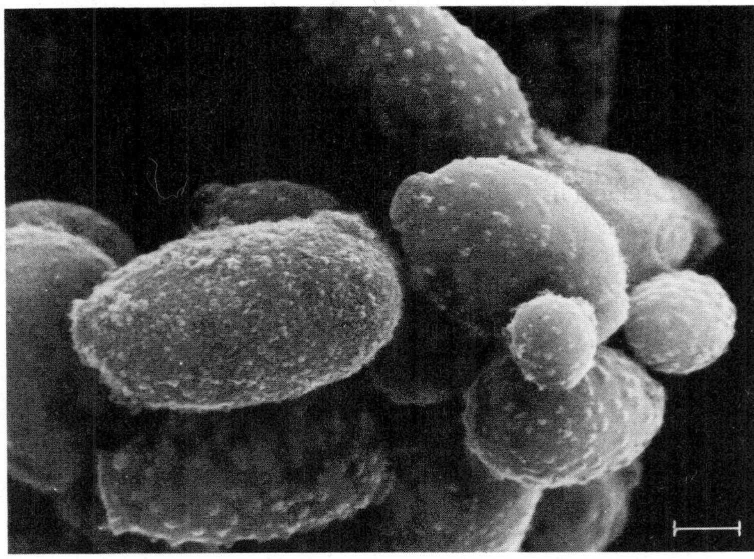

Fig. 2 Surface view of *C. tropicalis* grown on alkanes[6]. Many protrusions are observed on the cell surface. Bar, 1 μm

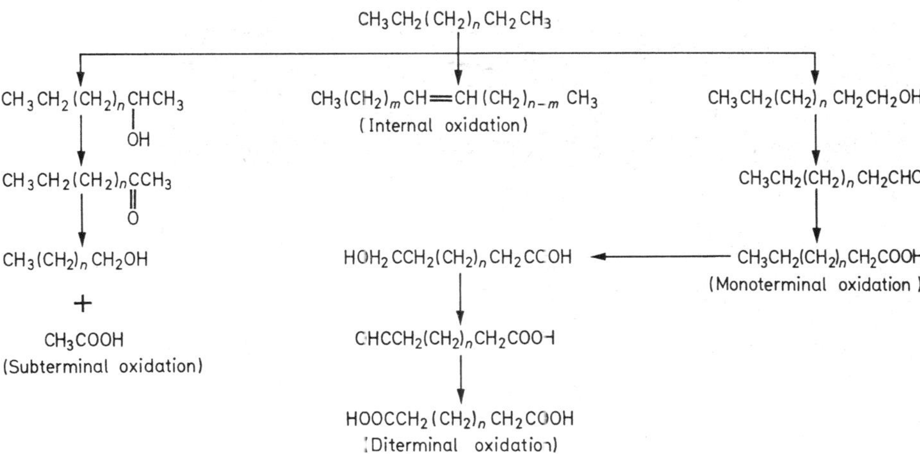

Fig. 3 Microbial oxidation pathways of alkanes

(cytochrome P-450) reductase and a heat-stable lipid fraction[23]. Although the catalytic mechanism of the yeast hydroxylation system has not yet been elucidated, it is presumed to be analogous to the liver microsomal system consisting of cytochrome P-450, reductase and a phospholipid, which is proposed by Strobel and Coon (Fig. 5)[24].

Another mechanism involving dehydrogenation of alkanes to alkenes and subsequent hydration to alcohols has been proposed. For example, Iida and Iizuka have reported the dehydrogenation system in *C. rugosa*[25]. However, biochemically conclusive evidence for this system has not been presented.

Recently, Finnerty[26] has proposed a system of a certain bacterial strain in which alkane peroxide is involved. This system has not yet been elucidated biochemically, too.

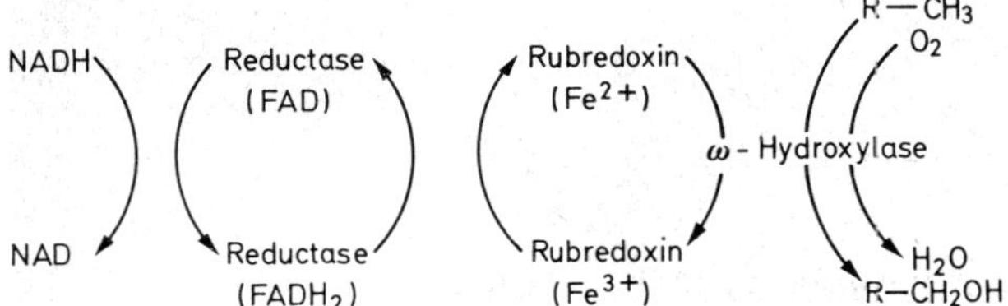

Fig. 4 Rubredoxin-dependent hydroxylation system of *Pseudomonas oleovorans*[11)]

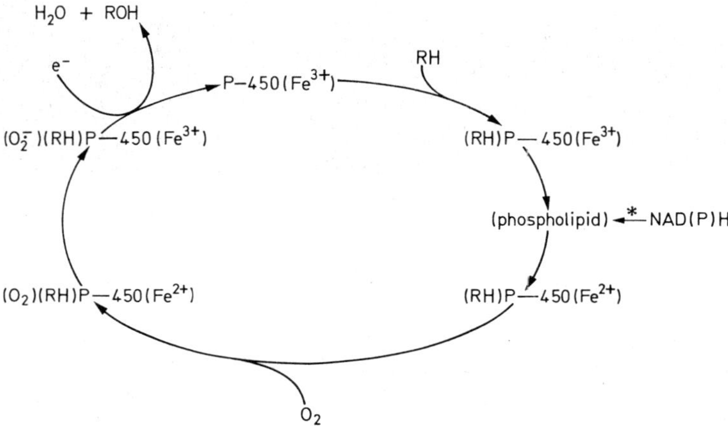

*NAD(P)H-cytochrome P—450 reductase

Fig. 5 Proposed mechanism of hydroxylation in liver microsomes[24)]

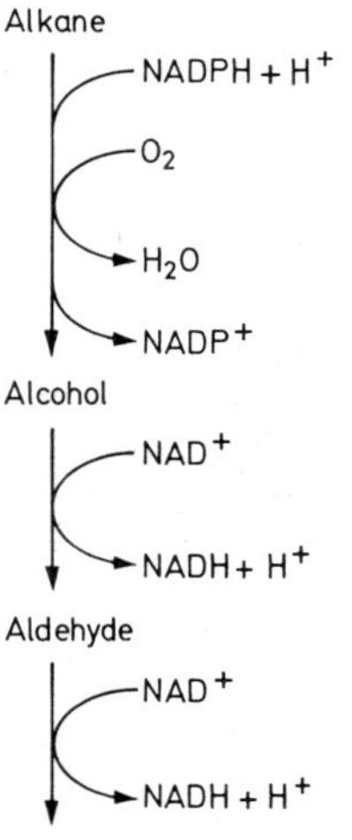

Fig. 6 Oxidation pathway of alkanes to the corresponding fatty acids

Of the alkane oxidation system mentioned above, the hydroxylation system involving cytochrome P-450, which operates under aerobic conditions, is most plausible in alkane-utilizing yeasts.

4 Oxidation of Higher Alcohols to Fatty Acids

Higher alcohols, derived from alkanes by hydroxylation in microsomes, are oxidized to the corresponding fatty acids *via* aldehydes (Fig. 6). NAD-linked alcohol dehydrogenase and aldehyde dehydrogenase, which are specific to long-chain substrates, participate in these reactions[27,28]. Both enzymes are inducible by alkanes, long-chain alcohols or aldehydes. The substrate specificity of alcohol dehydrogenase is shown in Table 4[29]. Aldehyde dehydrogenase also reveals a similar substrate specificity[27]. Interestingly, these enzymes of *C. tropicalis* have no or only weak activities on substrates with a chain length of C-15 or more, although yeast could assimilate C_{15} to C_{19} alkanes. This may be due to the difficulty of conducting experiments with water-insoluble substrates, or to the occurrence of other enzymes such as oxidases or dehydrogenases depending on an unknown hydrogen acceptor.

Azoulay's group has also investigated the localization of long-chain alcohol dehydrogenase and aldehyde dehydrogenase in alkane-grown *C. tropicalis* cells. However, the results reported are rather confusing. Thus in one paper both enzymes are said to be present in microsomes in a small portion and in mitochondria in a large portion[18] and, according to other papers, alcohol dehydrogenase is either localized in cytoplasm and mitochondria[29] or in mitochondria[30]. Furthermore, their mitochondrial fraction might include peroxisomes, since these were not recognized in alkane-grown yeasts at that time. Recently, the authors have detected the activities of long-chain alcohol dehydrogenase and aldehyde dehydrogenase among microsomes, mitochondria and peroxisomes of alkane-grown *C. tropicalis* cells[31].

5 Appearance of Peroxisomes

Before studying the further metabolism of alkanes, it is necessary to describe the appearance of peroxisomes in alkane-grown yeast cells, because peroxisomes play important roles in alkane assimilation.

When cultivated on alkanes, yeast cells exhibit a special fine structure — conspicuous appearance of specific organelles (Fig. 1A) [32,33]. Such organelles can be rarely

Table 4. Substrate specificity of soluble alcohol dehydrogenase from *C. tropicalis*[29]

Substrate	Relative activity (%)	Substrate	Relative activity (%)
Ethanol	0	1-Nonanol	75
n-Propanol	0	1-Decanol	100
n-Butanol	3	1-Undecanol	85
1-Hexanol	25	1-Dodecanol	120
1-Heptanol	40	1-Tridecanol	90
1-Octanol	70	1-Tetradecanol	90

Activity on 1-decanol = 100%

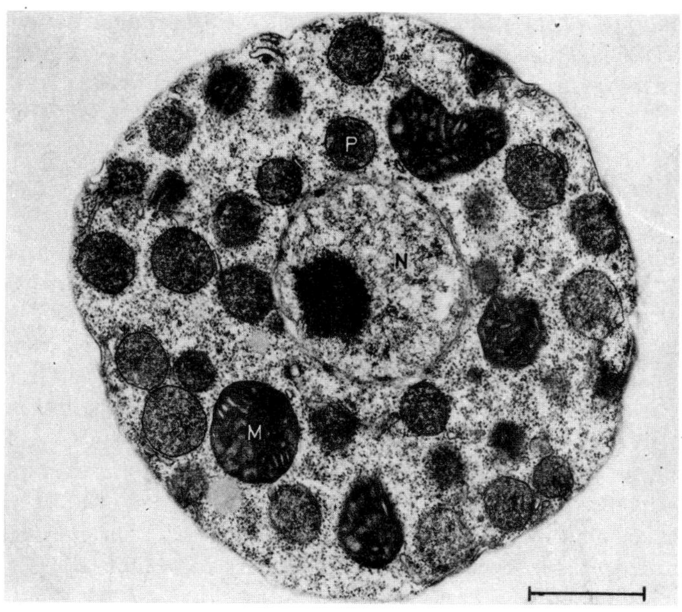

Fig. 7 Protoplast of *C. tropicalis* grown on alkanes. Abbreviations: M = mitochondrion; N = nucleus; P = peroxisome. Bar, 1 μm

observed in glucose-grown cells. These organelles have been identified as peroxisomes by a cytochemical technique. The development of peroxisomes in yeasts is closely correlated to the increase in catalase activity and the assimilation of alkanes or higher fatty acids[34,35], indicating the participation of the organelles and catalase in the alkane or fatty acid metabolism. The peroxisomes have been isolated intact from protoplasts of alkane-grown *C. tropicalis* (Fig. 7) [36,37] and from protoplasts of fatty acid-grown *C. lipolytica*[38] by means of differential and sucrose density gradient centrifugations (Fig. 8). Enzymes essential for fatty acid metabolism are located in peroxisomes as described below. The localization of long-chain alcohol dehydrogenase and aldehyde dehydrogenase in the peroxisomes was mentioned previously. The presence of DNA different from nuclear and mitochondrial DNAs has also been reported[39,40], although the role of peroxisomal DNA remains to be clarified.

Cytology and metabolic functions of yeast peroxisomes are summarized by the authors[41,42].

6 Activation of Fatty Acids to the Corresponding Coenzyme A Esters

Fatty acids must be activated to the corresponding CoA esters prior to further metabolism.

Trust and Millis[43] have reported the presence of acyl-CoA synthetase, which is active on heptanoate to tetradecanoate (maximum activity on decanoate and undecanoate), in the cell extract of alkane-grown *Torulopsis* sp. Duvnjak et al.[44] have also described the presence of acyl-CoA synthetase, which

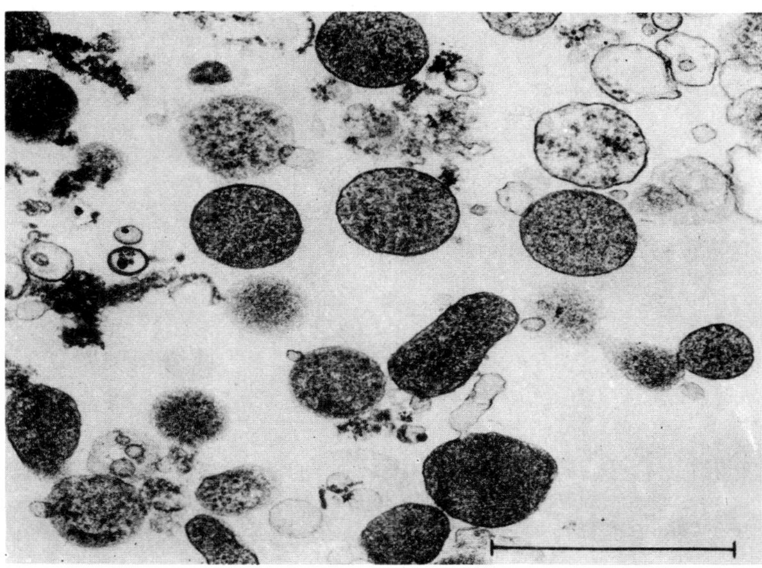

Fig. 8 Peroxisomes isolated from alkane-grown *C. tropicalis*. The peroxisomes are surrounded by a single unit membrane and have a granular matrix. Bar, 1 μm

Table 5. Comparison of the properties of acyl-CoA synthetases from *C. lipolytica*[38,47]

Properties	Synthetase I	Synthetase II
Induction by fatty acid	No	Yes
Phosphatidylcholine dependency	No	Yes
Stability	High	Low
Substrate specificity	Narrow	Wide
Solubilization by Triton X-100	Easy	Difficult
Subcellular localization	Microsomes, Mitochondria, *etc.*	Peroxisomes
Function	Lipid synthesis	Fatty acid degradation

is active on undecanoate to stearate (maximum activity on pentadecanoate), in the particulate fraction of alkane-grown *C. tropicalis*. The enzyme of *C. tropicalis* differs from that of *Torulopsis* sp. in the substrate specificity. We have found that peroxisomes isolated from alkane-grown *C. tropicalis* are able to degrade palmitate to acetyl-CoA in the presence of CoA and ATP, indicating the presence of acyl-CoA synthetase in these organelles[45].

Recently, Numa and his coworkers have reported the presence of two acyl-CoA synthetases differing in subcellular localization, control mechanism and metabolic functions in oleate-grown *C. lipolytica* (Table 5). They have isolated mutant strains of *C. lipolytica*, which cannot grow on glucose in the presence of cerulenin, an anti-lipogenic antibiotic, even when exogenous fatty acids are supplied[46]. These mutants cannot incorporate exogenous fatty acids into cellular lipids as a whole, suggesting the deficiency of acyl-CoA synthetase. Although the mutants do fail to

Table 6. Substrate specificities of acyl-CoA synthetases from *C. lipolytica*[47)]

Substrate	Relative activity (%)	
	Synthetase I	Synthetase II
Decanoic acid	28	19
Undecanoic acid	42	49
Lauric acid	48	61
Tridecanoic acid	58	61
Myristic acid	80	87
Pentadecanoic acid	100	100
Palmitic acid	104	90
Heptadecanoic acid	74	87
Stearic acid	20	53
Arachidic acid	0	16
Dodecanedioic acid	0	9
Hexadecanedioic acid	0	33
Eicosanedioic acid	0	16

Activities on pentadecanoic acid = 100%
Substrate concentration = 2 mM

show the activity of acyl-CoA synthetase when assayed under ordinary conditions, these strains can assimilate alkanes or fatty acids as the sole carbon source. The results suggest the presence of another acyl-CoA synthetase, whose activity cannot be detected under the conditions employed. In fact, they succeeded in isolating two distinct acyl-CoA synthetases from the particulate fractions of *C. lipolytica*[47)]. One (acyl-CoA synthetase I) has been observed in both oleic acid-grown and glucose-grown cells and is independent of phosphatidylcholine. The other (acyl-CoA synthetase II) has been detected only in oleic acid-grown cells and depends on phosphatidylcholine. Synthetase I is constitutive and lacking in the mutants mentioned above, while synthetase II is inducible by oleate. These enzymes can be distinguished immunochemically[48)]. Interestingly, synthetase I is widely distributed among subcellular fractions including microsomes and mitochondria where glycerophosphate acyltransferase is located, while synthetase II is exclusively localized in peroxisomes, the site of β-oxidation[38)]. Furthermore, synthetase II displays wider substrate specificity for fatty acids than synthetase I (Table 6)[47)]. These facts clearly indicate that synthetase II is responsible for the supply of acyl-CoAs to be degraded by the β-oxidation system yielding acetyl-CoA, while synthetase I plays a role in producing acyl-CoAs being incorporated into cellular lipids (Fig. 9).

A similar subcellular localization of acyl-CoA synthetase and enzymes participating in a further metabolism of acyl-CoA has also been observed in alkane-grown *C. tropicalis*, although acyl-CoA synthetase in *C. tropicalis* peroxisomes fails to reveal a phosphatidylcholine dependency[31)].

The distinct roles of acyl-CoA synthetases in different cellular organelles are shown in Fig. 10 together with those of alcohol dehydrogenase and aldehyde dehydrogenase.

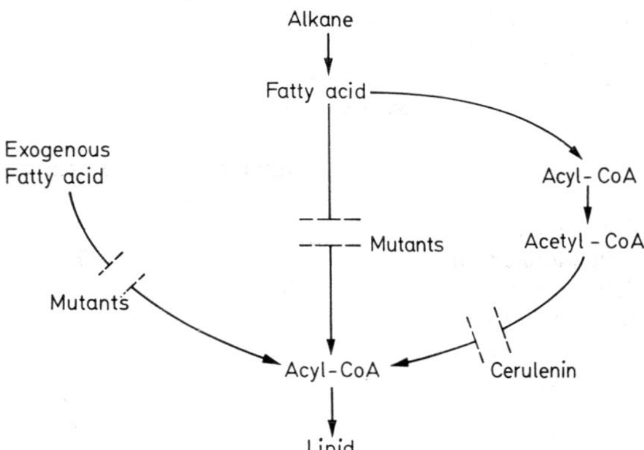

Fig. 9 Pathway of incorporation of fatty acids into lipids of *C. lipolytica*

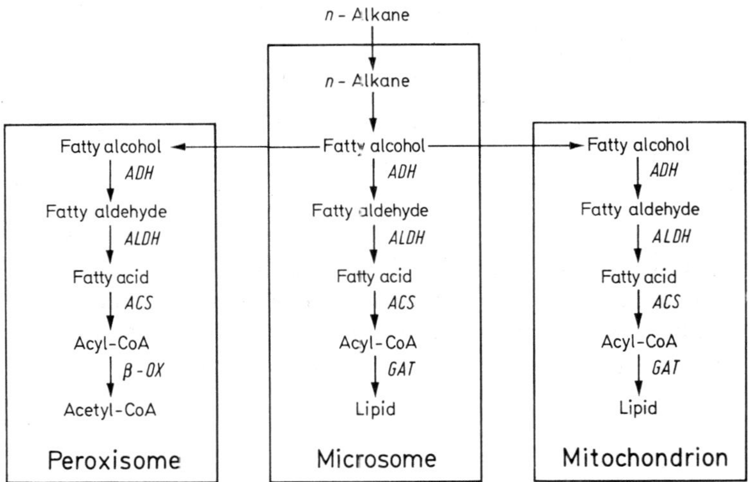

Fig. 10 Possible functions of microsomes, mitochondria and peroxisomes in alkane-utilizing yeasts[31]. Abbreviations: *ACS* = acyl-CoA synthetase; *ADH* = alcohol dehydrogenase; *ALDH* = aldehyde dehydrogenase; *GAT* = glycerophosphate acyltransferase; *β-OX* = fatty acid β-oxidation system

7 Degradation of Acyl-Coenzyme A

Although many investigators have suggested that fatty acids derived from alkanes might be metabolized through β-oxidation, experimental evidence of the presence of a β-oxidation system in yeasts has been lacking.

The conspicuous appearance of peroxisomes in yeasts associated with alkane utilization has stimulated us to investigate the presence of the β-oxidation system in yeast peroxisomes as the cases of castor bean[49] and rat liver[50]. Peroxisomes isolated from alkane-grown *C. tropicalis* reveal palmitate-dependent activities of NAD reduction, acetyl-CoA formation and oxygen consumption[45]. The activities also depend on the presence of CoA and ATP. When sodium azide, an inhibitor of catalase, is added to the system, a palmitate-dependent formation of hydrogen

Table 7. Stoichiometry of β-oxidation in *C. tropicalis* peroxisomes[45)]

Conditions	NAD reduced	Acetyl-CoA formed	O_2 consumed	H_2O_2 formed
		nmol min^{-1} mg $protein^{-1}$		
$-$ NaN_3	351	—	162	—
+ NaN_3	301	—	313	—
+ NaN_3	220	—	—	250
$-$ NaN_3	363	366	—	—

peroxide is observed. From stoichiometric studies (Table 7), the β-oxidation system in yeast peroxisomes has been found to be similar to that of castor bean[49)] and rat liver[50)] (Fig. 11). This means that, acyl-CoA is oxidized by acyl-CoA oxidase, an FAD-containing enzyme as described below, to enoyl-CoA with concomitant consumption of molecular oxygen and formation of hydrogen peroxide. Catalase induced by alkanes participates in the degradation of the hydrogen peroxide thus formed. Enoyl-CoA is further metabolized to yield acetyl-CoA in the presence of NAD and CoA. The β-oxidation system is exclusively localized in peroxisomes of yeasts[38, 45)] and is inducible by alkanes or fatty acids.

Acyl-CoA oxidase have been purified from oleate-grown *C. lipolytica*[51)] and crystallized from alkane-grown *C. tropicalis*[52)]. These enzymes are FAD-proteins and oxidize acyl-CoAs of C_4—C_{18} (*C. lipolytica*) and of C_4—C_{20} (*C. tropicalis*, Table 8). In both cases, lauroyl-CoA is the most effective substrate. The reaction is oxygen-dependent and the formation of hydrogen peroxide is observed.

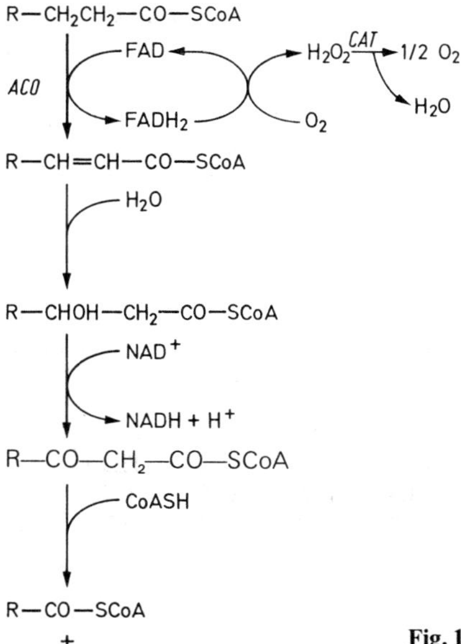

Fig. 11 Fatty acid β-oxidation system in yeast peroxisomes. Abbreviations: *ACO* = acyl-CoA oxidase; *CAT* = catalase

Table 8. Substrate specificity of acyl-CoA oxidase from *C. tropicalis*[52)]

Substrate	Relative activity (%)	Substrate	Relative activity (%)
Butyryl-CoA	5	Myristoyl-CoA	50
Hexanoyl-CoA	2	Palmitoyl-CoA	37
Octanoyl-CoA	16	Stearoyl-CoA	17
Decanoyl-CoA	97	Oleoyl-CoA	64
Lauroyl-CoA	100	Arachidoyl-CoA	7

Activity on lauroyl-CoA = 100%

NADH formed at the stage of β-hydroxyacyl-CoA dehydrogenation seems to be reoxidized to NAD by the glycerol-3-phosphate/dihydroxyacetone phosphate shuttle involving NAD-dependent glycerol-3-phosphate dehydrogenase in peroxisomes and FAD-dependent glycerol-3-phosphate dehydrogenase in mitochondria[53)]. The reducing power transferred to FAD from NADH may be used to yield energy.

8 Synthesis of Cellular Fatty Acids

The analysis of cellular fatty acids of alkane-grown yeast cells strongly suggests that these fatty acids are derived *via de novo* synthesis from acetyl-CoA, chain elongation of the substrate carbon skeleton with one or more acetyl units, and/or intact incorporation of the substrate carbon chain. It has been demonstrated that the biosynthetic routes of cellular fatty acids are determined by the chain-length of alkane substrates (Table 9)[54, 55)].

De novo fatty acid synthesis plays an important role for supplying a portion of cellular fatty acids from alkanes, because even-chain fatty acids are synthesized

Table 9. Fatty acid compositions of *C. lipolytica* and *C. tropicalis* grown on odd-chain alkanes[54)]

Fatty acid	*C. lipolytica* grown on				*C. tropicalis* grown on			
	C_{11}	C_{13}	C_{15}	C_{17}	C_{11}	C_{13}	C_{15}	C_{17}
	in %				in %			
C_{11}	1.1	Trace	—	—	—	—	—	—
C_{12}	—	—	—	—	—	—	—	—
C_{13}	0.3	8.5	—	0.2	—	2.7	—	—
C_{14}	Trace	0.3	0.3	0.4	—	—	0.2	Trace
C_{15}	2.5	18.4	22.9	2.9	1.9	11.9	19.2	0.5
C_{16}	28.0	8.1	1.7	1.3	7.9	4.3	2.2	0.8
C_{17}	5.7	33.1	72.3	94.9	38.6	45.2	54.3	83.4
C_{18}	62.4	31.7	2.7	0.5	50.2	25.8	24.0	15.1
Odd-chain	9.6	60.0	95.2	97.8	40.5	67.1	73.5	83.9

from odd-chain alkanes, especially from short-chain ones. Furthermore, the growth of *C. lipolytica* is inhibited by cerulenin, an anti-lipogenic antibiotic[56,57], when cultivated on glucose, undecane or dodecane[58]. When tridecane is the substrate, cerulenin partially inhibits the yeast growth, but not the growth on C_{14}–C_{18} alkanes. This inhibition by cerulenin is completely eliminated by the addition of oleate and palmitate, indicating that cerulenin inhibits the *de novo* synthesis of cellular fatty acids. In the meantime, it has been observed that the synthesis of acetyl-CoA carboxylase in *C. lipolytica*, the key enzyme of *de novo* fatty acid synthesis, is repressed by alkanes but a certain level of the enzyme activity is still retained in the cells, especially those grown on short-chain alkanes (Table 10) [59]. From these results it may be concluded that the synthesis of fatty acids *de novo* is one of the routes providing cellular fatty acids in yeasts from alkane substrates, although Gill and Ratledge[60] reject this possibility. They might miss the activity of acetyl-CoA carboxylase, because the enzyme of *Candida* yeast is very unstable[61]. This conclusion is also supported by the fact that *C. lipolytica* mutants lacking acyl-CoA synthetase I[46] could grow of alkanes or higher fatty acids without incorporation of exogenous fatty acids or of those derived from alkane substrates as a whole. Even when the mutants are grown on odd-chain alkanes, most of the cellular fatty acids have even-carbon-atom chains (Table 11) [62].

Chain elongation of the substrate carbon skeleton has been clearly demonstrated in yeast cells grown on pentadecane, tridecane and in some cases on undecane. This implies that a large part of cellular fatty acids in cells grown on these substrates contains a C_{17} carbon chain, indicating the addition of one or more C_2 units to the fatty acids derived from odd-chain alkanes[54,58]. Although the details of the chain elongation system in alkane-utilizing yeasts are still unclear, the system seems to be different from the *de novo* synthesis system, because the chain elongation is not inhibited by cerulenin[58]. The mutants of *C. lipolytica* defective in acyl-CoA synthetase I do not reveal the activity of chain elongation[62].

The intact incorporation of the substrate carbon skeleton is demonstrated when yeasts are grown on long-chain alkanes[54,58]. For example, C_{17} acids make up the greatest portion of cellular fatty acids in yeast cells grown on heptadecane. The intact incorporation system is not inhibited by cerulenin[58], but diminishes in the *C. lipolytica* mutants mentioned above[62].

Table 10. Acetyl-CoA carboxylase level of *C. lipolytica* grown on different carbon sources

Carbon source	Relative enzyme activity (%)	Carbon source	Relative enzyme activity (%)
Glucose	100	Fatty acid ($C_{14:0}$)	61
n-Alkane (C_{10})	36	($C_{16:0}$)	29
(C_{12})	37	($C_{18:0}$)	35
(C_{15})	25	($C_{18:1}$)	19
(C_{16})	21	($C_{18:2}$)	18
(C_{17})	12		
(C_{18})	15		

Enzyme activity of glucose-grown cells = 100%

Table 11. Ratios of odd-chain fatty acids to total cellular fatty acids in *C. lipolytica* wild and mutant strains[62)]

Carbon source	Proportion of odd-chain fatty acids (%)	
	Wild	Mutants
Glucose	3.3	0.7– 0.8
n-Undecane	9.2	1.5– 1.7
n-Tridecane	62.0	2.1– 2.4
n-Pentadecane	97.8	8.9–11.4
n-Heptadecane	98.6	11.7–12.4
Oleic acid	0.8	0.2– 0.3

From the results presented here, the synthetic pathways of cellular fatty acids in yeasts from alkanes can be summarized as in Fig. 12. Acyl-CoA synthetases of different subcellular localization play the important role to supply acyl-CoAs, which will be utilized for chain elongation and intact incorportation, or for degradation yielding acetyl-CoA, the substrate of the *de novo* synthesis system.

9 Synthesis of Tricarboxylic Acid Cycle Intermediates

As described above, a fraction of the fatty acids derived from alkanes is incorporated into cellular lipids without degradation. However, the largest portion of them are degraded by β-oxidation to acetyl-CoA (from even- and odd-chain alkanes) and propionyl-CoA (from odd-chain alkanes) (Fig. 13).

In the case of yeasts growing under gluconeogenic conditions, the glyoxylate cycle plays an important role in the biosyntheses of cellular components including carbohydrates and amino acids. In addition to the TCA cycle which produces CO_2 and the reducing power linked to a respiratory system to yield energy, the

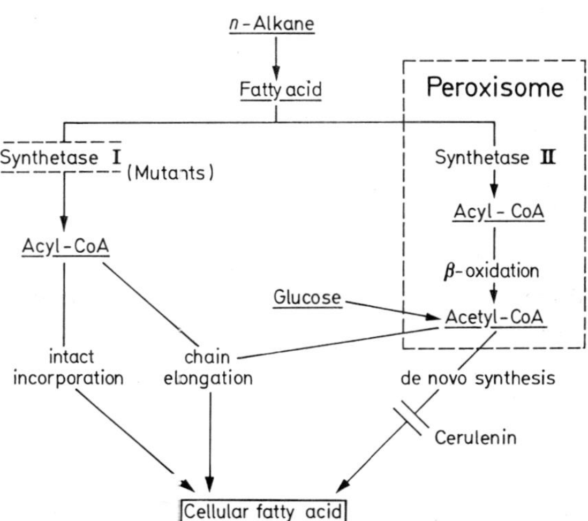

Fig. 12 Proposed scheme of fatty acid metabolism in alkane-utilizing yeasts. The mutants of *C. lipolytica* lack acyl-CoA synthetase I (see text)

$$CH_3(CH_2)_{2n}CH_3 \longrightarrow CH_3(CH_2)_{2n}COOH \xrightarrow{(n+1)\,CoASH} (n+1)\,CH_3CO{-}SCoA$$

$$CH_3(CH_2)_{2n+1}CH_3 \longrightarrow CH_3(CH_2)_{2n+1}COOH \xrightarrow{(n+1)\,CoASH} n\,CH_3CO{-}SCoA + CH_3CH_2CO{-}SCoA$$

Fig. 13 Formation of acetyl-CoA and propionyl-CoA from alkanes

glyoxylate cycle is known to produce one molecule of C_4-compounds, such as malate and succinate, from two molecules of acetyl-CoA by condensation with oxalacetate and glyoxylate (Fig. 14). An important role of this cycle in alkane-utilizing yeasts has also been demonstrated[63]. The key enzymes of the glyoxylate cycle, isocitrate lyase and malate synthase, are induced significantly in alkane-grown yeasts[64]. While the level of isocitrate lyase is high, that of NAD-dependent isocitrate dehydrogenase is low in alkane-grown *C. tropicalis* compared with glucose-grown cells[65,66]. These results suggest that the glyoxylate cycle will be more actively working than the TCA cycle in alkane-utilizing cells. A large part of isocitrate lyase thus induced is recovered in a particulate fraction from alkane-grown *C. tropicalis*[67]. Subcellular fractionation of alkane-grown *C. tropicalis* provides the interesting result that, of the enzymes belonging to glyoxylate cycle, only isocitrate lyase and malate synthase are peroxisomal enzymes. On the other hand, citrate synthase, aconitase and malate dehydrogenase, which are common to the TCA cycle, are localized in mitochondria, but not in peroxisomes[36]. These facts indicate that peroxisomes must cooperate with mitochondria in the metabolism of alkanes.

Since fatty acid β-oxidation activity could not be detected in mitochondria of *C. tropicalis*[45] and in *C. lipolytica*[38], acetyl-CoA required for the citrate synthesis must be transported to mitochondria from peroxisomes. Carnitine acetyltransferase in peroxisomes and mitochondria might be responsible for this transportation[68].

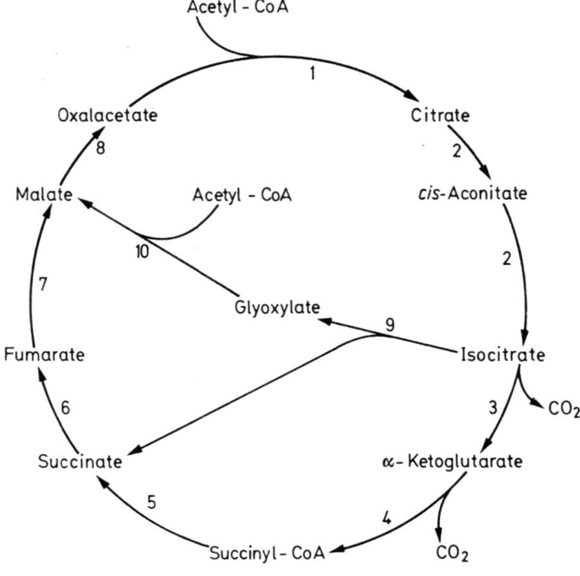

Fig. 14 TCA and glyoxylate cycles. Enzymes: 1 = citrate synthase; 2 = aconitase; 3 = isocitrate dehydrogenase; 4 = α-ketoglutarate dehydrogenase; 5 = succinyl-CoA synthetase; 6 = succinate dehydrogenase; 7 = fumarase; 8 = malate dehydrogenase; 9 = isocitrate lyase; 10 = malate synthase

Fig. 15 illustrates possible roles of microsomes, mitochondria and peroxisomes in alkane-utilizing yeasts. Thus, the metabolic significance of yeast peroxisomes in alkane or fatty acid assimilation will be easily understood on the basis of the currently available knowledge.

Propionyl-CoA derived from odd-chain alkanes is metabolized differently from acetyl-CoA. Tabuchi et al. propose a cyclic pathway of propionyl-CoA metabolism based on the accumulation of pyruvic acid and C_7 tricarboxylic acids in the culture broth of *C. lipolytica* grown on odd-chain alkanes[69]. They have detected the enzyme activities responsible for the methylcitrate synthesis and the methylisocitrate cleavage in *C. lipolytica*, and called the cycle "methylcitric acid cycle" (Fig. 16) [70]. In analogy to the isocitrate lyase reaction linking to the branching point between the TCA cycle and glyoxylate cycle, pyruvate instead of glyoxylate is formed from methylisocitrate in the cycle. The key enzymes of the methylcitric acid cycle are methylcitrate synthase (methylcitrate-synthesizing enzyme) [71] and methylisocitrate

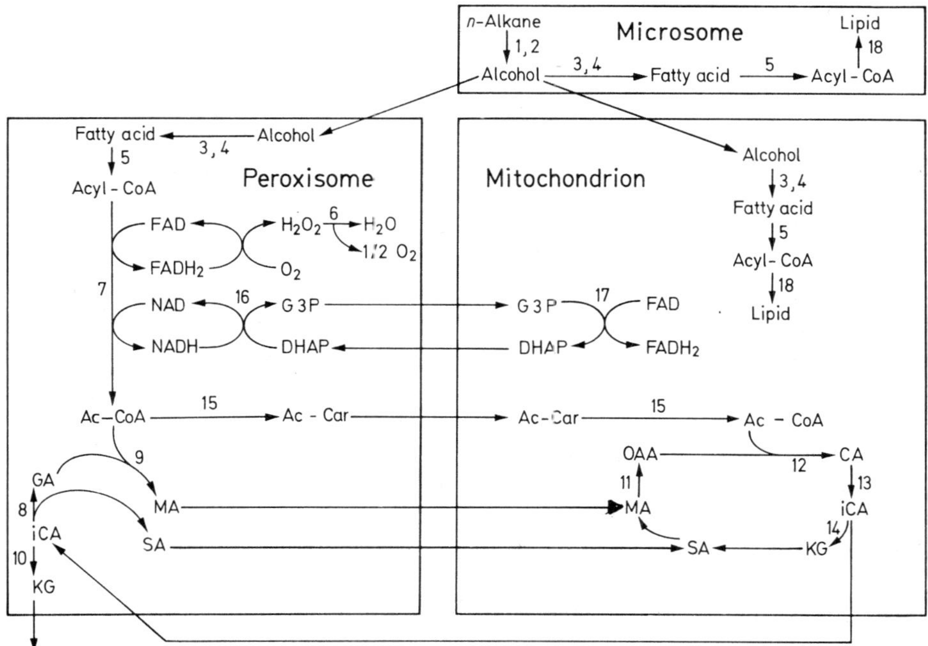

Fig. 15 Possible roles of peroxisomes in connection with those of mitochondria and microsomes in alkane assimilation by yeasts. Abbreviations: Ac-CoA = acetyl-CoA; Ac-Car = acetylcarnitine; CA = citrate; DHAP = dihydroxyacetone phosphate; GA = glyoxylate; G3P = glycerol-3-phosphate; iCA = isocitrate; KG = α-ketoglutarate; MA = malate; OAA = oxalacetate; SA = succinate. Enzymes: 1 = cytochrome P-450; 2 = NADPH-cytochrome P-450 (cytochrome c) reductase; 3 = alcohol dehydrogenase; 4 = aldehyde dehydrogenase; 5 = acyl-CoA synthetase; 6 = catalase; 7 = β-oxidation system; 8 = isocitrate lyase; 9 = malate synthase; 10 = NADP-dependent isocitrate dehydrogenase; 11 = malate dehydrogenase; 12 = citrate synthase; 13 = aconitase; 14 = NAD-dependent isocitrate dehydrogenase; 15 = carnitine acetyltransferase; 16 = NAD-dependent glycerol-3-phosphate dehydrogenase; 17 = FAD-dependent glycerol-3-phosphate dehydrogenase; 18 = glycerophosphate acyltransferase

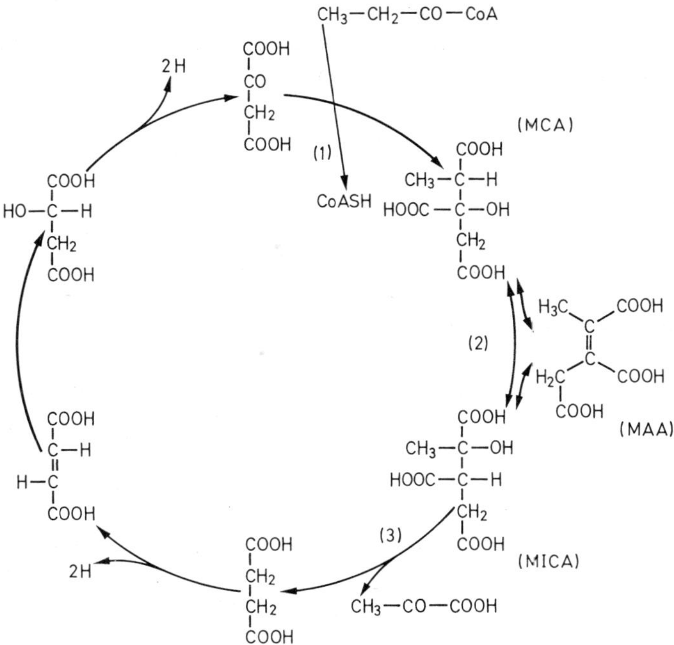

Fig. 16 Methylcitric acid cycle in *C. lipolytica*[70]. Abbreviations: MAA = methylaconitate; MCA = methylcitrate; MICA = methylisocitrate. Enzymes: 1 = methylcitrate synthase; 2 = aconitase; 3 = methylisocitrate lyase

lyase (methylisocitrate-cleaving enzyme) [72,73], these clearly differing from citrate synthase and isocitrate lyase, respectively. Methylcitrate synthase and methylisocitrate lyase seem to be constitutive, akin to the enzymes of the TCA cycle, while the key enzymes of the glyoxylate cycle are inducible[74]. Subcellular localization of methylcitric acid cycle enzymes is not known at present.

C_4-compounds such as succinate, formed in the glyoxylate cycle or methylcitric acid cycle are susceptible to further metabolism to synthesize various cellular constituents. However, no information is available about the regulation of cell component syntheses in alkane-utilizing yeasts.

10 Miscellaneous

As described previously, the main carbon flow from alkanes leading to biosyntheses of cellular lipids has been elucidated to an appreciable extent. Apart from these results, the conversion of secondary metabolites, namely conversions of citric acid to the polyol and of erythritol to mannitol, have been investigated on a biochemical basis[75,76], and applied successfully to the production of these compounds. Furthermore, productivities of several metabolites such as citric acid[77] and malic acid[78] have been discussed in connection with the levels of enzymes participating in their metabolism.

Another line of investigations has been carried out on the respiratory features of alkane-utilizing yeasts[79-81] and of mitochondria isolated from alkane-utilizable yeasts[82-86]. Such studies will offer interesting information on the energetics of alkane-utilizing yeast cells.

11 Future Prospects

Although biochemical and physiological studies on alkane-utilizing yeasts have been carried out for the last ten or more years, these are far inferior in quantity and in quality to those with *Saccharomyces* yeasts.

Discovery of peroxisomes in alkane-utilizing yeasts has made it easy to understand the carbon flow from alkanes to the TCA cycle and glyoxylate cycle leading to biosyntheses of cellular components. However, the regulations of gluconeogenesis, amino acid synthesis, ATP synthesis and so on still remain to be studied in more detail. Furthermore, it cannot be said that all the roles of peroxisomes have been elucidated.

Candida yeasts, typical alkane-utilizable yeasts, have specific characteristics in metabolic regulation mechanisms which are different from those of *Saccharomyces* yeasts. In addition, there will be considerable differences in the regulation mechanisms between fermentative and oxidative *Candida* yeasts. Such examples have been observed in the control of glucose-phosphorylating enzymes[87] and pyruvate kinase[88] among *C. tropicalis* (fermentative), *C. lipolytica* (oxidative) and *Saccharomyces* yeasts.

These facts strongly suggest that physiological studies of alkane-utilizing yeasts will provide new findings on the general physiology of living organisms.

12 References

1. Abbott, B. J., Gledhill, W. E.: Adv. Appl. Microbiol. *14*, 249 (1971)
2. Fukui, S., Tanaka, A.: Advan. Biochem. Eng. *17*, 1 (1980)
3. Iguchi, T., Takeda, I., Ohsawa, H.: Agric. Biol. Chem. *33*, 1657 (1969)
4. Meissel, M. N. et al.: Proc. Third Internatl. Spec. Symp. Yeasts (Otaniemi/Helsinki), p. 149 (1973)
5. Meissel, M. N., Medvedeva, G. A., Kozlova, T. M.: Mikrobiologiya *45*, 844 (1976)
6. Osumi, M. et al.: J. Ferment. Technol. *53*, 244 (1975)
7. Einsele, A., Schneider, H., Fiechter, A.: J. Ferment. Technol. *53*, 241 (1975)
8. Kaeppeli, O., Fiechter, A.: Biotechnol. Bioeng. *18*, 967 (1976)
9. Käppeli, O., Fiechter, A.: J. Bacteriol. *131*, 917 (1977)
10. Käppeli, O., Müller, M., Fiechter, A.: J. Bacteriol. *133*, 952 (1978)
11. Peterson, J. A. et al.: J. Biol. Chem. *242*, 4334 (1967)
12. Cardini, G., Jurtshuk, P.: J. Biol. Chem. *245*, 2789 (1970)
13. Liu, C.-M., Johnson, M. J.: J. Bacteriol. *106*, 830 (1971)
14. Lebeault, J. M., Lode, E. T., Coon, M. J.: Biochem. Biophys. Res. Commun. *42*, 413 (1971)
15. Gallo, M., Bertrand, J. C., Azoulay, E.: FEBS Lett. *19*, 45 (1971)
16. Müller, H. G. et al.: Acta Biol. Med. Germ. *38*, 345 (1979)
17. Gallo, M. et al.: Biochim. Biophys. Acta *296*, 624 (1973)
18. Gallo, M. et al.: Biochimie *55*, 195 (1973)
19. Gallo, M., Roche, B., Azoulay, E.: Biochim. Biophys. Acta *419*, 425 (1976)
20. Gilewicz, M. et al.: Can. J. Microbiol. *25*, 201 (1979)

21. Bertrand, J. C. et al.: FEBS Lett. *105*, 143 (1979)
22. Bertrand, J. C. et al.: Eur. J. Biochem. *93*, 237 (1979)
23. Duppel, W., Lebeault, J. M., Coon, M. J.: Eur. J. Biochem. *36*, 583 (1973)
24. Strobel, H. W., Coon, M. J.: J. Biol. Chem. *246*, 7826 (1971)
25. Iida, M., Iizuka, H.: J. Ferment. Technol. *47*, 442 (1969)
26. Finnerty, W. R.: Trends in Biochem. Sci. *2*, 73 (1977)
27. Lebeault, J. M. et al.: Biochim. Biophys. Acta *220*, 373 (1970)
28. Lebeault, J. M. et al.: Biochim. Biophys. Acta *220*, 386 (1970)
29. Lebeault, J. M., Azoulay, E.: Lipids *6*, 444 (1971)
30. Gallo, M., Roche-Penverne, B., Azoulay, E.: FEBS Lett. *46*, 78 (1974)
31. Yamada, T. et al.: Arch. Microbiol. *128*, 145 (1980)
32. Osumi, M. et al.: Arch. Microbiol. *99*, 181 (1974)
33. Teranishi, Y. et al.: Agric. Biol. Chem. *38*, 1213 (1974)
34. Teranishi, Y. et al.: Agric. Biol. Chem. *38*, 1221 (1974)
35. Osumi, M. et al.: Arch. Microbiol. *103*, 1 (1975)
36. Kawamoto, S. et al.: Arch. Microbiol. *112*, 1 (1977)
37. Osumi, M. et al.: J. Gen. Appl. Microbiol. *21*, 375 (1975)
38. Mishina, M. et al.: Eur. J. Biochem. *89*, 321 (1978)
39. Osumi, M.: J. Electron Microsc. *25*, 43 (1976)
40. Osumi, M., Kazama, H., Sato, S.: FEBS Lett. *90*, 309 (1978)
41. Fukui, S., Tanaka, A.: J. Appl. Biochem. *1*, 171 (1979)
42. Fukui, S., Tanaka, A.: Trends in Biochem. Sci. *4*, 246 (1979)
43. Trust, T. J., Millis, N. F.: J. Bacteriol. *104*, 1397 (1970)
44. Duvnjak, Z. et al.: Biochim. Biophys. Acta *202*, 447 (1970)
45. Kawamoto, S., Nozaki, C., Tanaka, A., Fukui, S.: Eur. J. Biochem. *83*, 609 (1978)
46. Kamiryo, T. et al.: Proc. Nat. Acad. Sci. USA *74*, 4947 (1977)
47. Mishina, M. et al.: Eur. J. Biochem. *82*, 347 (1978)
48. Hosaka, K. et al.: Eur. J. Biochem. *93*, 197 (1979)
49. Cooper, T. G., Beevers, H.: J. Biol. Chem. *244*, 3514 (1969)
50. Lazarow, P. B., de Duve, C.: Proc. Nat. Acad. Sci. USA *73*, 2043 (1976)
51. Kikuchi, T. et al.: Abstr. Res. Meeting Lipid Biochem., p. 144, Toyama, Japan 1979
52. Shimizu, S. et al.: Biochem. Biophys. Res. Commun. *91*, 108 (1979)
53. Kawamoto, S. et al.: FEBS Lett. *97*, 253 (1979)
54. Mishina, M. et al.: Agric. Biol. Chem. *37*, 863 (1973)
55. Rattray, J. M. B., Schibeci, A., Kidby, D. K.: Bacteriol. Rev. *39*, 197 (1975)
56. Nomura, S. et al.: J. Antibiotics *25*, 365 (1972)
57. Vance, D. et al.: Biochem. Biophys. Res. Commun. *48*, 649 (1972)
58. Tanaka, A. et al.: Europ. J. Appl. Microbiol. *3*, 115 (1976)
59. Mishina, M. et al.: Eur. J. Biochem. *71*, 301 (1976)
60. Gill, C. O. Ratledge, C.: J. Gen. Microbiol. *78*, 337 (1973)
61. Mishina, M. et al.: Eur. J. Biochem. *71*, 295 (1976)
62. Tanaka, A. et al.: Europ. J. Appl. Microbiol. Biotechnol. *5*, 79 (1978)
63. Hildebrandt, W., Weide, H.: Z. Allg. Mikrobiol. *14*, 47 (1974)
64. Nabeshima, S., Tanaka, A., Fukui, S.: Agric. Biol. Chem. *41*, 275 (1977)
65. Hirai, M. et al.: Agric. Biol. Chem. *40*, 1819 (1976)
66. Tanaka, A. et al.: Agric. Biol. Chem. *41*, 795 (1977)
67. Hirai, M. et al.: Agric. Biol. Chem. *40*, 1979 (1976)
68. Kawamoto, S. et al.: FEBS Lett. *96*, 37 (1978)
69. Tabuchi, T., Serizawa, N.: Agric. Biol. Chem. *39*, 1055 (1975)
70. Tabuchi, T., Uchiyama, H.: Agric. Biol. Chem. *39*, 2035 (1975)
71. Uchiyama, H., Tabuchi, T.: Agric. Biol. Chem. *40*, 1411 (1976)
72. Tabuchi, T., Satoh, T.: Agric. Biol. Chem. *40*, 1863 (1976)
73. Tabuchi, T., Satoh, T.: Agric. Biol. Chem. *41*, 169 (1977)
74. Tabuchi, T., Igoshi, K.: Agric. Biol. Chem. *42*, 2381 (1978)
75. Hatton, K., Suzuki, T.: Agric. Biol. Chem. *38*, 2419 (1974)
76. Hattori, K., Suzuki, T.: Agric. Biol. Chem. *39*, 57 (1975)
77. Akiyama, S. et al.: Agric. Biol. Chem. *37*, 885 (1973)

78. Sato, S., Nakahara, T., Minoda, Y.: Agric. Biol. Chem. *41*, 1903 (1977)
79. Teranishi, Y. et al.: Agric. Biol. Chem. *38*, 1581 (1974)
80. Teranishi, Y., Tanaka, A., Fukui, S.: Agric. Biol. Chem. *38*, 1779 (1974)
81. Yamamura, M. et al.: Agric. Biol. Chem. *39*, 13 (1975)
82. Akimenko, V. K., Medentsev, A. G., Golovchenko, N. P.: FEBS Lett. *45*, 22 (1974)
83. Gallo, M., Azoulay, E.: Biochimie *56*, 1129 (1974)
84. Mitsushima, K., Shinmyo, A., Enatsu, T.: J. Ferment. Technol. *54*, 863 (1976)
85. Mitsushima, K., Shinmyo, A., Enatsu, T.: J. Ferment. Technol. *55*, 84 (1977)
86. Mitsushima, K., Shinmyo, A., Enatsu, T.: Biochim. Biophys. Acta *538*, 481 (1978)
87. Hirai, M. et al.: Biochim. Biophys. Acta *480*, 357 (1977)
88. Hirai, M., Tanaka, A., Fukui, S.: Biochim. Biophys. Acta *391*, 282 (1975)

Protoplasts in Genetic Modifications of Plants

Oluf L. Gamborg*
Prairie Regional Laboratory, National Research Council, Saskatoon, Saskatchewan S7N OW9 Canada

Paul J. Bottino
Department of Botany, University of Maryland, College Park, Maryland, USA

Rapid advances have been made in recent years in plant protoplast technology and research applications. Isolated protoplasts have been used successfully in virology and cell genetics. Protoplast fusion is feasible and a variety of hybrid cells of different plant genera have been produced. Somatic hybridization and hybrid plant formation have been achieved. The hybrids between several plant species and genera possess diversity in chromosome number and reflect the parental species in isoenzyme patterns. Genetic complementation occurs and cytoplasmic inherited factors such as male sterility are transferred through somatic hybridization. The protoplast fusion methods are being implemented in plant breeding procedures to achieve gene transfer between plants which cannot be crossed sexually and for the purpose of achieving more effective disease resistance. Transfer of genetic information using protoplasts also has been investigated by DNA uptake, which may require a vector such as plasmids or viral DNA. Moreover, organelle and single cell uptake/fusion is being pursued with plant protoplasts. Technological and biological obstacles at present restrict the application of recombinant DNA and organelle transplant procedures in plant research and development but new and significant advances can be anticipated in the next few years.

1 Introduction

Vegetative or somatic plant cells are totipotent and possess the capability for regenerating into complete plants. This property has been utilized extensively in the multiplication of plants from tissues of leaf, stems, tubers or other plant parts.

* Present address: International Plant Research Institute, Inc., San Carlos, California.

Plant regeneration also is feasible from shoot meristems (0.4 mm) and from tissue cells grown in culture on a nutrient medium containing a plant growth hormone. These procedures require aseptic plant tissue culture methods. An extensive technology of plant cell and tissue culture has been developed during the last 15—20 years[1,2].

The technology encompasses a range of methods for growing plant organs, tissues, and cells aseptically in a controlled environment. Cells of most plant species can be grown for indefinite periods on chemically defined media containing plant growth hormones. A culture is initiated from a sterile plant tissue on agar media. The cell mass (callus) can be maintained by subculturing. Large populations of cells can be grown in liquid suspensions in flasks and fermentors; the cells have a generation time which varies from 20—36 h or longer. Partial synchrony of cell growth has been achieved in fermentors. The liquid suspension culture consists of cells clusters and single cells. Plating and cloning of cell lines have become routine procedures and utilized in selection. Plant cells in culture are grown on a carbon source of sucrose or glucose and generally lack chlorophyll. However, autotrophic cell lines which carry out photosynthesis have been obtained recently.

Under suitable conditions, the cells can be induced to differentiate, undergo morphogenesis and regenerate complete plants. The process of morphogenesis can occur via the formation of embryos or by hormone-induced organogenesis. Plants can be regenerated from tissues of leaves, seedlings, embryos, microspores as well as cultured cells[1,3]. The relative ease with which morphogenesis occurs vary with the plant species and the origin of the tissue being used. Plant regeneration is possible from meristems of many species and the procedure has been adopted and implemented in methods of disease control, propagation and freeze-preservation of crop plant species[4].

The culturing of anthers (microspores) is used in the production of haploid cells and plants. By employing populations of haploid cells it is possible to apply mutagenic treatment followed by chromosome doubling[5]. On subsequent plating of the homozygous cells the mutants with biochemical lesions can be selected. In species in which plant regeneration can be achieved it becomes possible to examine in complete plants the modifications induced in the single cells in culture[6].

Parallel with progress in the use of tissue culture has come the development of a plant protoplast technology[7,8,9]. Large populations of protoplasts can be obtained from plant tissues by the use of wall-degrading enzymes. Under appropriate condtions the protoplasts can be cultured, the cell wall reforms and cell division and sometimes plant regeneration can occur. Isolated protoplasts have been used extensively in investigations on plant virus[10]. The major reason for the rapidly expanding interest in protoplasts is their use in genetic manipulation in plants. Because they lack the rigid cell walls, plant protoplasts are materials of choice for somatic cell hybridization by cell fusion[4,11]. They are also being utilized in attempts to achieve genetic transformation by DNA uptake as well as in attempts at implantation of organelles for the purpose of transfer of genetic information[1].

2 Plant Protoplast Isolation

During the last decade a very extensive technology on plant protoplasts has been developed. The introduction in the 1960's of methods for isolation of viable protoplasts by enzyme treatment made it possible to obtain sufficient quantities for experimental purposes[12]. Considerable progress has also been made in methods of handling and manipulating isolated protoplasts. In suitable culture media and environmental conditions, the protoplasts regenerate a cell wall, undergo cell division and may grow into complete plants[2,7,13].

Protoplasts are usually isolated by treating tissues with a mixture of wall degrading enzymes in solutions which contain osmotic stabilizers to preserve the structure and viability of the protoplasts[14,15]. The relative ease with which protoplast isolation can be achieved depends upon a variety of factors. The most important of these are the physiological state of the tissues and cell materials, the choice of enzymes, the composition of the solutions and the concentration and type of osmotic stabilizer[16].

Table 1. Species from which protoplasts have been isolated and cell/plant regeneration observed

Plant species	Tissue origin	Plant regeneration	Ref.
Datura species	Leaf	P.	17, 18)
Datura innoxia	Cell culture	P.	19)
Petunia species	Leaf	P.	20, 21)
Solanum species, potato	Leaf	P.	23, 24, 25)
Nicotiana species, tobacco	Leaf	P.	26, 27)
N. tabacum albino mutant (Su/Su)	Cell culture	P.	28)
Lycopersicon esculentum, tomato	Leaf	P.	29)
Hyoscyamus spp	Leaf	P.	30)
Atropa belladonna	Cell culture		31)
Bromus inermis, bromegrass	Cell culture	P.	32)
Hordeum vulgare, barley	Cell culture		33)
Zea mays, corn	Shoot		34)
Pennisetum americanum, millet	Cell culture		22)
Oryza sativa, rice	Callus		35)
Sorghum bicolor, sorghum	Cell culture		36)
Triticum spp, wheat	Cell culture		37)
Asparagus officinalis	Callus, cladodes	P.	38)
Brassica napus, rape	Leaf	P.	39, 40)
Daucus carota, carrot	Cell culture	P.	41, 42)
Pisum sativum, pea	Leaf, shoot tip		43, 44, 45)
Medicago sativa, alfalfa	Leaf	P.	46, 47)
Melilotus spp, sweet clover	Leaf		46)
Caragana arborescens, caragana	Leaf		48)
Glycine max., soybean	Cell culture		49)
Phaseolus vulgaris, bean	Leaf		50)
Vicia faba, fababean	Leaf		51)
Vigna sinensis, cow pea	Leaf		52)
Linum usitatissimum, flax	Hypocotyl		53)
Gossypium hirsutum, cotton	Callus		54)
Cucumis sativus, cucumber	Leaf		55)
Citrus sinensis, orange	Callus	P.	56)
Pseudotsuga menziessii, douglas fir	Cotyledons		57)
Pinus pinaster, pine	Cotyledons		58)
Ranunculus sceleratus	Leaf	P.	59)
Amni visnaga	Cell culture		60)
Vicia hajastana	Cell culture		61)

Protoplasts have been isolated from tissues of a large number of plant species (Table 1). A convenient and suitable source of protoplasts is leaf mesophyll tissue. The most satisfactory results have been obtained with fully expanded leaves from young plants or new shoots. Protoplast yield and viability (quality) are influenced substantially by the age of the leaf as well as the environmental conditions of light, temperature, nutrition, and humidity under which the plants are grown[15, 16]. Since the physiological state of the tissue is critical, the plants are usually grown in environmental growth chambers or greenhouses where light and temperature can be carefully controlled. With respect to light intensity, a range of 0.3 to 1.0 Wcm^2 has proven suitable, although in special cases a much more intense illumination was beneficial[62]. A photoperiod of 18:6 or 16:8 h of light:dark is the most common. In some cases the yield of protoplasts has been greatly improved by placing the tissue in the dark for 24—72 h prior to isolation[43]. During the dark period degradation of starch occurs. The presence of starch grains have been reported to have adverse effects on protoplast viability. The choice of temperature and humidity is generally dictated by the requirements to ensure vigorous plant growth. Maintaining the plants at a high level of fertility and particularly of nitrogen fertilizers can be beneficial[25].

In addition to leaf tissue other plant materials have been used. These include shoot tips, cotyledons, flower petals and microspores[7, 16]. Callus tissue and cell suspension cultures are frequently used as sources of protoplasts. Since callus tissue often have a slow growth rate and have a broad diversity in cell age and physiological state these may be less suitable. Cells grown in liquid suspension when fully established can be expected to have a generation time of 24–30 h and consist of smaller cell clusters and single cells[63, 85]. For best production of protoplasts such cell cultures are subcultured at 3–4 day intervals to maintain maximum growth rate and uniformity of the cell population. Cells taken at the early log phase are generally the most suitable. In established cultures this would correspond to ca. two days after subculturing. Cultures which have been newly established in liquid suspension often consist of mixtures of cell size and age as well as a proportion of dead cells and are usually unsuitable as protoplast source.

2.1 Enzymes and Isolation Medium

The enzymes used in the isolation of protoplasts comprise three general classes; cellulases, hemicellulases and pectinases[14]. Commercial preparations of the enzymes are usually used. The range of available sources is limited. The preparations are not pure and contain deleterious enzymes such as proteases, lipases and nucleases which affect protoplast viability. Some purification of the preparations is achieved by gel filtration[13].

Both the composition and the concentration of enzymes affect the yield of protoplasts from a given tissue[29].

The incubation mixture used in protoplast isolation consists of the enzymes dissolved in a solution containing a few salts or a medium, a buffer and an osmotic stabilizer[14]. The process of isolation is carried out aseptically in petri dishes. Leaf or other tissues are sterilized in alcohol and solutions of hypochlorite. The tissues are sometimes placed in a plasmolyzing or nutrient hormone solution for a period prior to use[62,47]. The tissues are then cut into sections and when possible the epidermis is removed and incubated in the enzyme mixture. Vacuum infiltration of the incubation mixture may facilitate enzyme penetration. Following incubation for 4—18 h at 25–27 °C the mixture is filtered by gravity and the protoplasts carefully washed and collected by centrifugation[62,65].

2.2 Protoplast Culture

After isolation the protoplasts are suspended in nutrient media with adjusted osmolarity to maintain viability and structural integrity. The culture solution contains mineral nutrients, vitamins, carbon source and growth hormone as well as osmotic stabilizers[14,65]. For survival the protoplasts also require organic nitrogen sources and possibly coconut milk and various metabolites[61]. Protoplasts are usually cultured in liquid media in droplets until cell clusters have formed after which they are plated on nutrient agar[15,49].

2.3 Protoplast Development

Protoplasts when newly formed are completely free of cell walls. Upon culture in the absence of the enzymes, cell wall regeneration is initiated immediately[66]. Microfibrils are formed and within a few hours a structure can be detected. The cytology of protoplast development has been discussed in detail in recent reviews[67,68]. During wall regeneration, the mitotic division process also takes place. The percentage of divisions varies widely. Nevertheless, protoplasts of many plant species have been observed to undergo sustained division (Table 1). The success has recently been extended to protoplasts of conifers[57,58]. Protoplasts from plant parts of cereals and grasses remain difficult to induce into division, but those from cell cultures readily undergo sustained division[36]. Mitosis in protoplasts appears to be similar to the process in plant cells. A high percentage division is often difficult to achieve, but the consistently high frequency observed in some protoplast sources suggest that in due time the problems limiting division rates will be resolved.

The protoplasts retain the capacity for morphogenesis and plant regeneration residing in the original tissues. The regenerated cells can be placed on a medium which is conducive to plant regeneration. If protoplasts originate from cells of embryo-forming species such as carrot, asparagus, citrus, the regenerated cells produce complete plants by embryogenesis[35,41,56]. In most species shoot initiation requires the use of cytokinins such as benzyladenine, zeatin or isopentenyladenosine[17,25,39]. However, even in the presence of these compounds, the regenerated cells may form only roots. The success of shoot and plant regeneration depends primarily upon the ease with which the process can be achieved in the original tissue of the particular plant species. The list in Table 1 indicate the species in which plant regeneration from protoplasts have been reported. Many of the species belong to the Family *Solanacea*. It is significant that in none of the cereal grain or seed legume species has it been possible to regenerate plants from protoplasts. The production of differentiated structures from protoplasts also has been achieved in species of liverwort[69,70] and moss[71].

3 Uses of Protoplasts

The absence of the rigid cellulosic walls and the complete exposure of the plasma membrane makes protoplasts a particularly useful material for investigations on uptake and transport phenomena.

Protoplasts have been employed extensively in studies on plant virus uptake[10]. Much of the research has been performed with leaf protoplasts of *N. tabacum*. New insights have been gained on the uptake/infection process. One end of the virus attaches itself onto the membrane[72]. The uptake then appears to proceed by an endocytosis process which is aided by poly-L-ornithine, although there is not a unanimous agreement on the details of the process[10]. In a population of protoplasts up to 90% become infected and replication occurs nearly exponentially after a delay of 6—8 h. The protoplast system has also enabled investigation on virus species specificity and strain interactions as well as aspects of molecular biology which was not feasable with plant tissue cells.

Another promising area in which protoplasts would be the material of choice entail the elucidation of the specificity and mode of action of fungal and bacterial plant pathogens[73, 74]. Some species of fungal pathogens produce toxins which destroy plant cells. The toxins have a distinct deleterious effect on plant protoplasts and cause disruption of the membrane[75]. Detailed studies with protoplasts of sugar cane have shown that specific binding proteins for the toxin compound are located on the membrane[74]. Protoplasts of resistant plant tissues are not adversely affected and they lack the specific binding protein.

Protoplasts also have been used to advantage in photosynthetic studies of C_4 plants, which have a relatively efficient photosynthetic process[76, 77]. The availability of protoplasts from tissues of the C_4 plants has made it possible to elucidate the apparent compartmentation of enzymes and to study the mechanism and regulation of metabolism in C_4 plant species. Of particular interest is the observation that Fraction 1 protein (RUDP carboxylase) is present in bundle sheet cells and not in the leaf mesophyll (protoplasts), while in leaf mesophyll cells of C_3 plants the fraction 1 protein makes up close to fifty percent of the soluble proteins. Protoplasts also have proven very valuable material as a source of intact chloroplasts and nuclei[78].

In addition to the catagories of application already discussed, isolated protoplasts are now becoming widely used in studies on ion uptake, membrane surface properties as well as on elucidating the process of cell wall synthesis[67].

4 Plant Cell Mutants

4.1 Metabolic and Auxotrophic Mutants

A wide variety of metabolic mutants have been produced within prokaryotic organisms. The mutants have been extremely valuable in the elucidation of problems in genetics and biochemistry. In eukaryotes and in plants in particular it is only very recently that attempts have been successful in the production of auxotrophic mutant cell lines[79, 80, 81]. The production of complete auxotrophs has been successful by using mosses and liverwort. Spores which are haploid single cells can be exposed

Table 2. Auxotrophs obtained by mutagenesis of haploid plant cells in culture

Plant species	Required metabolites	Ref.
Sphaerocarpus donnellii	Arginine	80)
Sphaerocarpus donnellii	Arginine, ornithine, citrulline	80)
Sphaerocarpus donnellii	Choline	80)
Sphaerocarpus donnellii	Glucose	80)
Sphaerocarpus donnellii	Glucose & yeast extract	80)
Sphaerocarpus donnellii	Nicotinic acid	80)
Physcomitrella patens	Adenine	79)
Physcomitrella patens	p-Amino benzoic acid	79)
Physomitrella patens	Arginine or proline	79)
Physcomitrella patens	Nicotinic acid	79)
Physcomitrella patens	Ammonium (reduced nitrogen)	79)
Datura innoxia	Panthothenate	81)
Datura innoxia	Adenine	81)
Nicotiana tabacum	Reduced nitrogen	88)

to mutagenic treatment and subsequently grown into plants. Several auxotrophs were obtained after x-irradiation treatment of spores of *Sphaerocarpus donnellii*[80, 87]. These included four lines requiring arginine, four requiring nicotinic acid, two requiring choline and three requiring glucose (Table 2). Ashton and Cove[79] reported on 18 mutants isolated after chemical mutagen treatment of spores of *Physcomitrella patens*. They obtained two which required arginine, four requiring adenine, four requiring p-aminobenzoate, five requiring nicotinate and three which lacked nitrate reductase and required amonium salts.

The mutants in both plants were isolated by a non-selective procedure. The method consists of testing growth of tissue arising from single spores on supplemented and on minimal media.

The results obtained with the liverwort and moss suggested that the production of auxotrophic mutants might also be feasible with cells of higher plants. Using predominantly haploid cell suspension cultures of *Datura innoxia*, Savage et al.[64, 81] have isolated a mutant which is auxotrophic for pantothenate. A fraction consisting of single cells obtained by filtration of the cultures was treated with a chemical mutagen. The surviving cells were cultured, plated at low density and the callus arising from the cells subjected to non-selective testing. The procedure consists of hand replication and comparative growth of samples of the callus on supplemented medium and on minimal medium. Those failing to grow on minimal media were further tested on the same medium containing selected supplements. The pantothenate-requiring cell mutant of *Datura* is the first metabolite-requring plant cell line to be reported. The success has demonstrated the feasiblity of producing auxotrophic mutants in cells of higher plants. The *Datura* cell culture appears to be a convenient material for somatic cell genetics. The ploidy of the cells remain stable and haploid (predominantly) and they can be cultured for indefinite periods in defined media[19]. Other types of mutants of higher plants isolated recently are chlorate-resistant and lack nitrate reductase and associated enzymes[88]. After mutagenesis of allodihaploid cells of *Nicotiana tabacum* several chlorate-resistant

Table 3. Metabolic mutants in plants

Plant species	Nature of mutation	Materials	Ref.
Sphaerocarpus donnellii	Require vitamins, amino Acids, glucose	Plants	80)
Physomitrella patens	Require vitamins, amino acids, adenine, reduced nitrogen	Plants	79)
Arabidopsis thaliana	Require thiamine	Plants	89)
Zea mays	Require proline	Plants	90)
Nicotiana tabacum	Lack nitrate reductase	Cells	88)
Datura innoxia	Require panthothenate, adenine	Cells	64, 81)
Glycine max.	Utilize maltose	Cells	92)
Saccharum spp.	Utilize galactose	Cells	91)
Nicotiana tabacum	Utilize glycerol	Cells	94)

cell lines were isolated and characterized. Seven lines lacked nitrate reductase and two of those also lacked xanthine dehydrogenase.

The mutants discussed above appear to have biochemical lesions and consequently the cells have defects in metabolic pathways. A number of thiamine-requiring mutants of *Arabidopsis thaliana* plants[89)] and a proline-requiring mutant of maize[90)] have also been reported, but it is not clear if all cells of the plants have the biochemical defect (Table 3).

The usual carbon source of higher plant cells is sucrose or glucose. Cell lines have been reported recently which can utilize other compounds. These include sugar cane cells growing on galactose[91)], soybean cells growing on maltose[92)] and tobacco cells growing on glycerol[94)]. All of these cell lines have retained the capability to grow on sucrose or glucose. The exact biochemical nature of the property which enable them to utilize effectively other sugars has not been elucidated, but some of them could be membrane (uptake) mutants.

4.2 Chlorophyll Deficient Mutants

A class of mutants in higher plants are the chlorophyll deficient lines which are relatively common. The mutants range in appearance from shades of yellow green to white and are either genome (nuclear) or plastome (plastid) mutants.

The chlorophyll deficient mutants amongst higher plants have generally been considered a novelty. In some instances, however, chlorophyll mutants have been investigated extensively[82)]. Albino mutants exist in tobacco[28)], carrot[83)], potato[93)], tomato, corn, alfalfa and Brome grass[84)]. Chlorophyll deficient mutants have been observed in certain crosses[85)]. They have also been produced by mutagenesis[86)]. Potentially these mutants would be valuable in metabolic and cell regulation investigations, but relatively little information is available on the biochemical reactions directly or indirectly affected by the genetic defects. Albino mutants are auxotrophic in the absence of a carbon and energy source. Consequently, they survive only in a culture medium containing sugar and under aseptic condtions[28)].

The usefulness of albino mutants has been recognized recently and they have been employed for selection of somatic hybrid plants produced by protoplast fusion.

5 Somatic Hybridization

The concept of somatic hybridization was originally formulated and presented as a potential method to overcome some of the natural limitations inherent in producing wide crosses for plant breeding[95, 96, 97]. Hybrids between widely different plants are sought for the purpose of transferring resistance to diseases and pests, tolerance to stress conditions and to improve product quality and growth characteristics.

However, existing natural barriers in the sexual fertilization process or lack of adequate embryo development generally prevents the sucessful crossing of different plant genera[98]. The fusion of asexual cells with the formation of hybrids and subsequent plant regeneration has therefore been proposed as a method for the production of wide crosses in plants.

An extensive technology has been developed with cell fusion and somatic hybridization of animal and human cells[99]. The method has been incorporated as a valuable technique for biochemical and genetic analyses. A wide variety of animal hybrid cells have been produced including combinations with those of human origin. Animal cells can be fused directly and require no prior treatment. Plant cells, on the other hand, contain a rigid, cellulose wall and are adjoined by pectin and hemicellulose materials. These restrictions to cell membrane contact must be removed and the cells converted to protoplasts before fusion can be achieved. Somatic hybridization by protoplast fusion in plants involves several interrelated procedures which include: protoplast isolation, fusion, growth of fusion products, selection of hybrid cells and regeneration of hybrid plants. The isolation of protoplasts was discussed earlier and together with recent reviews provide an up-to-date account of this area[3, 7, 16].

5.1 Protoplast Fusion

Protoplast fusion can occur spontaneously during isolation. The process appears to take place when the plasmadesmata between adjoining cells expand rather than break during cell wall removal. Spontaneous fusion can occur between two or more adjacent protoplasts. The phenomenon tend to occur at greater frequency when protoplasts are isolated from cell cultures rather than from leaves or other plant tissues. The fusion products (homopolykaryocytes) regenerate a cell wall and may undergo mitosis[100]. Cytokinesis in such homopolykaryocytes has been reported. There is supportive evidence for nuclear fusion followed by division in di- and trinucleated homopolykaryocytes[101, 102], but information on the further development of the cell progeny has not become available. Protoplasts have been isolated from meiocytes of *Lilium* and other species of the *Liliaceae*. The isolated protoplasts readily fuse upon contact. The plasma membrane surface structure and properties apparently facilitate adhesion and contact and fusion may occur within a few minutes[103]. The meiotic process continues in the isolated protoplasts and reach the tetrad stage in the di- and trinucleated fusion products. Fusion of protoplasts from different sources requires a fusogenic agent. The isolated protoplasts are entirely spherical and the area of contact very small. To establish close contact between plant membranes, it is essential to introduce a treatment. Several agents such as antibodies, lectins and poly-L-lysine effectively agglutinate protoplasts, but fail to cause fusion. Kuster in the earliest recorded experiments on protoplast fusion in 1909 introduced the use of soium nitrate[104)]. Some degree of success was achieved more recently with salt mixtures including nitrate in fusion protoplasts of different species, but the rates were low[105, 106)]. A procedure involving high calcium levels and a pH near 10 was used in the fusion of tobacco protoplasts and subsequent formation of hybrid plants[107)]. Similar conditions also were used in fusion and hybridization of moss[71)].

Effective and reproducible procedures for protoplast fusion involves the use of polyethylene glycol (PEG)[108, 109)]. Addition of high molecular weight (1500—6000) PEG at high concentrations cause immediate adhesion of protoplasts and aggregates are formed. Fusion may be initiated at this point and goes to completion when the PEG is removed by dilution and washing.

The PEG treatment is efficient and the fusion rates may reach up to 50% of the protoplast population. Various views have been expressed on the function of PEG in the fusion process[108, 109)]. The highly polar nature of the compound as well as a weak ionic charge may facilitate integration of groupings on the proteins and lipids of

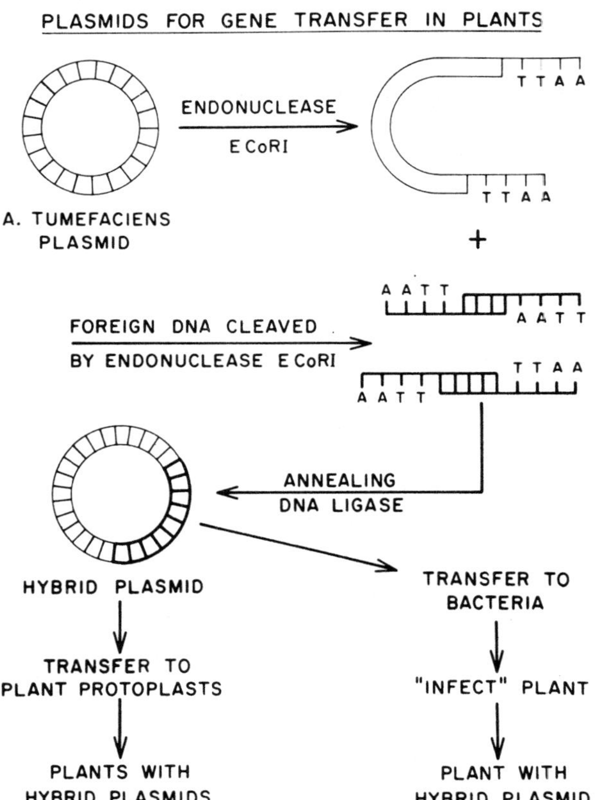

Fig. 1 Diagram projecting the use of plasmids (*A. tumefaciens*) as vectors in gene transfer in plants

opposing membranes[110]. PEG may function as a molecular bridge aided by calcium and facilitate the molecular dissociations of the plasmalemnas. Both the adhesion process and fusion appear to require critical but different concentration of PEG and of calcium. The enhancement of PEG-induced fusion by calcium ions and alkaline pH suggest that ionic groups are implicated[108, 111]. Adhesion requires a PEG concentration of 26—28% and occurs very rapidly. The membrane contact extends to large areas but may be discontinuous and forming intervening spaces. Upon dilution of the PEG the opposing plasmalemmas appear to erupt at several points and cytoplasmic continuity between adjacent protoplasts becomes established. At the outer edges the membranes of the fusing protoplasts join and the intervening sections form vesicles which gradually degrade[112, 113]. Initially the two cytoplasms remain separated but mixing occurs within 12 h. The comparatively rapid dissociation and reassociation of the plasma membranes occurs without apparent permanent damage to the fusion products (heterokaryocytes) which can regenerate a cell wall and divide[108, 48, 114]. The discovery of PEG as an efficient fusogenic agent has prompted studies on the use of the polymer in fusion of other biological systems. Apparently the compound is non-specific and induce fusion of protoplasts of bacteria[115], fungi[116], yeast[117], alga[119], He La cells with carrot protoplasts as well as carrot protoplasts with those of alga[118, 120].

5.2 Hybrid Development

When fusion products are cultured in a suitable medium they reform a cell wall. The identification of the fusion products is based on differences between the parental cells with respect to pigmentation, presence of chloroplasts and other cytoplasmic markers[108, 114, 121]. A system which is frequently used consists of fusing green leaf mesophyll protoplasts with those from a cell culture, which lack chloroplasts. The fusion products contain chloroplasts and can be distinguished from unfused protoplasts. Employing this procedure as well as a differential staining technique it has been possible to monitor the frequency of heteroplasmic fusion and heterkaryocyte formation[111, 122]. The fusion frequency varies widely and is affected by protoplast quality as well as by the fusion conditions.

The heterokaryocytes contain one or more nuclei from each parental protoplast. Constabel et al.[122] using a differential staining method monitored the heterokaryons of pea + soybean. The ratio of 1:1 nucleus of each parent occurred most frequently followed by those with a 2:1 ratio. When a larger number of nuclei were present, the heterokaryocytes deteriorated. During the first days in culture several developments may occur. In several intergeneric heterokaryons the fusion of interphase nuclei have been observed[122, 123]. The events may be detected one day after fusion and appear to require several hours to complete. The observations by Fowke et al. suggest that fusion of interphase nuclei may occur through the formation of nuclear membrane bridges[124]. Such phenomena have been observed in plant homopolykaryons[101] as well as in plant heteropolykaryons[124]. However, genome integration frequently occurs during mitoses. Heterokaryocytes can undergo division if the protoplasts of at least one of the parental species is able to divide. Division has been observed in heterokaryons arising from fusion of protoplasts from a wide variety of plants (Table 4). These include the intergeneric hybrids of soybean with

Table 4. Intergeneric protoplast fusion and heterokaryocyte division

Source of protoplast	Ref.
Barley (*Hordeum vulgare*) + Soybean (*Glycine max*)	108)
Corn (*Zea mays*) + *Soybean* (*Glycine max*)	108)
Pea (*Pisum sativum*) + *Vicia hajastana*	108)
Pea (*Pisum sativum*) + Soybean (*Glycine max*)	108)
Sweet clover (*Melilotus alba*) + Soybean (*Glycine max*)	122)
Alfalfa (*Medicago sativa*) + Soybean (*Glycine max*)	46)
Caragana (*Caragana arborescens*) + Soybean (*Glycine max*)	46)
Rapeseed (*Brassica napus*) + Soybean (*Glycine max*)	121)
Barley (*Hordeum vulgare*) + Carrot (*Daucus carota*)	123)
Tobacco (*Nicotiana tabacum*) + Soybean (*Glycine max*)	48)
Nicotiana glauca + Soybean (*Glycine max*)	111)
Nicotiana rustica + Soybean (*Glycine max*)	48)
Colchicum autumnale + Soybean (*Glycine max*)	48)
Nicotiana langsdorffii + Soybean (*Glycine max*)	48)
Soybean (*Glycine max*) + *Vicia hajastana*	125)
Fababean (*Vicia faba*) + Petunia (*Petunia hybrida*)	126)
Arabidopsis thaliana + *Brassica campestris*	127)
Carrot (*Daucus carota*) + Petunia (*Petunia hybrida*)	31)
Carrot (*Daucus carota*) + Tobacco (*Nicotiana tabacum*)	31)
Atrope belladonna + Petunia (*P. hybrida*)	31)
Tomato (*Lycopersicon esculentum*) + Petunia (*P. hybrida*)	128)
Sorghum (*S. bicolor*) + Corn (*Zea mays*)	36)

Vicia, pea, and other genera. The production of cell hybrids between plant families of *N. glauca* + soybean, carrot + barley, soybean + *B. napus* suggest the absence of an apparent somatic cell incompatibility. The hybrid nature of the cell progeny has been established on the basis of ultrastructural examination[124] chromosome identification[111, 122], isoenzymes and polypeptide patterns of the Fraction 1 protein[129, 130, 131]. In the initial stages the cells contain chloroplasts originating from leaf protoplasts as well as leucoplasts contributed by the cultured cells. In the nuclei the heterochromatin reflect that of the two parents and chromosomes of both parental species can be recognized.

5.3 Hybrid Selection and Plant Development

Various approaches have been implemented to permit isolation of hybrids. In most fusion experiments the division rates are relatively low. Moreover, one or both parental protoplast species also may divide and within a short period the hybrid cells cannot be distinguished from parental cells. Several selection methods have been successful (Table 5). A special plating procedure was used to isolate hybrids of *N. glauca* + soybean[111] and *Arabidopsis thaliana* + *Brassica campestris*[127, 132]. After allowing the hybrid cells to undergo a few divisions, the cell mixture is diluted. Microdroplets (ca 500 nl) are placed in Cuprak petri dishes designed with numerous small wells. Each droplet will contain a single or a few cell clusters. By scanning under a light microscope the wells containing single hybrid cell clusters can be identified. This procedure has been used to obtain several hybrid cell lines of *N. tabaccum* + soybean and of *A. thaliana* + *B. campestris*. In the latter example mature plants were also regenerated[133)].

Attempts have been made to take advantage of the species difference in resistance to uncommon amino acids such as canavanine[48)]. The compound inhibited division of soybean and pea cells but those of sweet clover were unaffected. Heterokaryons obtained by fusion of protoplasts from soybean (sensitive) with those from any one of the resistant plants failed to divide in the presence of canavanine. The sensitivity to canavanine appeared to be a dominant characteristic and was thus expressed in the hybrids.

A number of other types of potential chemical selection procedures may involve herbicides, phytotoxins or antibiotics. Plants differ in their capacity to metabolize and thus tolerate herbicides. As an example Rice cells can metabolize propanil (3,4-dichloropropionanilide) which accounts for their resistance to the compound[134)].

Phytotoxins produced by plant pathogens are metabolic analogs and have been shown to be species specific[135, 73)]. The compounds exert their effects at relatively low concentrations. Using the toxin of *Helminthosporium maydis*, Gengenbach and Green[136)] have selected cells and plants which are resistant to the compound. The same toxin has been used in protoplast fusion experiments and there are indications that toxin resistance is expressed in the fusion products[73)]. A variety of antibiotic resistant cells of different plant species have become available[137)]. Kanamycin resistant cells of *Nicotiana sylvestris* were used in hybridization of *N. sylvestris* + *N. knightiana*[138)]. The resistance to kanamycin was used in conjunction with the capacity for shoot induction in the selection of the somatic hybrids. Power et al.[139)] utilized the differential resistance of *Petunia parodii* and *P. hybrida* to actinomycin D in conjunction with culture medium adjustments in the selection of hybrids between these species. Any one of these methods based on differential sensitivity to chemicals have proven valuable when used in conjunction with other selective factors.

Selection systems based on genetic complementation are used routinely in microorganisms. The principle of genetic complementation is likely the most reliable and effective also for the recovery of somatic hybrids in plants. In the first reported experiments on somatic hybridization in plants two tobacco species were used. It was known that sexual hybrids of these species (*Nicotiana glauca* × *N. langsdorffii*) were oncogenic and produced genetic tumors. Cells from plant tumors can grow in culture in the absence of growth hormones (auxins). When protoplasts from these species were fused a proportion of the regenerated cells grew in the absence of hormones[106)]. The plants obtained from the cells were hybrids when compared with the sexual hybrids and produced tumors[106, 141)].

In later experiments, Melchers and Labib[140)] fused protoplasts of two chlorophyll-deficient, light-sensitive mutants of *Nicotiana tabacum*. The hybrid plants obtained by fusion had normal leaf color and had normal reaction to light. Chlorophyll-deficient (non-allelic) mutants have now been used in selection of both interspecies and intergeneric somatic hybrids (Table 5). In all cases complete plants were obtained. Nutritional mutants would provide a very desirable material. Until recently with the report of a pantothenate-requiring mutant in *Datura*, such

Table 5. Chromosome number and selection methods in somatic hybrids

Parental species	Selection method	No. of plants	Chromosome number (range)		Ref.
			expected	observed	
Nicotiana glauca + *N. langsdorffii*	Selective media	23	42	56–64	141)
Nicotiana tabacum + *N. tabacum*	Chlorophyll-deficient Light-sensitive	20+	48	48, 72, 96	140)
N. tabacum + *N. knightiana*	Albino, organogenesis	8	72	44–126	138)
N. tabacum + *N. rustica*	Chlorophyll-deficient	10	60	60–91	142)
N. tabacum + *N. glauca*	Albino	20	72	72	131)
Petunia hybrida + *P. parodii*	Albino	1+2	28	26, 28	21)
Datura innoxia + *D. innoxia*	Chlorophyll-deficient	17	24	24–108	86)
D. innoxia + *Atropa belladonna*	Albino		96	84–175	143)
D. innoxia + *D. stramonium*	Albino			48–72	
Daucus carota + *D. capillifolius*	Albino	12	36	34–54	83)
D. carota + *Aegopodium podagraria*	Albinism		60	18	144)
Arabidopsis thaliana + *Brassica campestris*	Mechanical, albino		60	55–60	133)
N. tabacum + *N. sylvestris*			72	76–80	145)

mutants were not availabe in higher plants[64]. Schieder[69] has used nutritional mutants in somatic hybridization in liverwort (*Sphaerocarpus donnellii*). Hybrids obtained by fusion of protoplasts from nicotinic acid and glucoserequiring mutants were selected on minimal media. The hybrid plants were identified on the basis of morphology and karyotype. Similar nutritional mutants also were used in selection of somatic hybrids of moss (*Physcomitrella patens*[71]).

Glimelius et al. have reported recently on the use of nitrate reductase deficient mutants of *Nicotiana tabacum* in selection of hybrids. The mutant cells had a requirement for reduced nitrogen[150]. The hybrids obtained by fusion regained the ability to grow on media in which nitrate was the sole nitrogen source.
greater tendency for multiples of the allotetraploid chromosome number[86,140].

6 Evidence for Gene Expression in Somatic Hybrids

The majority of hybrid plants which have been produced by somatic hybridization can also be obtained by sexual crossing. Direct comparison could then be made between the two types of plant hybrids, and the observations were used to confirm that hybridization had occurred[21,106,131]. As expected, the morphological, features of foliage, floral structure and color were intermediate and distinct from either of the parent plants. Somatic hybrization has yielded progeny of greater variety than is possible by sexual means because of the merging of the cytoplasm of the parent cells[147].

6.1 Ploidy

Somatic hybrid plants may vary considerably in ploidy and deviate from that expected by adding the parental chromosome numbers. Table 5 shows several hybrids and the number of plants on which chromosome counts were performed. The *N. glauca* + *N. langsdorffii* hybrids analyzed by Smith et al.[141] had chromosome numbers close to 60. The authors accounts for this observation by suggesting that hybrids arose by fusion of two protoplasts of *N. langsdorffii* $2n = 18$ and one of *N. glauca* ($2n = 24$). In the majority of interspecies hybrids, the chromosome numbers did not deviate substantially from the total of the two parents, but aneuploids were common[21,131,142,83,145]. Relatively few observations have been made on intergeneric hybrids and no generalization is possible[133,143,144]. In plants arising after fusion of protoplasts from different mutants of the same species there appears to be a greater tendency for multiples of the allotetraploid chromosome number[86,140].

6.2 Biochemical Changes

Zymograms of constitutive enzymes has proven a useful method to confirm that hybridization has occured. In Table 6 are listed examples of the zymograms of a number of enzymes from somatic hybrids. As indicated by these observations, the following can happen: (a) the isoenzyme bands are additive or the total of that of the parents[129,131], (b) some bands present in parent tissues are missing in the hybrid[138,146], (c) or new bands occurs in the hybrids[127,131,138]. In the hybrid of

Table 6. Isoenzymes of Constitutive enzymes used in identification of Somatic Hybrids

Parental species	Enzymes	Bands in hybrids	Ref.
N. glauca + *N. langsdorffii*	Alcohol dehydrogenase	— bands missing	146)
	lactate dehydrogenase	— some bands missing	
	esterase	— not sum of parental bands	
N. sylvestris + *N. knightiana*	Alcohol dehydrogenase	Bands non-additive	138)
	glucose-6-PO_4 dehydrogenase	New bands	
	esterase	In hybrids	
N. glauca + *Glycine max*	Alcohol dehydrogenase	— additive	129)
	aspartate Aminotransferase	— additive	
Arabidopsis thaliana + *Brassica campestris*	Esterase	— bands non-additive	127)
	alcohol dehydrogenase	— additive	
	lactate dehydrogenase	— new bands	
	peroxidase	— additive	
	myrosinase of *B. campestris*	— present	
N. glauca + *N. tabacum*	Alanyl aminopeptidase	— additive	131)
	aspartate aminotransferase	— additive & hybrid band	

N. glauca and *N. tabacum*, a distinct isoenzyme of aspartate aminotransferase occurs in the tissues of both the sexual and the somatic hybrid[131] indicating the production of new polypeptides which makes up the enzyme protein. It is fairly common that some isoenzymes shown to be present in parent tissues are missing in the hybrids[127, 129, 146]. The reason for that could be chromosome elimination[146] or other more subtle gene deletion or repression.

When hybrids are selected on the basis of genetic complementation and particularly through the use of albino mutants there is an extensive range of gene expression or correction. No report has appeared on which biochemical and structural corrections have occurred. The most visible changes are those associated with grana structure and chlorphyll formation[28]. The somatic hybridization method could be used to good advantage to elucidate the biochemical events underlying chlorphyll deficient and lethal, albino mutations, some of which are often ascribed to single gene mutants.

Green plants possess a unique enzyme, ribulose biphosphate carboxylase, which adds CO_2 to ribulose 1,5 biphosphate during photosynthesis[151]. The enzyme is referred to as Fraction 1 protein and consists of large subunits (LS) encoded for and synthesized in the chloroplasts, and smaller subunits (SS) which are genome encoded and synthesized outside the plastids. The protein can be obtained crystalline or separated from other proteins by using immunological procedure[151]. Through isoelectric focussing, the polypeptides of the subunits can be separated, forming band patterns which are specific for each plant species[153]. The procedure has been used in the identification of plant species. Bacause of the specificity of the patterns and sensitivity of the methods, the fraction 1 proteins also have been used in the verification of hybridization[151]. (Table 7) There is evidence that genome hybridization occurs and the hybrids contain SS coded for by both species[151, 154]. The LS of the hybrids generally resemble that of one or the other parent[154, 130]. In the somatic hybrids of tomato + potato three plants had the LS of tomato and one plant had those of the potato[130].

Other examples of gene expressions in somatic hybrids include those obtained by Glimelius et al.[150], which regained the capability to grow on nitrate media. Presumably, the hybridization resulted in the restoration of the nitrate reductase complex of proteins and enzymes[88]. In both liverwort[69] and moss[71], there are examples of corrections of metabolic auxotrophs by somatic hybridization (Table 7).

In somatic hybridization, as opposed to sexual crosses, the cytoplasms of the parental types become integrated[156]. Because of this unique feature of somatic hybridization, certain heritable factors associated with the cytoplasm can be transmitted. Thus it has been possible to transfer the male sterility property from one species to another by protoplast fusion[145, 148, 149] (Table 7). Male sterility can be a desirable feature of economic importance in plant breeding. The transfer can be made within compatible as well as non-compatible species. In the latter species, sexual crosses are not feasible. Restriction enzyme degradation and "finger printing" on polyacrylamide gels has become a useful method to analyze DNA sequences. The method has been employed in analyzing and mapping chloroplast DNA[155]. Belliard et al.[148] using a restriction enzyme digest compared the plastid DNA patterns of parents and hybrids and observed that the plastid DNA was that of either parent and neither a mixture nor a recombinant DNA.

Table 7. Biochemical gene expression in cytoplasm and organelles of somatic hybrids

Parental species	Hybrids			Other observations	Ref.
	Cytoplasmic male-sterility	Fraction 1 LS	Protein SS		
Nicotiana tabacum (*N. debneyi* cytoplasm) + *N. tabacum* (m-fertile)	M-sterile or M-fertile			C-DNA as either parent	148)
N. tabacum (*N. suaveolens*, cytoplasm) + *N. sylvestris* (m-fertile)	M-sterile or				145)
P. hybrida + *P. axillaris* (m-fertile)	M-sterile				149)
N. tabacum line nia-63 (NR^-) + *N. tabacum* line cnx-68 (NR^-)				Nitrate reductase	150)
Solanum tuberosum + *Lycopersicon esculentum*		3 as L.e. 1 as s.t.	Mixture		130)
Nicotiana glauca + *N. langsdorffii*		12 as N.L. 7 as N.g.	Mixture		154)
Sphaerocarpus donnellii-mutant nic-2 + mutant pal-2				Regained ability to synthesize nicotinic acid and glucose	69)
Physcomitrella patens mutant pab-3 + mutant thi-1, nic 10				Regained ability to synthesize nicotinic acid, thiamine and p-amino benzoic acid	71)

7 Other Gene Transfer Methods

Genetic modification in plants is also being considered through uptake of DNA and organelles and single cell uptake into protoplasts. Genetic transformation through DNA uptake implies that DNA from one source is taken up, incorporated into the recipient cell in a stable form and that genetic information encoded in the foreign DNA is expressed as new stable characteristics. The uptake of organelles such as chloroplasts or cells of bacteria and algae into protoplasts may provide new and effective approaches to study nuclear-organelle and nuclear cytoplasmic interactions as well as serving as a method of intergenetic transfer of such processes as nitrogen fixation.

7.1 DNA Uptake

There are several approaches which have been followed to accomplish genetic transformation with plants and plant cells. These include: a) uptake of DNA by plant tissues, b) uptake of DNA by plant cells and pollen, and c) uptake of DNA by plant protoplasts. Several recent reviews eleaborate on the details of these studies[157, 158, 159]. Most work so far involves studies which measure integration of foreign DNA by associating radioactively labeled foreign DNA with host cell DNA. A subsequent denaturation experiment shows that donor and host DNA cannot be separated suggesting they are complementary partners of a DNA double helix. Other types of experiments provide data which suggest replication of the donorhost DNA complex. However, after a very careful analysis of these studies, it has been pointed out that the fate of the exogenous donor DNA seems to be degradation and reutilization rather than integration and replication.

The apparent degradation of "naked" DNA in any form by nucleases from plant cells seems to be a major barrier to the successful insertion and expression of foreign genes in plant cells. In addition to nucleases of plant origin, enzyme preparations used in protoplast isolation may also contain high nuclease activity[159]. The solution presently being explored is to, in some way, provide protection for the donor DNA to eliminate susceptibility to nuclease attack. The approach so far has centered around the use of so-called gene vectors to facilitate both uptake and stabilization of foreign DNA inside the plant cell.

Two types of vectors have been used so far: bacterial plasmids and plant viruses. Plasmids are double-stranded, closed-circular, extrachromosomal DNA found in bacteria[160, 161]. In bacteria plasmids replicate independently, and can integrate into the bacterial chromosome. A series of endonucleases (restriction enzymes) from bacteria can be used to degrade any DNA at points of specified nucleotide sequences[160]. These restriction enzymes cleave the plasmid DNA into linear, double-stranded sections with overlapping, complementary nucleotide sequences. Foreign DNA can be inserted into a plasmid by co-digestion of plasmid and foreign DNA with the same restriction enzyme, mixing the linear fragments and then reconstituting the plasmid. The plasmid is taken up into the cell where the genes are expressed. According to Ohyama et al.[158] it appears that significant amounts of plasmid DNA uptake into plant protoplasts can occur. In the best work so far Fernandez et al.[162] showed that the pBR 313 bacterial plasmid could be incorporated into turnip protoplasts, in which the DNA retained its original molecular size in the nuclear fraction for up to 45.5 h. Other plasmids used were cleaved into smaller linear molecules very rapidly. These results, at least, suggest that plasmids may survive nuclease digestion in some protoplasts long enough to enable integration and possibly provide new information on gene transfer in these systems.

One unique plasmid system which is being extensively studied is that of the plant tumor inducing bacterium, *Agrobacterium tumefaciens*. The mechanism of this neoplasmic transformation has been recently elucidated and summarized[162]. Tumor inducing strains of the bacterium possess a large extrachromosomal DNA plasmid (the Ti plasmid) which is responsible for the oncogenic properties of the bacterium. The genes on these plasmids are only expressed in transformed plants. Since there is a mechanism for these plasmids to enter plant cells and to integrate into the DNA of the cell they might be used as vector to carry foreign genes into plant cells[159]. The proposed mechanism by which the process could be achieved is outlined in Fig. 1.

Recently another approach has been followed to incorporate foreign nucleic acids into plant protoplasts. Matthews et al.[164] reported using lipid vesicles (liposomes) to sequester bacterial RNA, and to incorporate the RNA into carrot protoplasts. Incubation medium which had been treated with liposome-sequestered [^{3}H] RNA contained large amounts of 16 s and 23 s RNA which was still intact. However, RNA extracted from the supernatant fluid of protoplasts which had been treated with naked RNA showed extensive degradation. Furthermore, protoplasts which had been treated with the liposome sesquestered RNA contained some intact RNA. The authors concluded that the RNA breakdown which occurred after lysosome mediated uptake into the protoplasts might have taken place because uptake occurred through endocytosis of the intact liposomes rather than fusion. This could result in the liposomes being rapidly degraded by the lysosomal apparatus. When uptake occurs through fusion the entrapped molecules are extruded into the cytoplasm, thus escaping degradation by the lysosomes. Although the liposome sequestered RNA is not totally protected inside the plant protoplasts, this approach does seem to present a gene transfer method worthy of further study.

The use of plant viruses as vectors for foreign DNA is also an area which is being considered. The DNA plant viruses may be the group with the greatest potential[165]. The virus DNA has the power of autonomous replication in the cytoplasm of plant cells without any apparent requirement for integration into the host chromosome. Presumably any additional foreign DNA attached to the viral chromosome by the previously recombinant DNA techniques would also be replicated and expressed. The present need here seems to be to develop recombinant DNA approaches with these viruses, so that specific foreign DNA molecules may be attached to the viral genome.

A slightly different gene transfer approach using plant viruses combined plasmid DNA with Tobacco Mosaic Virus (TMS) capsid protein[166]. The two plasmids, pBR 313 from *E.* coli, and pCK 135 of *A. tumefaciens* were reconstituted with the capsid protein of TMV. The DNA polymerized with the protein in the form of short rods with DNA "tails". Cowpea protoplasts were incubated with this DNA protein mixture. The *A. tumefaciens* DNA entered the nuclei of the cells where it was extensively degraded. The pBR 313 DNA, however, remained intact to a higher degree and it appeared that this DNA was cleaved into only three pieces with a molecular weight average of 2×10^6 Daltons. These results show that plasmid DNA coated with TMV protein is sufficiently protected from nuclease digestion, and can in some cases, survive inside the plant cell. The present approach here, as well as in all DNA uptake studies with DNA protoplasts, is to identify, if any biological function foreign genes carried by the plasmid DNA is expressed. This still remains to be adequately demonstrated.

7.2 Uptake of Organelles and Single Cells

Plant protoplasts have also made it possible to consider uptake of cell organelles and whole cells as a gene transfer method. These subjects have been recently reviewed by Davey[167)] and Giles[168)]. Most of the emphasis has been in using organelle transfer to create new cytoplasmic combinations which might be of value to plant breeders and to use cell uptake to explore the transfer of nitrogen fixation to species where this process is absent. As far as organelle transfer is concerned, little more than successful uptake has been reported to date. According to Thomas et al.[169)] convincing evidence of integration and expression of organelles in foreign protoplasts has not been established.

Beyond uptake, success has been reported on the transfer of nitrogen fixing ability by establishing cells of *Azotobacter vinelandii* in the protoplasts of the mycorrhizal fungus *Rhizopogon*[170) 171)]. The nitrogen fixing capacity seemed to be localized in stable spherical bodies indentified as L-forms of the bacteria in the cytoplasm of the fungal hyphae[172)]. Several reports have appeared on using protoplasts of green plants. In the first study Meeks et al.[172)] induced the uptake of cells of an auxotrophic N_2-fixing mutant of *Anabaena variabilis* into protoplasts of tobacco using PEG. The main question was the fate of the algal cells and the protoplasts over time. About 7.6% of the protoplasts contained at least one algal cell. After five days of culture, less than 1.0% of the cells contained algal cells. During a two week period of culture no protoplast containing an algal cell had divided and 59% of the protoplasts deteriorated and died.

A novel approach to uptake of blue-green algae into protoplasts was recently reported by Bradley and Leith[173)]. They were able to successfully induce the uptake of blue-green alga cells into onion protoplasts, when the algal cells were contained in oil drops. This seems to be an approach similar to the liposome-mediate uptake of foreign DNA into protoplasts.

Finally, Fowke et al.[120)] were successful in inducing fusion between carrot protoplasts and the walless mutant of *Chlamydomonas reinhardii*. The fusion products were cultured and observed with light and electron microscopy. The frequency of fusions was usually 10—20% with most protoplast containing 3 or less algal cells. Upon fusion the algal organelles were released into the cytoplasm of the protoplasts. Protoplasts containing algal organelles had regenerated walls and had divided in 3—5 days. However, mitochondria and golgi bodies were never observed and it was assumed that they disintegrate. Chloroplasts were recognized up to 10 days after fusion and wall regeneration.

The results of these studies provided valuable information on the relative merits of fusion vs. uptake as a means of gene transfer. The previously discussed work on protoplast fusion combined with the work of Fowke et al.[120] suggest that cell fusion may be a more productive approach than uptake. This was mentioned earlier in the context of liposome mediated gene transfer and may have additional relevance here. When uptake occurs through endocytosis into intracellular vessicles, these vessicles can fuse with the lysosomal apparatus resulting in breakdown of the foreign material. However, when fusion occurs, the contents of the cells intermix and are less susceptible to lyosomal degradation. This hypothesis still remains to be tested but might suggest the direction of further thought and experimentation.

8 Acknowledgement

I would like to express my gratitude (OLG) to former colleagues in Saskatoon and noteably Dr. L. Fowke, University of Saskatchewan, and Dr. F. Constabel of the Prairie Regional Laboratory, for helpful discussions and reviewing of the manuscript. Our sincere thanks to Donna Fairwell for her competent typing as well as her patience in the preparation of the manuscript.

9 References

1. Reinert, J., Bajaj, Y. P. S.: Plant Cell, Tissue and Organ Culture, p. 803. New York: Springer 1977
2. Street, H. E.: Plant Tissue and Cell Culture. Botanical Monographs. Vol. 11, Second Ed. London: Blackwell Scientific Publications 1977
3. Thorpe, T. A. (ed.): Frontiers of Plant Tissue Culture, 1978. Univ. of Calgary, Canada 1978
4. Sharp, W. R. et al. (eds.): Plant Cell and Tissue Culture, 4th Biosciences Collog. Columbus, Ohio: Ohio State University Press 1979
5. Kasha, K. J.: Haploids in Higher Plants. p. 421. Univ. of Guelph, Guelph, Canada 1974
6. Nabors, Murray, W.: Bioscience *26*, 761 (1976)
7. Vasil, I. K.: Adv. Agronom. *28*, 199 (1976)
8. Cocking, E. C.: An. Rev. Plant Physiol. *23*, 29 (1972)
9. Gamborg, O. L. et al.: In: Coll. Int. Cent. Nat. Rech. Sci. *212*, 155 (1973)
10. Takebe, I.: Compr. Virol. *11*, 237 (1977)
11. Gamborg, O. L. et al.: Can. J. Genet. Cytol *16*, 737 (1974)
12. Cocking, E. C.: Nature *187*, 962 (1960)
13. Gamborg, O. L., Wetter, L. R. (eds.): Plant Tissue Culture Methods. p. 109. Saskatoon: National Research Council of Canada, Prairie Regional Laboratory 1975
14. Gamborg, O. L.: Culture Media for Plant Protoplasts, In: CRC Handbook in Nutrition & Food. M. Recheigl, Jr. (ed.), p. 415. Cleveland: CRC Press, Inc. 1977
15. Gamborg, O. L.: In: Cell Genetics in Higher Plants, Duduts, D., Farkas, G. L., Maliga, P. (eds.), p. 107. Budapest: Akademiai Kiado 1976
16. Eriksson, T.: In: Plant Tissue Culture and Its Bio-technological Application, p. 313. Berlin, New York: Springer 1977
17. Schieder, O.: Z. Pflanzenphysiol. *76*, 462 (1975)
18. Schieder, O.: Mol. gen. Genet. *162*, 113 (1978)
19. Furner, I. J., J. King, O. Gamborg: Plant Sci. Lett. *11*, 169 (1978)
20. Binding, H.: Z. Pflanzenphysiol. *74*, 327 (1975)
21. Cocking, E. C. et al.: Plant Sci. Lett. *10*, 7 (1977)
22. Vasil, V., Vasil, I. K.: Z. Pflanzenphysiol. *92*, 379 (1979)
23. Butenko, R. G., Vitenko, A. A., Avetisov, V. A.: Soviet Plant Physiol. (Moscow) *24*, 541 (1977)
24. Grun, P., Chu, L. J.: Amer. J. Bot. *65*, 538 (1978)
25. Shepard, J. F., Totten, R. E.: Plant Physiol. *60*, 313 (1977)

26. Takebe, I., Labib, G., Melchers, G.: Naturwiss. *58*, 318 (1971)
27. Bourgim, J., Chupeau, Y., Missonier, C.: Physiol. Plant *45*, 288 (1979)
28. Gamborg, O. L. et al.: Z. Pflanzenphysiol. *95*, 255 (1979)
29. Cassells, A. C., Barlass, M.: Physiol. Plant *42*, 236 (1978)
30. Lorz, H., Wernicke, W., Potrykus, I.: Planta Medica 21 (1979)
31. Gosch, G., Reinert, J.: Protoplasma *96*, 23 (1978)
32. Kao, K. N. et al.: Colloq. Int. C.N.R.S. *212*, 207 (1973)
33. Koblitz, H.: Biochem. Physiol. Pflanzen *170*, 287 (1976)
34. Potrykus, I. et al.: Mol. gen. Genet. *156*, 347 (1977)
35. Cai, Qi-qui et al.: Acta Bot. Sinica *20*, 97 (1978)
36. Brar, D. S. et al.: Z. Pflanzenphysiol. *96*, 269 (1980)
37. Dudits, D., In: Cell Genetics in Higher Plants, D. Dudits et al. (eds.), p. 153. Budapest: Akademiai Kiado 1976
38. Bui-Dang-Ha, D., I. A. MacKenzie: Protoplasma *78*, 215 (1973)
39. Kartha, K. K. et al.: Plant Sci. Lett. *3*, 265 (1974)
40. Thomas, E. et al.: Molec. gen. Genet. *145*, 245 (1976)
41. Grambow, H. J. et al.: Planta *103*, 348 (1972)
42. Dudits, D. et al.: Can. J. Bot. *54*, 1063 (1976)
43. Constabel, F., Kirkpatrick, J. W., Gamborg, O. L.: Can. J. Bot. *51*, 2105 (1973)
44. Arnold, S. V., Eriksson, T.: Physiol. Plant. *36*, 193 (1976)
45. Gamborg, O. L., Shyluk, J., Kartha, K. K.: L. Plant Sci. Letters *4*, 285 (1975)
46. Constabel, F. et al.: Biochem. Physiol. Pflanzen *168*, 319 (1975)
47. Kao, K. N., M. R. Michayluk: Z. Pflanzenphysiol. *96*, 135 (1980)
48. Constabel, F. et al.: Z. Pflanzenphysiol. *79*, 1 (1976)
49. Kao, K. N., Keller, W. A., Miller, R. A.: Exp. Cell. Res. *62*, 388 (1970)
50. Pelcher, L. E., Gamborg, O. L., Kao, K. N.: Plant Sci. Lett. *3*, 107 (1974)
51. Binding, H., Nehls, R.: Z. Pflanzenphysiol. *88*, 327 (1978)
52. Davey, M. R., Bush, E., Power, J. B.: Plant Sci. Lett. *3*, 127 (1974)
53. Gamborg, O. L., Shyluk, J. P.: Bot. Gaz. *137*, 301 (1976)
54. Bhojwani, S. S., Power, J. B., Cocking, E. C.: Plant Sci. Lett. *8*, 85 (1977)
55. Coutts, R. H. A., Wood, K. R.: Plant Sci. Lett. *4*, 189 (1975)
56. Vardi, A., Spiegel-Roy, P., Galun, E.: Plant Sci. *4*, 231 (1975)
57. Kirby, E. G., Cheng, T. Y.: Plant Sci. Lett. *14*, 145 (1976)
58. David, A., David, H.: Z. Pflanzenphysiol. *94*, 173 (1979)
59. Dorion, N., Chupeau, Y., Bourgim, J. P.: Plant. Sci. Lett. *5*, 325 (1975)
60. Fowke, L. C., Bech-Hansen, C. W., Gamborg, O. L.: Protoplasma *79*, 235 (1974)
61. Kao, K. N., Michayluk, M. R.: Planta *126*, 105 (1975)
62. Shepard, J. F.: Amer. Potato *54*, 486 (1977)
63. Chu, Yaw-en, Lark, K. G.: Planta *132*, 259 (1976)
64. Savage, A. D., King, J., Gamborg, O. L.: Plant Sci. Lett. *16*, 367 (1979)
65. Constabel, F.: Isolation and culture of plant protoplasts. In: Plant Tissue Culture Methods, Gamborg, O. L., Wetter, L. R. (eds.). Ottawa: National Research Council of Canada 1975
66. Williamson, F. A. et al.: Microfibril deposition on cultured protoplasts of *Vicia hajastana*, Protoplasma *91*, 213 (1977)
67. Fowke, L. C., Gamborg, O. L.: Application of Protoplasts to the Study of Plant Cells, Int. Rev. Cytology *68*, 9 (1980)
68. Fowke, L. C.: Ultrastructure of isolated and cultured protoplasts, In: Proc. Symp. 4th Int. Congr. Plant Tissue and Cell Culture (T. Thorpe, ed.), p. 223. Calgary 1979
69. Schieder, O.: Z. Pflanzenphysiol. *74*, 357 (1974)
70. Ono, K., Ohyama, K., Gamborg, O. L.: Plant Sci. Lett. *14*, 225 (1979)
71. Grimsley, N. H., Ashton, D. J. Cove: Mol. Gen. Genet. *154*, 97 (1977)
72. Burgess, J., Linstead, P. H., Harndeu, J. M.: Micron *8*, 181 (1977)
73. Earle, E. D., In: Frontiers of Plant Tissue Culture, p. 363. Univ. Calgary, Canada 1978
74. Strobel, G. A., Steiner, G. W., Byther, R.: Biochm. Gen. *13*, 557 (1975)
75. Pelcher, L. E. et al.: Can. J. Bot. *73*, 427 (1975)
76. Edwards, G. E., Huber, S. C., In: Proc. 4th Int. Congr. on Photosynthesis, p. 95. Univ. of Wisconsin, Madison 1977

77. Edwards, G. E. et al.: Plant Physiol. *63*, 821 (1979)
78. Ohyama, K., Pelcher, L. E., Horn, D.: Plant Physiol. *60*, 179 (1977)
79. Ashton, N. W., Cove, D. J.: Mol. Gen. Genet. *154*, 87 (1977)
80. Schieder, O.: Molec. gen. Genet. *144*, 63 (1976)
81. Savage, A. D.: Thesis, Faculty of Graduate Studies and Research, Univers. Saskatchewan, Saskatoon, Canada 1979
82. von Wettstein, D. et al.: Science *184*, 800 (1974)
83. Dudits, D. et al.: Theoret. Appl. Genet. *51*, 127 (1977)
84. Gamborg, O. L., Constabel, F., Miller, R. A.: Planta *95*, 355 (1970)
85. Burk, L. G., Menser, H. A.: Tobacco Sci. *8*, 101 (1977)
86. Schieder, O.: Planta *137*, 253 (1977)
87. Schieder, O.: In: Frontiers of Plant Tissues Culture, T. A. Thorpe (ed.), p. 393. Univ. Calgary, Calgary, Canada 1978
88. Muller, A. J., Grafe, R.: Molec. Gen. Genet. *161*, 67 (1978)
89. Redei, G. P., Acedo, G., In: Cell Genetics in Higher Plants (Dudits, D., Farkas, G. L., Maliga, P., eds.), p. 39. Budapest: Akademiai Kiado 1976
90. Racchi, M. L. et al.: Plant Sci. Lett *13*, 357 (1978)
91. Maretzki, A., Thom, M.: Plant Phyiol. *61*, 544 (1978)
92. Limberg, M., Cress, D., Lock, K. G.: Plant Physiol. *63*, 718 (1979)
93. Lam, Shue-Lock, Erickson, T.: J. Hered. *62*, 207 (1971)
94. Chaleff, R. S., Parson, M. R.: Genetics *89*, 723 (1978)
95. Borlaug, N. E.: Cereal Sci. Today *16*, 401 (1971)
96. Nickell, L. G., Torrey: J. G., Science *166*, 1068 (1969)
97. Holl, F. B.: Can. J. Genet. Cytol. *17*, 517 (1975)
98. Bates, L. S., Deyoe, C. W.: Econ. Bot. *27*, 401 (1973)
100. Miller, R. A. et al.: Can. J. Genet. Cytol *13*, 347 (1971)
101. Fowke, L. C. et al.: J. Cell Sci. *18*, 491 (1975)
102. Kao, K. N., Wetter, L. R.: Int. Cell Biol. Pap. p. 216. Int. Congr. 1974
103. Ito, M., Maeda, M.: Bot. Mag. (Tokyo) *87*, 209 (1974)
104. Kuster, E.: Ber. dtsch. Bot. Ges. *27*, 589 (1909)
105. Eriksson, T.: Coloq. Int. Centr. Nat. Rech. Sci. (Paris) *193*, 297 (1970)
106. Carlson, P. S., Smith, H. H., Dearing, R. D.: Proc. Nat. Acad. Sci. (USA) *69*, 2292 (1974)
107. Keller, W. A., Melchers, G.: Z. Naturforsch. *28c*, 737 (1973)
108. Kao, K. N. et al.: Planta *120*, 215 (1974)
109. Wallin, A., Glimelius, K. G., Eriksson, T.: Z. Pflanzenphysiol. *74*, 64 (1974)
110. Lucy, J. A.: Trends in Biochemical Sciences (TIBS) *2*, 17 (1977)
111. Kao, K. N.: Mol. Gen. Genet. *150*, 225 (1977)
112. Fowke, L. C. et al.: Can. J. Bot. *53*, 272 (1975)
113. Burgess, J., Fleming, E. N.: Planta (Berl.) *118*, 183 (1974)
114. Fowke, L. C., Constabel, F., Gamborg, O. L.: Planta *135*, 257 (1977)
115. Fodor, K., Alfoldi, L.: Mol. Gen. Genet. *168*, 55 (1979)
116. Anne, J., Eyssen, H., Desomer, P.: Nature *262*, 719 (1976)
117. Ferenczy, L., Maraz, A.: Nature *268*, 524 (1977)
118. Dudits, D. et al.: Hereditas *32*, 121 (1976)
119. Ohiwa, T.: Protoplasma *97*, 185 (1978)
120. Fowke, L. C., Gresshoff, P. M., Marchant, H. J.: Planta *144*, 341 (1979)
121. Kartha, K. K. et al.: Can. J. Botany *52*, 2435 (1974)
122. Constabel, F. et al.: Can. J. Bot. *53*, 2092 (1975)
123. Dudits, D. et al.: Can. J. Genet. Cytol. *18*, 263 (1976)
124. Fowke, L. C. et al.: Planta *130*, 39 (1976)
125. Constabel, F., Weber, G., Kirkpatrick, J. W.: C. R. Acad. Sci. (Paris) *285*, 319 (1977)
126. Binding, H., Nehls, R.: Molec. Gen. Genet. *164*, 137 (1978)
127. Gleba, Yury Yu, Hoffmann, F.: Molec. Gen. Genet. *164*, 137 (1978)
128. Binding, H.: Molec. gen. Genet. *144*, 171 (1976)
129. Wetter, L. R.: Mol. gen. Genet. *150*, 231 (1977)
130. Melchers, G., Sacristan, M. D., Holder, A. A.: Carlsb. Res. Comm. *43*, 203 (1978)
131. Evans, D. A., Wetter, L. R., Gamborg, O. L.: Physiol. Plant *48*, 225 (1980)

132. Gleba, Yury Yu: Naturwissenschaften *65*, 158 (1978)
133. Gleba, Yury Yu, Hoffmann, F.: Naturwissenschaften *66*, 547 (1979)
134. Still, G. G.: Plant Physiol. *43*, 543 (1968)
135. Strobel, G. A.: Ann. Rev. Plant Physiol. *25*, 541 (1974)
136. Gengenbach, B. G., C. E. Green: Crop. Sci. *15*, 645 (1975)
137. Maliga, P.: In: Frontiers of Plant Tissue Culture, p. 381. Univ. of Calgary, Calgary, Canada 1978
138. Maliga, P. et al.: Mol. gen. Genet. *157*, 291 (1977)
139. Power, J. E. et al.: Plant Sci. Lett. *10*, 1 (1977)
140. Melchers, G., Labib, G.: Mol. gen. Genet. *135*, 277 (1974)
141. Smith, H. H., Kao, K. N., Combatti, N. C.: J. Hered. *67*, 123 (1976)
142. Nagao, T.: Jap. Crop Sci. *47*, 491 (1978)
143. Schieder, D.: Planta *145*, 371 (1979)
144. Dudits, D. et al.: Plant Sci. Lett. *15*, 101 (1979)
145. Zelcer, A., Avid, D., Galun, E.: Z. Pflanzenphysiol. *90*, 397 (1978)
146. Wetter, L. R., Kao, K. N.: Z. Pflanzenphysiol. *80*, 455 (1976)
147. Gamborg, O. L.: In: Genetic Engineering for Nitrogen Fixation (A. Hollaender, ed.), Nat. Acad. Sci., Wash., D.C. 1977
148. Belliard, G. et al.: Mol. gen. Genet. *165*, 231 (1978)
149. Izhar, S., Power, J. B.: Plant Sci. Lett. *14*, 49 (1978)
150. Glimelius, K., Eriksson, T., Grafe, R.: Physiol. Plant *44*: 273 (1978)
151. Kung, Shain-Dow: Ann. Rev. Plant Physiol. *28*, 401 (1977)
152. Gray, J. C., Wildman, S. G.: Plant Sci. Lett. *6*, 91 (1976)
153. Chen, K. et al.: Plant Sci. Lett. *7*, 429 (1976)
154. Chen, K., Wildman, S. G., Smith, H. H.: Proc. Nat. Acad. Sci. *74*, 5109 (1977)
155. Bedbrook, Jr., Bogorad, L., Kolodner, R.: Cell *11*, 739 (1977)
156. Gleba, Y.: In: OSU Biosciences Colloquia (Sharp et al., eds.), Vol. 4, p. 775. Columbus, Ohio 1979
157. Kleinhofs, A., Behke, R.: Ann. Rev. Genet. *11*, 79 (1977)
158. Ohyama, K., Pelcher, L. E., Schaefer, A.: In: Proc. Symp. 4th Int. Congr. Plant Tissue and Cell Culture, T. Thorpe (ed.), p. 75. Calgary 1979
159. Gamborg, O. L. et al.: In: Plant Cell and Tissue Culture. Sharp, W. R., Larsen, P. O., Paddock, E. F., Raghaven, V. (eds.), 4th Biosciences Colloq. p. 371. Columbus: Ohio State Univ. Press 1979
160. Primrose, S. B.: Sci. Prog. *64*, 293 (1977)
161. Cohen, S. N.: Sci. American *233*, 24 (1975)
162. Fernandez, S. M., Lurquin, P. F., Kado, C. I.: FEBS Letters *87*, 277 (1978)
163. Schell, J. et al.: Proc. Roy. Soc. Lond. B. *204*, 251 (1979)
164. Matthews, B. et al.: Planta *145*, 37 (1979)
165. Shepherd, R. J.: Ann. Rev. Plant Physiol. *30*, 405 (1979)
166. Kado, C. I., Lurquin, P. F.: Microbiology, p. 231 (1978)
167. Davey, M. R.: In: Applied and fundamental aspects of plant cell, tissue, and organ culture (Reinert, J., Bajaj, Y. P. S., eds.), p. 551. Berlin: Springer 1977
168. Giles, K. L.: In: Applied and fundamental aspects of plant cell, tissue, and organ culture (Reinert, J., Bajaj, Y. P. S., eds.), p. 536. Berlin: Springer 1977
169. Thomas, E., King, P. J., Potrykus, I.: Z. Pflanzenzüchtg. *82*, 1 (1979)
170. Giles, K. L., Whitehead, H.: Science *193*, 1125 (1976)
171. Giles, K. L., Whitehead, H.: Plant Sci. Lett. *10*, 367 (1977)
172. Meeks, J. C., Malmberg, R. L., Wolk, C. P.: Planta *139*, 55 (1978)
173. Bradley, P. M., Leith, A.: Naturwissenschaften *66*, 111 (1979)

Author Index Vol. 1—19

Advances in Biochemical Engineering

Managing Editor: A. Fiechter

Distribution rights for all socialist countries: Akademie-Verlag

A Selection

Volume 7

Biotechnology

1977. 112 figures, 14 tables. V, 150 pages
ISBN 3-540-08397-9

Contents:
K. Schügerl, J. Lücke, U. Oels: Bubble Column Bioreactors. Tower Bioreactors without Mechanical Agitation. – *R. T. Acton, J. D. Lynn:* Description and Operation of a Large-Scale, Mammilian Cell, Suspensions Culture Facility. – *S. Aiba, M. Okabe:* A Complementary Approach to Scale-Up Simulation and Optimization of Microbial Processes. – *L. Kjaergaard:* The Redox Potential: Its Use and Control in Biotechnology.

Volume 8

Mass Transfer in Biotechnology

1978. 95 figures. V, 151 pages
ISBN 3-540-08557-2

Contents:
M. Charles: Technical Aspects of the Rheological Properties of Microbial Cultures. – *K. Schügerl, J. Lücke, J. Lehmann, F. Wagner:* Application of Tower Bioreactors in Cell Mass Production. *M. Zlokarnik:* Sorption Characteristics for Gas-Liquid Contacting in Mixing Vessels.

Volume 16

Plant Cell Cultures I

1980. 82 figures, 21 tables. IX, 148 pages
ISBN 3-540-09807-0

Contents:
G. Wilson: Continuous Culture of Plant Cells Using the Chemostat Principle. – *P. Spiegel-Roy, J. Kochba:* Embryogenesis in Citrus Tissue Cultures. – *E. Reinhard, A. W. Alfermann:* Biotransformation by Plant Cell Cultures. – *S. J. Stohs:* Metabolism of Steroids in Plant Tissue Cultures. – *S. S. Radwan, H. K. Mangold:* Biochemistry of Lipids in Plant Cell Cultures.

Volume 17

Products from Various Feedstocks

1980. 57 figures. VII, 172 pages
ISBN 3-540-09955-7

Contents:
S. Fukui, A. Tanaka: Production of Useful Compounds from Alkane Media in Japan. – *J. Jui:* Microbial Reactions in Postaglandin Chemistry. – *W. Sittig, U. Faust:* Analysis of a Loop Reactor for Methanol as a Carbon Source. – *Y.-H. Lee, L. T. Fan:* Properties and Mode of Action of Cellulase. – *L. T. Fan, Y.-H. Lee, L. S. Fan:* Kinetics of an Insoluble Cellulose Hydrolysis by Cellulase.

Volume 18

Plant Cell Cultures II

1980. 90 figures. VII, 193 pages
ISBN 3-540-09936-0

Contents:
P. J. King: Plant Tissue Culture and the Cell Cycle. – *K. Hahlbrock, J. Schröder, J. Vieregge:* Enzyme Regulation in Parsley and Soybean Cell Cultures. – *P. J. Wang, C. Y. Hu:* Regeneration of Virus Free Plants Through in Vitro Culture. – *L. A. Withers:* Low Temperature Storage of Plant Tissue Cultures. – *K. Tran-Thanh-Van:* Control of Morphogenesis or What Shapes a Group of Cells?

Springer-Verlag
Berlin
Heidelberg
New York

The Handbook of Environmental Chemistry

Editor: O. Hutzinger

This handbooks is the first advanced level compendium of environmental chemistry to appear to date. It covers the chemistry and physical behavior of compounds in the environment. Under the editorship of Prof. O. Hutzinger, director of the Laboratory of Environmental and Toxicological Chemistry at the University of Amsterdam, 37 international specialists have contributed to the first three volumes.
For a rapid publication of the material each volume will be divided into two parts. Part A of the first three volumes are now available, Part B will follow in the Spring of 1981. Each volume contains a subject index.

The Handbook of Environmental Chemistry is a critical and complete outline of our present knowledge in this field and will prove invaluable to environmental scientists, biologists, chemists (biochemists, agricultural and analytical chemists), medical scientists, occupational and environmental hygienists, research geologists, and meteorologists, and industry and administrative bodies.

Volume 1 (in 2 parts)
Part A

The Natural Environment and the Biogeochemical Cycles

With contributions by numerous experts
1980. 54 figures. XV, 258 pages
ISBN 3-540-09688-4

Contents:
The Atmosphere. – The Hydrosphere. – Chemical Oceanography. – Chemical Aspects of Soil. – The Oxygen Cycle. – The Sulfur Cycle. – The Phosphorus Cycle. – Metal Cycles and Biological Methylation. – Natural Organohalogen Compounds. – Subject Index.

Volume 2 (in 2 parts)
Part A

Reactions and Processes

With contributions by numerous experts
1980. 66 figures, 27 tables. XVIII, 307 pages
ISBN 3-540-09689-2

Contents:
Transport and Transformation of Chemicals: A Perspective. – Transport Processes in Air. – Solubility, Partition Coefficients, Volatility and Evaporation Rates. – Adsorption Processes in Soil. – Sedimentation Processes in the Sea. – Chemical and Photo Oxidation. – Atmospheric Photochemistry. – Photochemistry at Surfaces and Interphases. – Microbial Metabolism. – Plant Uptake, Transport and Metabolism. – Metabolism and Distribution by Aquatic Animals. – Laboratory Microecosystems. – Reaction Types in the Environment. – Subject Index.

Volume 3 (in 2 parts)
Part A

Anthropogenic Compounds

With contributions by numerous experts
1980. 61 figures. XV, 274 pages
ISBN 3-540-09690-6

Contents:
Mercury. – Cadmium. – Polycyclic Aromatic and Heteroaromatic Hydrocarbons. – Fluorocarbons. – Chlorinated Paraffins. – Chloroaromatic Compound Containing Oxygen. – Organic Dyes and Pigments. – Inorganic Pigments. – Radioactive Substances.

Springer-Verlag
Berlin
Heidelberg
New York